D1070056

The organic chemistry of tin

This is a volume in the series

THE CHEMISTRY OF ORGANOMETALLIC COMPOUNDS

THE CHEMISTRY OF ORGANOMETALLIC COMPOUNDS

A Series of Monographs

Dietmar Seyferth, *editor*

Department of Chemistry
Massachusetts Institute of Technology
Cambridge, Massachusetts

The organic chemistry of tin

WILHELM P. NEUMANN
University of Dortmund

Translated by WOLF MOSER
University of Aberdeen

Interscience Publishers
a division of
JOHN WILEY & SONS
London · New York · Sydney · Toronto

Library of Congress catalog card no. 74-111281

ISBN 0 471 63237 6

The original German edition published by Ferdinand Enke Verlag, Stuttgart, 1967

Printed by J. W. Arrowsmith Ltd., Bristol, England

Preface to the 1967 German Edition

In 1849, Edward Frankland reported that he had been able to synthesize diethyltin diiodide. Exploration of the organic chemistry of tin had begun. Exactly 100 years later, in 1949, after intervening periods of sometimes greater and sometimes lesser activity, exploration started to expand almost explosively, both in breadth and in depth. Since then new compounds, new syntheses, and new applications have appeared in rapid succession but with an increasing trend towards the study of reaction mechanisms and the bond forces responsible. Theoretical organic chemistry has provided both the premise for mechanisms and the impetus for their study. A start has thus been made in fitting the confusing multiplicity of experimental results into a clear pattern of chemical correlations. By far the greater part of this development belongs to the future, and many more scientists will have to contribute to it. C. K. Ingold's words apply as truly to the organic chemistry of tin as to all organo-metal chemistry, of which he said in 1964: "It is obvious that organo-metal chemistry will grow into a major domain of chemistry. At present it is a dark continent with far too few explorers" (reference 315a).

The rapid development of the last few years appeared to call for a summary and a critical review of present knowledge. The suggestion that I should write such a book, therefore, presented me with a pleasant task, to which I started to apply myself with enthusiasm. However, it soon became apparent that the material would be accommodated only with difficulty within the suggested format. Accordingly some fundamental aspects and topics of general importance were singled out and treated as far as possible in separate chapters. This is the origin of the chapters on, e.g., structure and bonding (2), reactions of the C–Sn bond (4), the Sn–H bond (8-4, 9, 10), the Sn–O and Sn–N bonds (19), and on analytical methods (22) and spectroscopy (23). The chapters and most sub-sections are prefaced by brief introductions to facilitate survey and study of the topic for the non-specialist reader. This has enabled detailed treatment of the various types of compound to be kept within the required limits. Results which could not be discussed at length have, as far as possible, been given at least a brief mention and a reference to provide the specialist reader with a source of further information. Unfortunately this has meant that a lot of material is sometimes packed into very little space. The selection of material is of course open to argument, and it is also bound to have been influenced by my own views. The literature has been covered to about April 1966, with many references to the end of 1966 (some to early 1967) added in proof. Criticisms and suggestions for improvement will be welcome.

I am truly grateful to my revered teacher Professor Karl Ziegler for his constant encouragement and his kindness in supplying a preface to (the 1967 German edition of) this book. It was he who originally awakened my interest in organometallic chemistry and later nurtured it during my stay at his Institute from 1955 to 1960.

I have received much help in the preparation of the manuscript, not only in Germany but also from abroad. I should like to mention particularly Professor F. Kröhnke, the Director of the Institute for Organic Chemistry of this University (at Giessen), who read the entire manuscript, and also Professor E. Habermann, Professor B. Kockel, and Professor H. Schmutterer (Giessen), Professor O. R. Klimmer (Bonn), Dr. W. J. Considine (New York, and Rahway, New Jersey), and Professor D. Seyferth (Cambridge, Massachusetts), whose critical reading of particular chapters, valuable suggestions, and other help contributed to the successful completion of this volume. My own colleagues Dr. E. Heymann, Dipl. Chem. F. Kleiner, Dr. R. Schick, Dr. B. Schneider, and Dr. R. Sommer worked on individual chapters. In addition, Dr. Sommer brought much skill and keenness to the laborious task of compiling literature references and other material. Miss Bärbel Hock assisted with the compilation of the subject index, and by tireless secretarial work. To all of these I herewith express my warmest thanks. Finally, I should like to express my gratitude to my wife Mechtild, who not only helped me with the proofs, but made a major contribution by her patience, consideration, and constant encouragement.

It is my hope that this book may help to further the interest in the organic chemistry of tin, and furthermore that it may focus attention on the wide gaps in our present knowledge of this field and thus induce others to participate in its further exploration.

WILHELM P. NEUMANN

Giessen,
March, 1967

PREFACE

WHEN Professor Dietmar Seyferth, as Editor, invited me to write the organotin volume for this series, I was just finishing the manuscript of my book "Die Organische Chemie des Zinns". With the agreement of the publishing companies concerned, we decided to prepare an English translation of the German volume.

Three years have now passed since the appearance of the German edition of the book. In the circumstances, it would have been unwise merely to translate the original text, especially as progress in organotin chemistry has been rather rapid since 1967. I therefore took the opportunity to re-examine the entire manuscript, to rectify a number of errors, and to take into account some of the suggestions sent to me by chemists from all over the world.

Alterations and additions have been extensive, and several chapters have been completely rewritten. About 160 new references have been included, most of them from 1967 and 1968, some (added in proof) from 1969. A further 40 references, most of them "in the press" or "private communication," have been replaced by ordinary ones, easily accessible and therefore more valuable for the reader. About 1180 references are now listed. To sum up, this book is published as a translation of an extensively revised German text.

The scope of the book has remained the same and has been described in the original German "Vorwort," of which a translation appears above. We have tried to take into account that organotin chemistry in recent years has become a topic in which structural and stereochemistry, applied, preparative, mechanistic, and theoretical organic chemistry meet and interact. A surprising expansion of both theoretical and preparative aspects has been the happy outcome of this interaction. For example, Sn–H, Sn–N, Sn–O, and even some Sn–C compounds have proved themselves as versatile tools in preparative chemistry, while strong connections have been demonstrated between organotin and free-radical chemistry. The growing industrial importance of organotin compounds should also be mentioned.

I am very grateful to Mr. W. Moser (Reader in Chemistry, University of Aberdeen), who translated the book, for his extremely able work and his patience with my many suggestions for additions and changes in the original text. It was a pleasure to cooperate with him, and also with the several departments of John Wiley and Sons Ltd. involved in preparing and producing this book. Many colleagues from several countries made valuable comments and suggestions for additions to the German book, and these have been evaluated for the English edition. Dr. Terence N. Mitchell, who is now with me at the University of Dortmund as post-doctoral research fellow, has spent much time on proof-reading and trying to improve my English. I owe very many thanks to all of them, and also to Professor Dietmar Seyferth for his creative activity, and to my wife Mechtild for her encouragement and patience during a rather busy time.

WILHELM P. NEUMANN

Dortmund,
January, 1970

CONTENTS

1 **INTRODUCTION AND NOMENCLATURE** **1**

Introduction 1
Nomenclature 4

2 **STRUCTURE OF ORGANOTIN COMPOUNDS** . . . **5**

Compounds containing Tetra-covalent Tin Atoms . . . 5
Compounds containing Penta-coordinate Tin 14
Compounds containing Hexa-coordinate Tin 18

3 **TETRAORGANOTINS** **20**

Preparation of Symmetrical Tetraorganotins 22
From Tin Alloys 22
From Inorganic Tin Compounds 22
Electrochemical Methods 24

Preparation of Unsymmetrical Tetraorganotins . . . 24
Tetraorganotins with Vinyl, Allyl, and Alkynyl Groups . 25
Vinyltin Compounds 25
Allyltin Compounds 26
Alkynyltins 27

Tetraorganotins with Functional Groups 32

4 **REACTIONS OF THE CARBON–TIN BOND** . . . **33**

Hydrolysis: Attack by Acids and Bases 34
Cleavage by Halogens 36
Cleavage by Hydrogenation 38
Cleavage by Free Radicals 38
Cleavage by Strongly Polar Alkyl Halides and Some Other Compounds 40
Alkylation by C–Sn Bonds (Transmetalation) 40
Grignard-like Reactions 43

5 ORGANOTIN HALIDES 45

Preparation of Organotin Halides 49

Starting from Tin Metal 49

By Partial Alkylation of Inorganic Tin Compounds . . 52

By Comproportionation 53

By Cleavage of C–Sn *or* Sn–Sn *Bonds* 57

Preparation of Individual Organotin Halides from Other Organotin Halides 58

6 ORGANOTIN PSEUDOHALIDES 59

7 STANNYL ESTERS 61

Esters of Carboxylic Acids 61

Esters of Inorganic Acids 64

8 ORGANOTIN HYDRIDES 66

Preparation of Alkyltin Hydrides R_3SnH, R_2SnH_2, and $RSnH_3$ 66

Preparation of Organotin Hydrides containing Negative Substituents 70

Tetraalkylditin Dihydrides 71

Reactions of Organotin Hydrides 72

Reactions with Radical Sources and Short-lived Radicals . 73

Reactions with Proton Donors 76

Exchange Reactions with Other Stannyl Compounds . . 77

Condensations leading to Sn–Sn *Bonds* 78

Reactions with Metal Alkyls 81

Other Reactions 82

9 HYDROSTANNATION OF UNSATURATED COMPOUNDS 85

Olefins 85

Use of Radical Catalysts 86

Use of Organometallic Catalysts 92

Addition of Alkylhalotin Hydrides 93

Conjugated Dienes 93

Allenes 95

Alkynes 96

Aldehydes and Ketones 99
Unsaturated Nitrogen Compounds 100
Azomethines 100
Azo-compounds 101
Nitroso-compounds 101
Isocyanates and Isothiocyanates 102
Carbodiimides, CO_2, and CS_2 103
Polar Conjugated Systems 103

10 ORGANOTIN HYDRIDES AS REDUCING AGENTS . 107
Aldehydes and Ketones 110
Alkyl and Aryl Halides 113
Acyl Halides 116
Other Compounds 117

11 ALKALI AND ALKALINE EARTH STANNYLMETALLICS 120
Stannyl-Alkali Compounds 120
Stannyl-Alkaline Earth Compounds 124

12 STANNYLBORON COMPOUNDS 125

13 STANNYL-SILICON, -GERMANIUM, AND -LEAD COMPOUNDS 126

14 COMPOUNDS WITH Sn–Sn BONDS 128
Hexaorganoditins 128
Substituted Organic Ditins 133
Linear and Branched Polytins 134
Cyclic Polytins 140
Aromatic Cyclic Polytins 141
Aliphatic Cyclic Polytins 144

15 STANNYLAMINES 148

16 STANNYL-PHOSPHINES, -ARSINES, -STIBINES, AND -BISMUTHINES 152

17 ORGANOTIN OXYGEN, SULFUR, SELENIUM, AND TELLURIUM COMPOUNDS (EXCEPT STANNYL ESTERS) . . . 155

Triorganotin Hydroxides and Stannoxanes $R_3SnOSnR_3$. 155
Diorganotin Oxides . . . 160
Stannonic Acids . . . 163
Organotin Alkoxides . . . 165
Organotin Peroxides . . . 167
Organotin Sulfur Compounds . . . 169
Organotin Selenium and Tellurium Compounds . . . 174

18 ORGANOTIN COMPOUNDS CONTAINING OTHER ELEMENTS . . . 176

19 REACTIONS OF STANNYLAMINES, STANNOXANES, AND ORGANOTIN ALKOXIDES . . . 179

Condensation Reactions . . . 179

Condensation with Proton Donor Compounds . . . 179
Condensations with Other Compounds . . . 183

Addition to Polar Multiple Bonds . . . 184

Primary Addition . . . 184
Secondary Reactions . . . 187

20 TIN HETEROCYCLES . . . 191

Ring Systems with One Heteroatom . . . 191
Ring Systems with Two or More Heteroatoms . . . 195

21 MACROMOLECULAR ORGANOTINS . . . 199

Polymers with Tin Atoms in the Main Chain . . . 199

Polymers with Tin inserted at Regular Intervals in a Carbon Chain . . . 199
Polymeric Stannoxanes and Stannyl Esters . . . 204

Polymers with Stannyl Substituents . . . 206

Polymers from Olefinic Organotin Compounds . . . 206
Other Organotin Polymers . . . 208

22 ANALYSIS OF ORGANOTIN COMPOUNDS . . . 209

Separation Techniques 209

Paper Chromatography 209
Thin-layer Chromatography 210
Gas–liquid Chromatography 210

Elemental Analysis 211
Determination of Functional Groups 212

23 SPECTROSCOPIC INVESTIGATION 214

Ultraviolet Spectroscopy 214
Infrared and Raman Spectroscopy 215
Nuclear Magnetic Resonance Spectroscopy. 216

Proton Magnetic Resonance Spectroscopy. 216
^{119}Sn *Magnetic Resonance Spectroscopy* 226

Mössbauer Spectroscopy 226
Mass Spectrometry 227

24 TOXICITY OF ORGANOTIN COMPOUNDS . . 230

25 TECHNOLOGY AND COMMERCIAL APPLICATIONS 238

Industrial Preparation 239
Organotins as Polymer Stabilizers 240
Organotins as Insecticides and Fungicides 242
Various Applications 245

References 247

Index 277

1 Introduction and nomenclature

1-1 INTRODUCTION

The first chemist to report an organotin compound seems to have been E. Frankland. The report occurs in his fundamental work (212, 213) on ethylzinc compounds (1849), but appears to have remained unknown to most of his contemporaries as well as to later authors. Alkyltin compounds are thus amongst the earliest known organometallic derivatives.

Frankland had a specimen of diethyltin diiodide as early as 1849, and unambiguous characterization followed some years later (214). In 1853 he also described diethyltin oxide and dichloride as well as an air-sensitive "diethyltin" which on heating above 150° deposited tin and yielded a new distillable liquid which must have been tetraethyltin. Frankland's intuition and experimental technique, which he adapted with surprising skill to these problems, must be considered remarkable even today. His primary interest was not in organometallic compounds as such but in testing the then topical "Theory of Radicals," an important link in the development of organic structure theory. This is also true of the work of his contemporaries. 1852 saw Cahours and Riche (120) extending Frankland's work of 1849, and C. Löwig reporting independent work on organotin compounds (486). Until now this last publication has usually been considered to represent the beginning of organotin chemistry. Löwig treated alkyl halides with a tin–sodium alloy, but worked up the product by such a complicated procedure that it unintentionally yielded yet further compounds and made separation even more difficult. Apart from the compounds described by him, there were others such as stannoxanes present, as shown already in 1858–1862 by A. Strecker (829, 830).

By investigation of organometallic compounds it was hoped to gain further knowledge of the structure of organic molecules. It is therefore not surprising that even A. Kekulé, that genius of organic structure theory, concerned himself with tin alkyls. In 1861 he was already propounding the following (348): "I cannot resist the opportunity of once again drawing the attention of chemists to the analogy between tin compounds and ... carbon compounds. ..."

This analogy is connected with the tetrahedral model of 4-covalent tin, with the covalent bonding of the ligands, and with the properties and reaction modes consequent upon these. It has recently received considerable attention. Naturally there are also appreciable differences arising from the more complicated structure of the tin atom, as compared with that of carbon, and from its size.

From about 1880 further investigation of tin alkyls fades into the background in comparison with the rapid progress of organic chemistry in general. Individual important results appeared at intervals, but intensive work is evident again only from about 1917. This is concerned with organotin compounds as such, and lasts until about 1931. Among the names associated with this chapter of development are particularly those of E. Krause (416) and his co-workers in Germany, C. A. Kraus in the United States, and K. A. Kozeshkov in Russia. About 200 publications on these materials had appeared by 1935. [For an excellent review, see E. Krause and A. von Grosse (416).]

Then came the discovery of industrial applications of organotin compounds, e.g. as stabilizers against photodecomposition for poly(vinyl chloride) and other polymers, and later as fungicides. These factors and an increased general scientific interest produced a striking renaissance of organotin chemistry starting about 1949 and continuing to the present day. The rapid progress was made possible largely by developments in other fields, e.g. by new preparative techniques and modern physico-chemical methods of analysis. By the end of 1965 the rapidly increasing number of groups working in many different countries had produced some 3000 publications. Figure 1-1 shows the development of organotin chemistry as measured by the number of publications. It should be mentioned that such numbers cannot imply any scale of merit.

The recent evolution of the subject is marked especially by two factors: the intense interest in bonding and structure theory, and the shift in preparative problems. Preparation of simple organotins R_4Sn, R_3SnX, R_2SnX_2, or $RSnX_3$ is not as predominant as it was formerly. On the other hand, a rapidly growing space is taken up by syntheses making use of such compounds, especially in reactions with complicated organic molecules. Some of the impetus has come from the realization that stannyl compounds containing Sn—H, Sn—N, and Sn—O groups are capable of far easier and more varied reactions than would have been suspected even a few years ago. The scope of organotin chemistry has thereby increased considerably, and the study of the relevant reaction mechanisms has become more important. Better understanding has been aided by the ideas of theoretical organic chemistry.

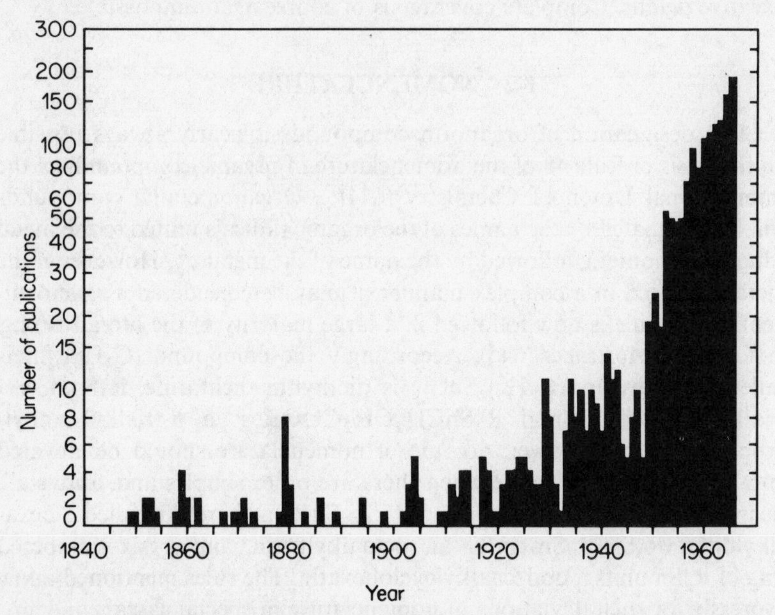

FIG. 1-1 Publications on organotin compounds, arranged by year of publication; most of the patent literature has been omitted

In the following pages an attempt will be made to describe the present state of organotin chemistry. It should be clear from the numbers mentioned that reference cannot be made to all published papers or all described substances.* On the other hand, use is made of unpublished results, in part communicated privately to the author, in part arising from his own laboratory. Whilst selection has been unavoidable, the reader is shown whenever possible where to find the original literature source. This is also true for the tables giving examples from the various groups of substances. The set scope of the work requires some limitation,

* Comprehensive tabulated summaries up to 1959 are given by R. K. Ingham, S. D. Rosenberg, and H. Gilman (315), and up to 1964 by M. Dub (183). Annual literature surveys (768) are available from 1964 onwards. Reviews covering parts of the subject are mentioned in the relevant chapters.

and the material has been reviewed critically. Values of melting points, boiling points, etc. are usually those which have proved reproducible from information supplied by colleagues or from experience in the author's own laboratory. Where literature sources are cited in connection with tables, these have been selected from the multiplicity of material in such a way that the reader is referred as far as possible to recommended preparative details. Complete coverage is of course again impossible.

1-2 NOMENCLATURE

Clear designation of organotin compounds is nearly always possible on the basis of Rule 48 of the nomenclature of organic compounds of the International Union of Chemistry (644): "Organometallic compounds will be designated by the names of the organic radicals united to the metal which they contain, followed by the name of the metal.... However, if the metal is united in a complex manner, it may be considered as a substituent." This rule is now followed in a large majority of the literature, e.g. in *Chemical Abstracts* (645). Accordingly the compound $(C_2H_5)_4Sn$ is called tetraethyltin, $(C_2H_5)_2SnCl_2$ is diethyltin dichloride, R_3SnH is a trialkyltin hydride, and $R_3SnCH_2CH_2COOR$ is a β-trialkylstannylpropionic ester. However, no rule of nomenclature should be elevated into an inviolable principle when there are other simpler and, above all, clearer designations. A compound $R_3Sn{-}SnR_3$ is therefore called a hexaalkylditin, $R_3Sn{-}R_2Sn{-}SnR_3$ an octaalkyltritin, and a six-membered ring of R_2Sn units a dodecaalkylcyclohexatin. The rules mentioned allow expressly for such deviations in nomenclature in special cases.

2 Structure of organotin compounds

2-1 COMPOUNDS CONTAINING TETRA-COVALENT TIN ATOMS

The fifty electrons of the tin atom are arranged as follows: $1s^2\ 2s^2\ 2p^6\ 3s^2\ 3p^6\ 3d^{10}\ 4s^2\ 4p^6\ 4d^{10}\ 5s^2\ 5p^2$. The valence electrons can undergo sp^3-hybridization, and tetra-covalent tin atoms are therefore tetrahedral. The valencies may be satisfied by R or X. Four series of compounds R_4Sn, R_3SnX, R_2SnX_2, and $RSnX_3$ are known, and these are discussed later in separate chapters.

R can be identical or different, substituted or unsubstituted, aliphatic or aromatic groups. X can be negative groups such as –OR, –SR, –OCOR, $–OSnR_3$, $–NR_2$, or halogen or some other acid radical, or neutral ligands such as –H, or electropositive ones such as Li or Na. The three series of organotin hydrides R_3SnH, R_2SnH_2, and $RSnH_3$ (Chapter 8) have recently assumed considerable importance.

Stannyl–metal compounds (Chapters 11, 13, 14, 16, and 18) should also be mentioned, amongst them compounds containing tin–tin bonds (Chapter 14). These include the so-called "tin dialkyls," derivatives of tetra-covalent tin containing normal tin–tin bonds (Chapter 14-4).

The steric arrangement around the tin atoms is tetrahedral. There is no measurable change in the tetrahedral angle when negative groups, e.g. –OR (881) or halogen (803), are present alongside carbon. Even the proximity of several such groups, e.g. in R_2SnHal_2 or $RSnHal_3$, has no measurable influence on the angles (294, 803). The angle Sn–Sn–Sn also agrees with that in the tetrahedral model (639), so that, e.g. for a cyclohexatin, the steric arrangement found in cyclohexane is largely valid.

It should thus be possible also to synthesize chiral organotins with an asymmetric tin atom as chiral center. The resolution of a methyl-ethylpropyltin halide was reported as early as 1900 (674b). Rotations obtained were, however, so small that one would regard them now with reservations (809a), especially in view of the successes in the synthesis of optically active alkyls of silicon and germanium.

Better insight into the behavior and strength of C–Sn bonds has suggested new approaches to the preparation of chiral organotins, e.g. (91b)

$$\text{cyclo-}C_6H_{11}SnMe_3 \xrightarrow{MeOH/Br_2} \text{cyclo-}C_6H_{11}SnMe_2Br \rightarrow$$

$$\text{cyclo-}C_6H_{11}SnMe_2\textit{i}Pr \xrightarrow{MeOH/Br_2} \text{cyclo-}C_6H_{11}SnMeBr\textit{i}Pr \rightarrow$$

$$\text{cyclo-}C_6H_{11}SnMeEt\textit{i}Pr \xrightarrow{MeOH/I_2} \text{cyclo-}C_6H_{11}SnIEt\textit{i}Pr$$

Another attempt has been made with $PhSn(Cl)(Me)CH_2CMe_2Ph$ (648a). However, no optically active compound has been isolated.

The covalent radius of the tin atom is 1.40 Å, surprisingly independent of the nature of the ligands. This value is always obtained (within experimental error) when one deducts the covalent radius of the ligand, which it shows attached to carbon, from the measured bond-length in the tin compound (see Table 2-1). Only when there is an accumulation of strongly

Table 2-1
Bond-lengths and Covalent Radius of Tin (Å)

BOND Sn–X[a]	LENGTH	MEASURED IN	METHOD[b]	REF.	COV. RADIUS OF X	COV. RADIUS OF Sn (BY DIFF.)
Sn–C	2.18 ± 0.03	Me_4Sn	a	95	0.77	1.41 ± 0.03
	2.17 ± 0.06	Et_3SnOAr	b	881	0.77	1.40 ± 0.06
	2.15 ± 0.06	Et_3SnOAr	b	881	0.77	1.38 ± 0.06
	2.21 ± 0.06	Et_3SnOAr	b	881	0.77	1.44 ± 0.06
	2.19 ± 0.03	Me_3SnCl	a	803	0.77	1.42 ± 0.03
	2.19 ± 0.03	$MeSnCl_3$	a	803	0.77	1.42 ± 0.03
	2.17 ± 0.03	Me_3SnBr	a	803	0.77	1.40 ± 0.03
Sn–H	1.700 ± 0.015	$MeSnH_3$	c	476	0.28	1.42 ± 0.02
	1.701 ± 0.001	SnD_3H	d	886	0.28	1.42 ± 0.02
Sn–O	2.08 ± 0.06	Et_3SnOAr	b	881	0.66	1.42 ± 0.06
Sn–S	2.35 ± 0.10	$(MeSn)_4S_6$	b	177a	1.04	1.31 ± 0.10
Sn–Cl	2.37 ± 0.03	Me_3SnCl	a	803	0.99	1.38 ± 0.03
Sn–Br	2.49 ± 0.03	Me_3SnBr	a	803	1.14	1.35 ± 0.03
Sn–I	2.72 ± 0.03	Me_3SnI	a	803	1.33	1.39 ± 0.03
Sn–Sn	2.77	$(Ph_2Sn)_6$	b	639		1.39
	2.78	$(Ph_2Sn)_6$	b	639		1.39
	2.80	grey Sn	b	639		1.40

[a] Accuracy and error limits are those in the original literature.
[b] Determined from (a) electron diffraction, (b) X-ray diffraction, (c) microwave spectra, (d) infrared spectra.

Table 2-2

Contraction of the Sn–Hal Bond (Å) *in Tin Halides* (803)

	X = Cl	Δ[a]	X = Br	Δ[a]	X = I	Δ[a]
Me_3SnX	2.37 ± 0.03	0.03	2.49 ± 0.03	0.01	2.72 ± 0.03	0.03
Me_2SnX_2	2.34 ± 0.03	0.02	2.48 ± 0.02	0.03	2.69 ± 0.03	0.01
$MeSnX_3$	2.32 ± 0.03	0.02	2.45 ± 0.02	0.01	2.68 ± 0.02	0.04
SnX_4	2.30 ± 0.03		2.44 ± 0.02		2.64 ± 0.02	

[a] It should be noted that Δ-values are of the same order as the error limits on the actual bond-lengths.

negative ligands round the tin is there some decrease in bond-lengths (Table 2-2).

The bonding of the tin is thus almost entirely covalent, at least in crystalline solids, in non-polar media, and in the vapor. This is true even for tin–halogen bonds. The polarity which should follow from the difference in electronegativities (see Table 2-5 and the related discussion) is not observed, probably because of the relatively large bond-lengths. This is particularly noticeable in many reactions! However, for the same reason the bonds are easily polarizable, and at the limit will undergo ionic dissociation especially in polar solvents:

$$R_3Sn\text{–}X \rightleftharpoons R_3Sn^+ + X^- \tag{2-1}$$

The extent of dissociation is generally very slight even with strongly negative ligands: e.g. the specific conductance of Et_3SnCl is 1.8×10^{-9} in $1M$ solution at 20° in benzene ($\kappa = 2.3$), and only rises to 1×10^{-6} in nitrobenzene ($\kappa = 36$) (736). Even trimethyltin iodide shows only very slight conductance, and this is not much higher in nitrobenzene or nitro-

Table 2-3

Specific Conductance of Tin Halides $\times 10^8$ (ohm^{-1} $cm.^{-1}$) (736)

COMPOUND	PURE SUBSTANCE[a] 20°	PURE SUBSTANCE[a] 90°	1*M* SOLUTION IN BENZENE, 20°
Et_4Sn	< 0.02	< 0.02	< 0.02
Et_3SnCl	460	3200	0.18
Et_2SnCl_2	—	1200	0.43
$EtSnCl_3$	10	170	0.22
$SnCl_4$	< 0.02	< 0.02	< 0.02

[a] The compounds here act as their own solvents, and differences in conductance are mainly due to different dielectric constants (see Table 2-5). Further purification did not lower the conductance.

methane ($\kappa = 89$) than in ether or benzene (408). By way of further example, Table 2-3 gives the specific conductance of tetraethyltin and the ethyltin halides (736). The overwhelmingly covalent character of the tin bonds is also confirmed by Mössbauer spectra (Chapter 23-3). Also, in dimethylformamide, triphenylmethyl chloride conducts about 40 times better than triphenyltin chloride, and as much as 80 times better than triphenyltin fluoride (850). There is thus no indication of "salt-like constitution" of the organotin halides, of which there is occasional mention in the literature.*

Therefore, while polar or ionic reaction mechanisms are known in organotin chemistry, they are not nearly as predominant as with silicon and germanium, although the original electronegativity values according to Pauling differ only by 0.1 or less from that of tin (see Table 2-6). Radical reactions, on the other hand, are common with tin, and will be discussed for the different series of compounds. All these circumstances account for the multiplicity of organotin compounds and for the surprisingly close parallel with carbon.

Within the Fourth Main Group the bond-lengths to carbon increase considerably: C–C 1.54, C–Si 1.94, C–Ge 1.99, C–Sn 2.17, C–Pb 2.29 Å.

The increased bond-length for tin is the cause of the increased reactivity and the related lower thermal stability of the tin alkyls compared with their C-, Si-, and Ge-analogs (Chapter 3). Long bonds naturally have low strength and also lessen the screening of the central atom by the ligands. Attacking reagents thus have easier access. In the above series this can be seen clearly in the organolead compounds (887).

These relationships are also shown clearly by the dissociation energies of the bonds between carbon and the Fourth Main Group elements in the tetraalkyls. The mean values are: C–C 87, C–Si 70, C–Ge 60, C–Sn 50, C–Pb 31–37 kcal./mole.

These values are of course further dependent on the nature of the alkyl group, i.e. on the stabilization of the corresponding alkyl radical by mesomerism. Because of the importance of bond-dissociation energies for the understanding of radical reactions in organotin chemistry, the presently available data (still rather incomplete) are set out in Table 2-4.

Enthalpies of combustion and formation of a number of tin alkyls are given in reviews (455a, 802).

The dipole moment of the alkyl–Sn bond, mostly estimated as 0.45—0.6 D (261, 482, 524) depends in both magnitude and direction on the

* Only pronounced electron-pair donors, such as pyridine, have a greater effect on conductance (736). However, this is not a case of simple dissociation in the sense discussed above; there is extensive rearrangement of the organotin molecule to give, at least in the case of R_3SnCl, penta-coordination around the tin (see Chapter 2-2).

nature of the alkyl group. If this contains more than one C atom, the further C atoms contribute to the moment, but only up to the fourth, as also shown for the alkyl halides (820). Because of the free rotation about the long Sn–C bond (836) there is, however, no longer a tetrahedral angle between the group moments. The angle and therefore the extent of compensation of the moments now depends on the preferred conformation of the molecules.

We therefore find that only tetramethyltin has zero dipole moment. For the higher tetraorganotins there is incomplete compensation of the bond dipoles, as shown in Table 2-5. The limit is reached in the butyl

Table 2-4

Dissociation Energies, D (kcal./mole), *for Sn–X Bonds in Organotins* Me_3SnX

X	CALCULATED FROM THERMOCHEMICAL VALUES ONLY[a]	CALCULATED FROM APPEARANCE POTENTIALS[b]	APPROXIMATION FROM CHEMICAL EVIDENCE[c]
H	—	—	~ 35
Me	52, 56[d]	55 (61 ± 5[e])	—
Et	46	50	—
Pr	47	50	—
Bu	47	—	—
$CH{=}CH_2$	—	60	—
Ph	61	63	—
$CH_2CH{=}CH_2$	—	—	~ 37
CH_2Ph	—	39	—
NR_2	—	—	65—70[f]
OCOPh	95 (ref. 681a)	—	—
Cl	—	—	~ 85
Br	—	76	—
I	—	62	—
$SnMe_3$	50 (ref. 681a)	57	—

[a] Unless stated otherwise, from H. A. Skinner (802) ($\bar{D}$-values).

[b] A. L. Yergey and F. W. Lampe (899a) have measured the appearance potentials of Me_3Sn^+ ions in the mass spectrometer. But, for calculation of *D*(Sn–X) they use a $\Delta H_f°$ value for $Me_3SnSnMe_3$, which is considered by other authors as not fully reliable (176a, 455a, 648b). The *D*-values given in ref. 899a are therefore too high—in the author's opinion by about 20 kcal.—and not really consistent with chemical evidence. If one uses the $\Delta H_f°$ value for $Me_3SnSnMe_3$ preferred by W. F. Lautsch et al. (455a), one arrives at the values given in this column (300a). These are, in the author's experience, in fair agreement with chemical evidence.

[c] Approximation from chemical behavior as studied in the author's laboratory (300a).

[d] Measured in the compound $MeSn(Me)Cl_2$ (676a).

[e] Independently measured and calculated by M. F. Lappert et al. (454c).

[f] Dependent rather strongly on the nature of R (300a).

compounds. It can be confirmed from structure models that for most of the possible conformations the centers of positive and negative charge do not coincide. Incidentally, tetraethyltin crystallizes in at least ten modifications, which again can be explained by rotation isomerism (see Chapter 3).

Rotation isomerism is to be expected not only at the Sn–C bond but also at the neighboring C–C bond, as can be seen from the Raman and IR bands at about 500 and 600 cm.$^{-1}$ (163, 836). The latter belongs to the *trans*-conformation (**1**) which exists at low temperature and in the solid. To the former is assigned the *gauche*-conformation (**2**), the proportion of which increases with rise in temperature.

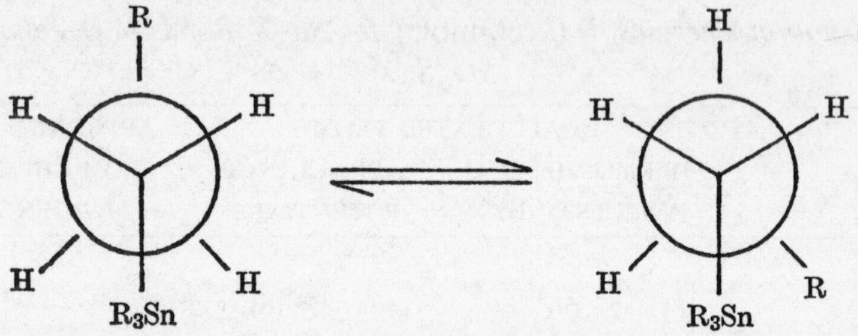

(**1**) ⇌ (**2**) (2-2)

It is assumed that these results, which apply to compounds R_2SnCl_2, are generally valid for R_4Sn, R_3SnX, R_2SnX_2, and $RSnX_3$, where R is an alkyl chain of at least three carbon atoms. For tributyltin mono- and di-chloroacetate, conformational analysis has been carried out with regard to the relative positions of carbonyl oxygen and halogen (164).

As has been mentioned already, an effect of further substituents on tin bond-lengths has so far been observed only in the case of polyhalogen

Table 2-5
Dipole Moments (Debye Units) of Tetraorganotins and Organotin Halides in Hexane at 20.0° (ref. 524 or 573 *unless stated otherwise*)

R = X =	Me Cl	Et Cl	*n*-Bu Cl	*i*-Bu Cl	*n*-C_8H_{17}	Ph Cl	Et Br
R_4Sn	0.0	0.41	1.00	1.00	0.82	—	0.41
R_3SnX	3.46[a]	3.50	3.53	3.42	—	3.16[b]	3.43
R_2SnX_2	4.14[a]	4.07	4.26	4.28	—	4.07	3.88
$RSnX_3$	3.77[a]	3.83	3.96	4.07	—	4.03	3.44

[a] In benzene at 25° (482). The values for R_2SnX_2 and $RSnX_3$ may be on the high side, since benzene affects the measurements; cf. other authors (524, 573).
[b] Poorly soluble in hexane at 20°, and therefore measured in benzene; ref. 482 gives 3.46 at 25°.

compounds [see Table 2-2 (803)]. Even there the effect is small and often lies within the limits of experimental error. It is doubtful whether any conclusions concerning partial double-bond character of the Sn–halogen bond should be drawn from such a contraction (878). As far as the long Sn–Br bond is concerned, there is no appreciable change in moment. Et_3SnBr and $EtSnBr_3$ have the dipole moment that would be expected for a tetrahedral arrangement provided that the Et–Sn moment remains constant (see Table 2-5 and pp. 8, 12). On the other hand, in the alkyltin chlorides the dipole moment rises as expected from R_3SnCl to R_2SnCl_2, and then falls to $RSnCl_3$ but does not drop back to the value for R_3SnCl (see Table 2-5). The Sn–Cl bond moment must thus increase in this series, and increased charge must outweigh decreased separation. No dipole measurements on alkyltin iodides have been reported.

It has not been established whether the bond dipole of the $C^{\delta-}$–$Sn^{\delta+}$ bond changes with increasing halogen substitution, in which the tin of course becomes more positive. NMR measurements on methyltin compounds, and the coupling constants J(Sn–H) calculated from these, suggest that this might be the case (207, 301).

The electronegativities of the elements of Group IV have been the subject of extensive investigations. Pauling's original values (646) have been modified repeatedly (see Table 2-6). The results always differ according to the method of measurement and the compounds selected (20). Electronegativity is not a constant atomic property. It is therefore not surprising that the values in Table 2-6 (columns 1—3), which are derived from different physical measurements, do not agree either amongst themselves or with the values in column 4 which are derived for sp^3-hybridization. One may therefore take the view that, in the series Si, Ge, Sn, Pb, differences in electronegativity are very slight, if there are any at all (discussion and further literature in refs. 65 and 677a). Other factors such as atomic size, bond-length, and nature of the substituent play a large, and often dominant, part (90). The influence of electronegativity is superimposed on these factors. The concept has therefore become somewhat blurred and difficult to grasp.* In practice one will generally have to work not with an "electronegativity of tin" but with a value for tin in a particular combination; i.e. one will have to allow for the influence of all ligands.

Closely connected with this is the inductive effect which tin atoms or stannyl groups exert on their surroundings. The bond polarization $C^{\delta-}$–$Sn^{\delta+}$, which is there in principle,† may be changed by substitution at C as well as at Sn. According to Eaborn et al. (90) polarization is

* There has, after all, never been an exact definition (see refs 677a and 708).

† For example, $SnMe_3$ groups have been shown to be electron-donating when introduced into the 2-, 3-, or 4-position of pyridine (measurement of pK values) (25a).

Table 2-6
Electronegativities[a]

	Pauling (REF. 646)	Sanderson (REF. 717)	Fineman–Daignault (REF. 205)	Allred–Rochow (REF. 20)
C	2.5	2.47	2.57	2.60
Si	1.8	1.74	1.90	1.90
Ge	1.8	2.31	2.02	2.00
Sn	1.8	2.02	2.47	1.93
Pb	1.8	—	—	2.45

[a] Since the authors use different reference systems, comparisons should only be made within the same column.

increased by electron-donor substituents R′ in the *p*-position in a phenyl group, as shown by the increase in the rate of scission of the C–Sn bond by iodine according to reaction 2-3. The reaction thus belongs to the class of electrophilic aromatic substitutions (90, 91):

$$R'C_6H_4SnR_3 + I_2 \rightarrow R'C_6H_4I + R_3SnI \qquad (2\text{-}3)$$

As expected, electron-attracting groups R′ decrease the rate by lowering the nucleophilicity of the ring carbon attached to tin. The sequence of relative reaction rates is for R′ = OMe > *t*-Bu > *i*-Pr > Et > Me > H > Br > F > Cl > COOH; in other words it is much as one would expect (262). Different substituents R at the tin atom act by raising or lowering the electron density at the tin, and produce the following sequence of reaction rates (90): cyclo-C_6H_{11} 5.4 > Et 5.1 > Me 1.0 > Ph 0.018.

Electron-donor properties increase in the series (R = Me): SiR_3 < GeR_3 < SnR_3, as may be seen from the relative rates of acidolysis of *p*-R_3MCH_2-C_6H_4-SnR_3. These are for M = Si 1.00, Ge 1.36, Sn 3.21 (89). (By interposing the –CH_2– group one avoids complications from *dπ–pπ* bonding; see below.) A similar sequence emerges from IR and NMR measurements (735), and also from IR studies on acetic acid esters $MeCOOMMe_3$, which give M = C < Si < Ge (818). Here one may, however, have to assume in addition to such inductive effects some back-donation to the tin (*dπ–pπ* bonds). Its contribution cannot yet be assessed quantitatively. A similar situation is found in the substituted benzoic acids, where a mildly decreasing acidity seems once again to

Table 2-7
Dissociation Constants of Substituted Benzoic Acids (133)

R IN p-R-C_6H_4COOH	$K \times 10^6$ (AT 25°)
H	1.05
Me_3C	0.70
Me_3Si	1.11
Me_3Ge	1.07
Me_3Sn	1.05

establish increasing electron-donor character in the series $SiR_3 < GeR_3 < SnR_3$ (see Table 2-7). But, whereas the acidity decreases, understandably, when H is replaced by CMe_3, it rises again for R = $SiMe_3$. This might be attributed to partial double-bond character of the aryl–metal bond, caused by electron-donation from π-orbitals of the aromatic system into low-lying empty d-orbitals of the metal atom. This is so for both Ge and Sn. The extent of partial double-bonding ($d\pi$–$p\pi$) appears to be practically the same; i.e. it does not depend on the size of the participating atom (133). Dipole measurements show (308) that electron-releasing substituents in the aromatic ligand promote electron-donation to the tin, most likely into one of its vacant $5d$-orbitals. More recent work seems to indicate that $d\pi$–$p\pi$ effects are more pronounced for Ge, and still more so for Si (90). This is supported by ESR measurements on Sn-, Ge-, or Si-substituted biphenyl radical-anions (168a). Anyway, some authors assume $d\pi$–$p\pi$ bonding for the tin, and support this with various arguments beyond those mentioned (133, 482) (see scheme 2-4).

(2-4)

The same conclusion has been drawn from UV spectra (266, 267). Back-donation can of course take place with other ligands with π-electrons, e.g. C≡C–R (Chapter 3-3-3), or may be considered e.g. for halogens. Force constants for Sn–Cl bonds, calculated from IR measurements, admittedly do not support this (419). It might be stressed that partial $d\pi$–$p\pi$ bonding should not be deduced on the sole evidence of bond-shortening (878). Appreciable $d\pi$–$p\pi$ bonding has also been postulated for the Sn–N group, but this is refuted by other authors (see Chapter 15).

2-2 COMPOUNDS CONTAINING PENTA-COORDINATE TIN*

Tetraalkyltins and tetraaryltins form no stable derivatives in which the coordination number is higher than 4. Moieties with higher coordination number must, however, be assumed as unstable transition states for many reactions of the C–Sn bond (see Chapter 4). Hitherto, these transition states have rarely been discussed more fully in the literature. One example with a penta-coordinate tin atom is found in the reaction of Ph_4Sn with KNH_2 (261b):

$$Ph_4Sn + NH_2^- \rightleftharpoons \left[Ph_3Sn \genfrac{}{}{0pt}{}{\cdot\cdot NH_2}{\cdot\cdot Ph} \right]^- \rightarrow \text{Products}$$

The conditions are distinctly reminiscent of the intermediate state (**3**) of an S_N2 reaction such as 2-5.

$$X^- + R_3C{-}Y \longrightarrow \left[X{-}{-}{-}CR_3{-}{-}{-}Y \right] \longrightarrow X{-}CR_3 + Y^- \qquad (2\text{-}5)$$

(**3**)

Compounds with penta-coordinate tin atoms become more stable or even fully stable if, first, the tin has become more positive by attachment of at least one negative ligand (e.g. in R_3SnCl) and, secondly, a powerful donor is present. Penta-coordination around tin was first clearly formulated by van der Kerk et al. (360, 490) (for reviews, see refs 635a and 669).

Many combinations fulfil these conditions, e.g. (**4**)—(**9**). In the imidazolyl compounds (**4**) and related cyclic derivatives with N atoms in positions 1 and 3, the ring system acts as negative ligand as well as donor. The penta-coordination and consequent association can be demonstrated in the solid state and also in not too dilute solution (319, 360, 490). In the stannyl esters (**5**) penta-coordination again leads to association as in (**6**), at least in the solid, just above the melting point or in concentrated solution in inert solvents (318, 490)† (see equation 2-6). From the IR data this holds for the entire homologous series from the formyl- right up to the stearyl-stannyl ester (166). In more dilute solution and in the vapor state (630, 631) the compounds are monomolecular and just like esters (**5**).

* For a review with ninety-eight references, see ref. 635a.

† For the liquid state this conclusion stems mainly from IR measurements. Against this it should be noted that NMR measurements do not support it. In a number of solvents the signals from acyl and alkyl protons are independent of concentration (175).

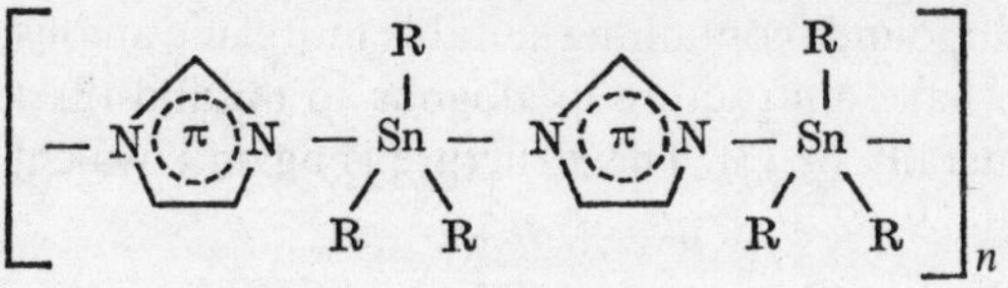

$$\left[-N\!\left(\pi\right)\!N - \underset{R\quad R}{\overset{R}{Sn}} - N\!\left(\pi\right)\!N - \underset{R\quad R}{\overset{R}{Sn}} - \right]_n$$

(4)

The extent of complexing depends on the nature of the carboxylic acid. For example, the IR spectrum of trimethylstannyl laurate in CCl_4 shows that at 2.3% concentration only the ester form (**5**) is present. At 8% this and the associated form (**6**) are found in about equal parts, and at 30% the latter predominates entirely (318). The analogous formyl derivative $HCOOSnMe_3$ shows stronger association; e.g. at 1.1% in cyclohexane it still shows an average degree of polymerization of 4.5. In these compounds (**6**) the trialkylstannyl group is planar or almost planar, analogous to the intermediate state (**3**) of an S_N2 reaction mentioned earlier. Evidence comes from the absence of the symmetric Me_3Sn– vibration at 510—520 cm.$^{-1}$

The repeat period of the linear chain (**6**) is 10.3—10.8 Å for both the formyl and the acetyl derivative (631). Analogous penta-coordination can be assumed also for the plumbyl esters $RCOOPbR_3$, but not for similar Si- and Ge-compounds (318).

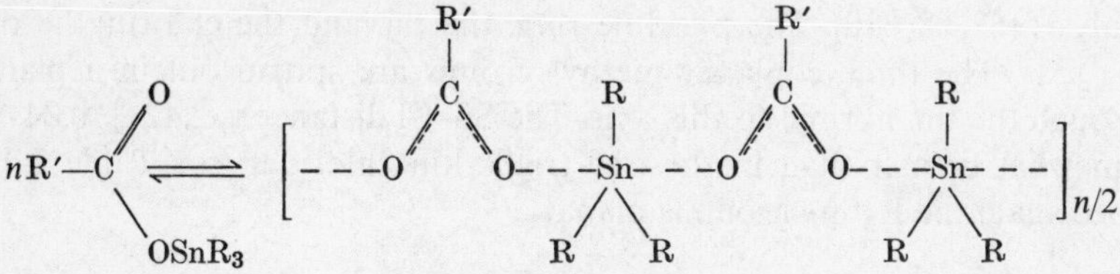

$$n\,R'\!-\!C\!\begin{smallmatrix}\nearrow O\\ \searrow OSnR_3\end{smallmatrix} \rightleftharpoons \left[\cdots O\!\overset{R'}{\overset{|}{C}}\!O \cdots \underset{R\ \ R}{\overset{R}{Sn}} \cdots O\!\overset{R'}{\overset{|}{C}}\!O \cdots \underset{R\ \ R}{\overset{R}{Sn}} \cdots \right]_{n/2} \qquad (2\text{-}6)$$

(5) (6)

Although the Sn–O bonds in (**6**) have all been made the same, there must be appreciable differences inasmuch as each acid group belongs to a particular stannyl group, whereas the second acid group is only attached by coordination. (There are as yet no X-ray data on the Sn–O bond-lengths.) In any case, when p-$MeC_6H_4COOSnEt_3$ and $PhCOOSnBu_3$ are melted together at 130°, the compounds can be recovered unchanged by fractional distillation; i.e. there is no exchange of stannyl groups (584).

The planar structure of the stannyl group led to earlier, erroneous conclusions that the stannyl esters (220, 636) and other alkyltin compounds with penta-coordination (66) were ionic. However, in this last case the suggestion was made with reservations.

Stannyl compounds containing suitable inorganic anions, or rather acid radicals, also have a structure analogous to (**4**) and (**5**), and can be represented generally by (**7**). The structure is again covalent and not built up from ions (139, 629).

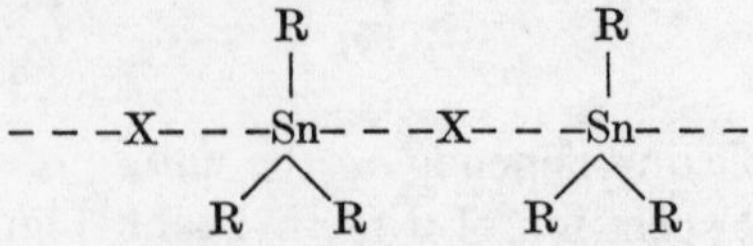

```
           R                    R
           |                    |
---X---Sn---X---Sn---
          / \                  / \
         R   R                R   R
```

(**7**)

X can be NO_3, ClO_4 (139, 628, 898), AsF_6, SbF_6, BF_4 (138, 287), $AlCl_4$, $AlBr_4$ (587), and even F (140). For the last case an X-ray structure determination has shown that the F–Sn···F bridges are slightly bent and the interatomic distances unequal. The R_3Sn groups (R = Me) are said to be slightly non-planar, as shown by the very weak symmetric Me_3Sn vibration at 515 $cm.^{-1}$.*

The electron donor need not of course form part of the molecular assembly, as it has in the examples given hitherto, but can be a separate molecular species. It has recently become clear that the well known molecular compounds formed by R_3SnHal with Lewis bases, e.g. $R_3SnCl \cdot NR_3$ (315, 339), also belong to this category. For $Me_3SnCl \cdot C_5H_5N$ a monomeric covalent structure (**8**) has been proved by IR and X-ray analysis (67, 68, 310). The pyridine ring, the tin, and the chlorine lie on one axis. The three coplanar methyl groups are spread out in a plane through the tin, normal to this axis. The Sn–Cl distance is 2.42 ± 0.04 Å, somewhat greater than in the free trialkyltin chlorides (see Table 2-1). Once again the R_3Sn– group is planar.

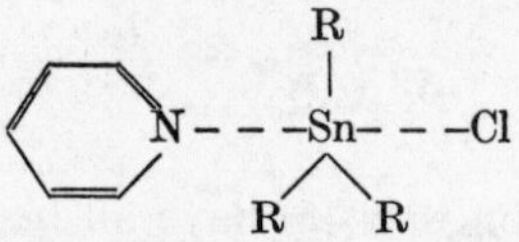

```
                 R
                 |
(C5H5)N---Sn---Cl
                / \
               R   R
```

(**8**)

Such compounds were formerly regarded as -onium salts $(R_3SnNR_3)^+$ Cl^- (410). It is now apparent that their higher conductance in solution, as compared with free trialkyltin chlorides, accords merely with an extreme, dissociated state of (**8**).

Analogous trigonal-bipyramidal structures have also been deduced for the molecular compounds formed by R_3SnCl or R_3PbCl with the Lewis

* In a more recent measurement this band was not detected. The Me_3Sn– group would then be planar, in agreement with the Raman spectrum (897).

bases tetramethylene sulfoxide, dimethylformamide, and dimethylacetamide (521).

Apart from the conditions just discussed, which lead to penta-coordinate tin, there is another arrangement known to give the same coordination. It is usually made up of two alkyl groups, two negative ligands, and one donor per penta-coordinate tin. The disubstituted stannoxanes $XR_2SnOSnR_2X$ (Chapter 14-2) are typical of this arrangement. In the solid state, and in not too dilute solution in benzene or cyclohexane, they are dimeric and have the structure (**9**). This was recognized simultaneously (151) by R. Okawara (627, 635) and by D. L. Alleston, A. G. Davies, et al. (17, 19). X-ray structure analysis shows that all four tin atoms lie in one plane, with one oxygen above and the other below. The exact position of the other components is not yet known. X can be OCOR, OH, OR, halogen, $OSiR_3$, and even OOR or N=C=S (174, 871). The dimeric form (**9**) ($R = Me$, $X = OSiMe_3$) is also present in solution in CCl_4 (15 %), as shown by NMR measurements (147). Dissociation according to equation 2-7 occurs with increase in temperature; at 85° only the monomeric form remains. ΔH for equation 2-7 is 9 ± 1 kcal./mole.

$$\left[{>}Sn{-}O{<}^{Sn}_{Sn}{>}O{-}Sn{<}\right] \rightleftharpoons 2\ {>}Sn{-}O{-}Sn{<} \qquad (2\text{-}7)$$

Stannoxanes containing R_3Sn– groups cannot undergo this association, and are therefore monomeric (Chapter 14-1). Apparently the tin atom is first rendered sufficiently electrophilic by the substituent X before it can form a stable donor–acceptor bond. It is notable that the bridge oxygens

R_2SnX | O | XR_2Sn — SnR_2X (2,3 Å; 2,2 Å) | O | R_2SnX

(**9**)

never donate electrons back to the tin atom of their own molecule, but always to that of a second molecule. One is reminded of an analogous situation in the dialkylaluminum-dialkylamides $(R_2AlNR'_2)_2$ (547, 552) and -alkoxides (299, 547).

Trimethyltin hydroxide also belongs to this category. In solution it is dimeric with a four-membered ring similar to that in (**9**); in the solid state it is however polymeric with planar Me_3Sn– groups. In either case it is always associated through Sn–O bridges and not through hydrogen-bonding (422, 638). Similar penta-coordination with Sn–O bridges has been suggested for dialkylchlorostannyl carboxylates, but this time by intramolecular bridging (872). Compounds of type (**9**) had been known for quite some time (276, 486, 829, 830), but their structure was interpreted incorrectly or not at all. Perhaps there was some reluctance to accept penta-coordination which is rare in coordination chemistry. Unambiguous proof by modern analytical techniques, as illustrated above, is only a recent development. This also accounts for the fact that penta-coordinate tin appeared quite early in formulae, e.g. with Harada (276), but that no conclusions were drawn in the text. It is quite likely that re-investigation of a number of long-known organotin complexes (315) will demonstrate penta-coordination.

2-3 COMPOUNDS CONTAINING HEXA-COORDINATE TIN

Hexa-coordination of tin, which has been found in compounds of type R_2SnX_2 and $RSnX_3$ (and of course also SnX_4) (67, 204, 669), is in every way compatible with current ideas in complex chemistry. It occurs when the substituents X are sufficiently electronegative. With decreasing electronegativity of X the tin becomes a worse acceptor with respect to its nucleophilic partner in the complex. Compound (**10**) thus forms no further stable adducts (667).

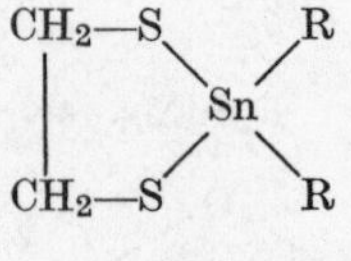

```
CH2—S       R
 |    \    /
 |     Sn
 |    /    \
CH2—S       R
```

(**10**)

Numerous complexes with hexa-coordinate tin are known, e.g. $Et_2SnBr_2{\cdot}2C_5H_5N$, $Pr_2SnI_2{\cdot}2PhNEt_2$, $PhSnCl_3{\cdot}2C_5H_5N$, and $SnBr_4{\cdot}$ 2THF. Apart from nitrogen, one finds hydroxy-, carbonyl-, or ether-oxygen as ligand (summaries in refs. 315, 669, and 736). Detailed investigations are extant, e.g. for $Bu_2SnCl_2{\cdot}2,2'$-bipyridyl, the 8-hydroxyquinolinate (**11**) (843) (R = Ph or Bu), and $Me_2SnCl_2{\cdot}M$ or $MeSnCl_3{\cdot}M$ (M = 1,10-phenanthroline or 2,2'-bipyridyl) (13, 67). For the R_2Sn-compounds one must, in most cases, assume a linear structure for this group. In the case of the acetylacetonate (**12**) this has been proved by IR and Raman measurements (263, 347).

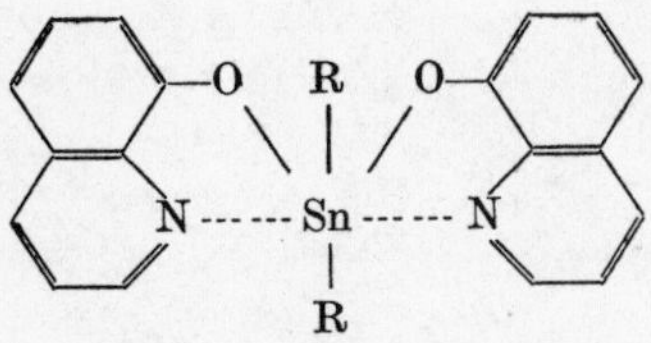

(**11**) 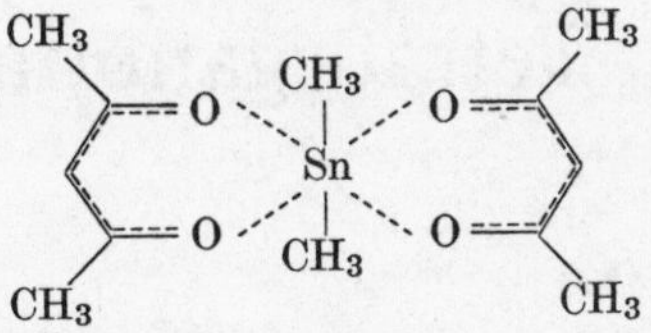 (**12**)

(The corresponding dialkyllead compounds are likewise octahedral.) Dimethyltin diacetate also has the stretched R–Sn–R arrangement,* proving hexa-coordination of the tin. The silicon in dimethylsilicon diacetate does not show this feature (636).

* This was earlier taken to indicate a salt-like character for this class of compound (636). This view has now been abandoned.

3 Tetraorganotins

Tetraalkyltins resemble the branched paraffins of similar molecular weight in appearance, boiling point, good solubility in organic solvents, and poor solubility in water and methanol (see Table 3-1).* When pure, they are colorless, almost odorless, monomolecular in solution, and generally stable towards water. Most of them are oxidized slowly by air in the presence of UV light. They decompose only above 200—250°; below 200° they can be distilled without decomposition, in certain cases with steam. The higher members of the series are waxy solids (tetra-*n*-dodecyltin melts at 15°) which dissolve easily only in certain liquids such as benzene, chloroform, and pyridine. Stereochemistry, dipole moments, and possible partial double-bond character of the C–Sn bond have been discussed in Chapter 2-1. Tetraethyltin (and almost certainly other aliphatic tetraorganotins also) shows an interesting type of isomerism in the solid state which should probably be ascribed to rotation isomerism of the alkyl groups. It crystallizes in at least ten modifications, all of which melt between -136 and $-125°$ (315).

The situation is similar with tetraethyllead, which occurs in at least six crystalline modifications, and with tetraethylgermanium, which is known in two solid forms; only one form of tetraethylsilicon has been found (143, 315). The corresponding tetramethyl derivatives are known in only one solid form each, as one would expect.

Tetraaryltins are colorless, stable against air and water, nicely crystalline, and melt above 100°, many even above 200° (Table 3-1). They are less soluble than the aliphatic derivatives, and dissolve readily only in hot benzene and chloroform. The compounds with substituted aromatic moieties often have lower melting points than those with such groups

* Comprehensive tables and literature review up to 1959 in ref. 315.

Table 3-1
Some Important Tetraalkyltins and Tetraaryltins

COMPOUND	B.P. (°C/mm.)	M.P. (°C)	n_D^{20}	METHOD OF PREPARATION
Me_4Sn	76—77	−54.2, −55.5	1.4415°	189, 832
Et_4Sn	180.5—181	−136 to −125[a]	1.4717	357a, 549, 662
n-Pr_4Sn	222—225, 116/13	−109.3	1.4748	720
i-Pr_4Sn	103/10	—	1.4851	355, 677
n-Bu_4Sn	145/11	—	1.4739	338, 549
i-Bu_4Sn	267, 130/10	—	1.4742	549
s-Bu_4Sn	148—150/10	—	1.4977	591
$(t\text{-}C_5H_{11})_4Sn$	157/19	—	—	511
$(C_6H_{13})_4Sn$	187—190/1.5	—	1.4706	358
$(cyclo\text{-}C_6H_{11})_4Sn$	—	263—264	—	417
$(PhCH_2)_4Sn$	—	42—43	—	806
$(C_8H_{17})_4Sn$	268/10, 250—255/5	—	1.4745	549
$(C_{12}H_{25})_4Sn$	—	15—16	1.4736	523
$(C_{16}H_{33})_4Sn$	—	41.5—42.5	—	523
Ph_4Sn	—	224—225, 229	— —	131, 662, 891
$(p\text{-}MeC_6H_4)_4Sn$	—	233—235, 238	—	50, 661
$(\alpha\text{-Naphthyl})_4Sn$	—	310—320[b]	—	55
$(9\text{-Phenanthryl})_4Sn$	—	360—370[b]	—	55
$(2\text{-Biphenylyl})_4Sn$	—	300—301	—	54

[a] Several modifications; see ref. 315. [b] Decomposition.

unsubstituted; they also usually show better solubility. Steric hindrance by bulky aromatic groups apparently does not occur with tin. (The covalent radius of tin is 1.40 Å; see Chapter 2-1.) Certainly, tetra-2-biphenylyltin (54) and even tetra-9-phenanthryltin (55) can be prepared successfully.*

Tetraphenyltin crystals are isomorphous with those of the analogous silicon and lead compounds (416, 922). Their melting point is hardly depressed by a number of other metal aryls (315).

The reactions of the tetraalkyltins and tetraaryltins are almost entirely those of the C–Sn bond, so they are discussed in Chapter 4 along with other properties of this linkage.

* Only via the aryllithium. The Grignard reagents do not produce complete arylation of the tin, even in large excess.

3-1 PREPARATION OF SYMMETRICAL TETRAORGANOTINS

Compounds of the type R_4Sn are important starting materials for almost all other organotin derivatives. They represent by far the commonest route from inorganic tin chemistry into organotin chemistry, and will therefore be discussed first. The difficulty of further alkylation increases with the number of organic moieties already attached to the tin. Depending on the inorganic starting material, different methods are used to arrive at the tetraalkyltin.

3-1-1 From tin alloys

When a tin–sodium alloy is allowed to react with an alkyl halide (a method used already by C. Löwig in 1852; see Chapter 1) the tetraalkyltin is formed by reaction 3-1 (X = halogen) (315) but is contaminated with the incompletely alkylated by-products R_3SnX and R_2SnX_2; their separation is difficult (see following section). Various minor additives to the tin alloy have been recommended, especially zinc, but also alloys with magnesium (353, 491) and lithium (474). The intermediate steps of the overall reaction

$$4\,RX + Sn + 4\,Na \rightarrow R_4Sn + 4\,NaX \qquad (3\text{-}1)$$

are still largely in doubt, although tentative suggestions have been made (315).

3-1-2 From inorganic tin compounds

Alkylation of tin halides, especially $SnCl_4$, can often be recommended. The appropriate Grignard reagent (315, 906) may be used as in reaction 3-2, and this is sometimes prepared in situ (683, 906).* Tetrahydrofuran has often proved a useful solvent (682).

$$4\,RMgX + SnCl_4 \rightarrow R_4Sn + 4\,MgXCl \qquad (3\text{-}2)$$

In spite of the difficulties and the need for handling large amounts of solvent, the Grignard synthesis is commonly used in the laboratory and even on the technical scale. There it is carried out in batteries of glass tanks, each of 2000 litre or even 10,000 litre capacity (see also Chapter 25).

Alkylation with alkyllithium is also a useful laboratory method (315):

$$4\,RLi + SnCl_4 \rightarrow R_4Sn + 4\,LiCl \qquad (3\text{-}3)$$

* Useful preparative details will be found in the references in Table 3-1. Hydrocarbon–ether mixtures have often been found advantageous.

The use of sodium in admixture with the alkyl halide (289, 352) should also be mentioned here. In place of $SnCl_4$, partially alkylated tin halides can of course be alkylated further, e.g. dibutyltin dichloride (289).

In all these processes (reactions 3-1—3-3) the pure tetraalkyltin is obtained only when the alkylating agent is present in considerable excess. With a slight excess, organotin halides are still present as impurities. Several methods have been proposed for their removal, e.g. precipitation of R_4Sn by methanol (718), precipitation of the complex $R_3SnCl \cdot 2NH_3$ by passing ammonia gas through the dry ethereal solution, and lastly transformation of the chloride contaminants into the corresponding poorly soluble fluorides by means of alkali fluoride. The halides can also be converted into the less volatile or non-volatile oxides by means of alkali (315) (see also Chapter 17). A nuisance in the laboratory-scale synthesis by reaction 3-2 is the complex $SnCl_4 \cdot 2Et_2O$. This is crystalline, poorly soluble in ether, and can block stopcocks and condensers. Suitable laboratory instructions can be found in the references in Table 3-1.

New possibilities arose when alkylaluminum compounds were made accessible through the work of K. Ziegler and his collaborators (925). Reaction 3-4 is, however, not quantitative (191, 330, 902, 904), just as in the Grignard synthesis. With stoichiometric quantities the reaction stops mainly at the trialkyltin chloride stage owing to the formation of stable

$$4\,R_3Al + 3\,SnCl_4 \longrightarrow 3\,R_4Sn + 4\,AlCl_3 \qquad (3\text{-}4)$$

complexes $R_2SnCl_2 \cdot AlCl_3$ and $R_3SnCl \cdot AlCl_3$ from the intermediate alkylation stages (see Chapter 2-2). If these are destroyed by introduction of a better complexing agent for the $AlCl_3$, alkylation proceeds smoothly and without an excess of R_3Al right to the tetraalkyltin as in reactions 3-5 or 3-6. As complexing agents one adds ether or tertiary amines in stoichiometric amount (549, 551, 578), e.g.

$$4\,R_3Al + 3\,SnCl_4 + 4\,R'_2O \longrightarrow 3\,R_4Sn + 4\,R'_2O \cdot AlCl_3 \qquad (3\text{-}5)$$

The etherate often separates as an insoluble, heavy phase which simplifies work-up. Dry, finely powdered alkali halides may also be used as the complexing agent (322, 324, 578), e.g.

$$4\,R_3Al + 3\,SnCl_4 + 4\,NaCl \longrightarrow 3\,R_4Sn + 4\,NaAlCl_4 \qquad (3\text{-}6)$$

In none of these cases is a solvent required. The advantages over Grignard-type alkylations are therefore obvious, and both reactions 3-5 and 3-6 have repeatedly proved useful as laboratory-scale methods.

The patent literature gives a number of variants on this alkylation technique, e.g. the use of sodium fluoride as complexing agent (538), and the use of sodium tetraalkylaluminates (323) or alkoxy-alkylaluminums

(320). The latter require no further complexing agents. Reaction 3-5 is not confined to the aliphatic series; phenylaluminum compounds react in the same way (891). However, aromatic aluminum compounds are not yet as easily accessible as aliphatic ones. Discussion of starting materials has so far been confined to the tin halides, but alkoxides $R_nSn(OR')_{4-n}$ (510), stannoxanes $R_3SnOSnR_3$ (788) or $(R_2SnO)_n$, and other tin compounds with negative ligands have also been used for special purposes. These can be alkylated successfully by Grignard reagents (reaction 3-7) or by alkylaluminums.

$$4\,RMgBr + Sn(OR')_4 \rightarrow R_4Sn + 4\,Mg(OR')Br \quad (R = Me, Et, n\text{-}Pr, i\text{-}Pr) \qquad (3\text{-}7)$$

3-1-3 Electrochemical methods

As a matter of interest it is sometimes possible to alkylate tin metal directly by electrochemical means.

Electrolysis in a suitable electrolyte, e.g. $NaAlEt_4$, transforms a tin anode into tetraethyltin (927). Electrolysis with a tin cathode in aqueous alkaline acrylonitrile or methacrylonitrile yields the tetraalkyltins $(NC–CHR–CH_2)_4Sn$, where R = H or Me (340). The same compound (R = H) and the corresponding ditin are formed in the electrolysis of β-iodopropionitrile in an acid or neutral aqueous medium with a tin cathode (854). Tetramethyltin is produced from methyl chloride and $SnCl_4$ by electrolysis in molten salts (831).

3-2 PREPARATION OF UNSYMMETRICAL TETRAORGANOTINS

Compounds having dissimilar organic residues attached to tin are perfectly stable, at least at room temperature and in the absence of catalysts (see Chapter 4). One usually starts with partially alkylated compounds as in reactions 3-8 and 3-9.

$$\begin{aligned} R_3SnCl + R'MgCl &\rightarrow R_3SnR' + MgCl_2 \\ R_2SnCl_2 + 2\,R'MgCl &\rightarrow R_2SnR'_2 + 2\,MgCl_2 \qquad (3\text{-}8) \\ RSnCl_3 + 3\,R'MgCl &\rightarrow RSnR'_3 + 3\,MgCl_2 \end{aligned}$$

The residues R and R′ can be aromatic, aliphatic, or alicyclic. As alkylating agents, alkyllithium and more recently alkylaluminums have in several instances proved superior to Grignard compounds, e.g. see reaction 3-9 (671).

$$(C_4H_9)_2SnCl_2 + R'\text{—}Al\begin{matrix} CH_2\text{—}CH_2 \\ CH_2\text{—}CH_2 \end{matrix}C\begin{matrix} CH_3 \\ CH_3 \end{matrix} + OR''_2 \longrightarrow (C_4H_9)_2Sn\begin{matrix} CH_2\text{—}CH_2 \\ CH_2\text{—}CH_2 \end{matrix}C\begin{matrix} CH_3 \\ CH_3 \end{matrix} + R'AlCl_2 \cdot OR''_2 \quad (3\text{-}9)$$

A large number of alkyltins of this type were prepared quite some time ago (315, 416), mainly for the purpose of studying the difference in reactivity of the various C–Sn bonds (see Chapter 4). For the same reason there is still much preparative activity in this field even today (187, 713, 741, 788), but the objects now include bond theory (Chapter 2), spectroscopy (Chapter 23), and other systematic comparative studies.

Stepwise attack by different Grignard reagents on a tin halide R_nSnX_{4-n} ($n = 0, 1, 2$) leads to mixtures and is therefore of little practical importance. Compounds such as $R_2SnR'R''$ are prepared from a suitable tetraalkyltin by removal of one residue (see Chapter 4) followed by realkylation. Competitive attack by different alkyl halides, in a few cases, yields 40—60% of the expected alkyltin $R_2SnR'_2$.

$$2\,EtI + 2\,MeI + 4\,Mg + SnCl_4 \rightarrow Et_2SnMe_2 + 4\,MgClI \quad (3\text{-}10)$$

Similarly for R_3SnR' and $RSnR'_3$.

Comproportionation of different tetraalkyltins, e.g. as in reaction 3-11, does not occur.

$$R_4Sn + R'_4Sn \rightarrow 2\,R_2SnR'_2 \quad (3\text{-}11)$$

Without a catalyst there is no reaction at all. With a Lewis acid such as $AlCl_3$ or AlR_3 as catalyst, one observes the redistribution reactions first discovered by Calingaert (123, 124), i.e. a statistical distribution of all available alkyl residues amongst all the tin atoms. Thus, equimolar amounts of tetraethyltin and tetramethyltin with 2.5 mole % $AlCl_3$ at 60° do not react according to equation 3-11 but yield a mixture of (in mole %) Et_4Sn 8.4, Et_3SnMe 28.6, Et_2SnMe_2 38.4, and $EtSnMe_3$ + $SnMe_4$ 24.6, i.e. fairly much the distribution predicted on statistical grounds. For preparative purposes the method must be of very limited use.

3-3 TETRAORGANOTINS WITH VINYL, ALLYL, AND ALKYNYL GROUPS

3-3-1 Vinyltin compounds

Numerous aliphatic and aromatic vinyltin compounds, including tetravinyltin, can be prepared successfully by Grignard synthesis (199,

292, 712, 772) as in reactions 3-12.

$$R_{4-n}SnCl_n + n\,BrMgCH{=}CH_2 \rightarrow R_{4-n}Sn(CH{=}CH_2)_n + n\,MgBrCl$$

$$SnCl_4 + 4\,ClMgCH{=}CH_2 \rightarrow (H_2C{=}CH)_4Sn + 4\,MgCl_2$$

$$R_{4-n}SnCl_n + n\,BrCF{=}CF_2 + n\,Mg \rightarrow R_{4-n}Sn(CF{=}CF_2)_n + n\,MgBrCl \quad \text{(ref. 769)}$$

$$Me_3SnCl + BrMgCH{=}CHMe \rightarrow Me_3SnCH{=}CHMe + MgBrCl \quad \text{(ref. 776)}$$

(3-12)

The Grignard reagents are prepared in tetrahydrofuran (622, 772, 778). Apart from the tin halides mentioned, stannoxanes and even dibutyltin oxide (712) have proved suitable. Alkenyllithium can also be used for vinylation, as in reaction 3-13 (776):

$$R_3SnCl + LiCH{=}CH_2 \rightarrow R_3SnCH{=}CH_2 + LiCl \qquad (3\text{-}13)$$

but in the aromatic series there are frequent complications from transmetalation (778). Some other preparative methods are of special importance, e.g. the addition of organotin hydrides to alkynes (Chapter 9).

Amongst the reactions of vinyltin compounds the cleavage of the C=C–Sn link is the most important, and this finds preparative application in the vinylation of other elements (see Chapter 4-6). On the other hand, vinyltin halides may be reduced to hydrides, alkylated further with Grignard reagents, hydrolyzed to stannoxanes or oxides, or transformed into stannyl esters, without the cleavage occurring as more than a side-reaction. Often it can be avoided entirely. A few addition reactions of the vinyl group are known, and these seem to be mostly of the radical type, e.g.

$$Et_3SnCH{=}CH_2 + HCCl_3 \xrightarrow[\text{peroxide}]{\text{Benzoyl}} Et_3SnCH_2CH_2CCl_3 \qquad (3\text{-}14)$$

Apart from $HCCl_3$, the reaction is known for $BrCCl_3$, CCl_4, $HSiCl_3$, and $HSnR_3$ (see Chapter 9).

Nothing certain is known about polar addition reactions. Reference should be made to recent detailed reviews of vinyltin compounds (343, 763).

3-3-2 Allyltin compounds

Allyltin compounds, including tetraallyltin (862) and methallyltins (758), can be prepared as for the vinyl compounds (241, 777), e.g. as in reaction 3-15. Diallyl-dialkyltins are also obtainable by alkylation of diallyltin dibromide (749) (for this compound see Chapter 5-1-1).

$$Ph_2SnCl_2 + BrMgCH_2CH{=}CH_2 \rightarrow Ph_2Sn(CH_2CH{=}CH_2)_2 + 2\,MgBrCl$$

$$Bu_3SnCl + LiCH_2CH{=}CH_2 \rightarrow Bu_3SnCH_2CH{=}CH_2 + LiCl \quad (3\text{-}15)$$

$$SnCl_4 + 4\,BrMgCH_2CH{=}CH_2 \rightarrow (H_2C{=}CHCH_2)_4Sn + 4\,MgBrCl$$

Many polar and radical reagents effect replacement of the allyl group, since the allyl–tin link is unusually reactive (see Table 2-4 and Chapter 4). Addition takes place with a number of unsaturated systems, e.g. with benzaldehyde (562) according to reaction 3-16 (see Chapter 4-7).

$$PhCHO + H_2C{=}CHCH_2SnEt_3 \rightarrow \underset{\displaystyle OSnEt_3}{PhCHCH_2CH{=}CH_2} \xrightarrow{H_2O} \underset{\displaystyle OH}{PhCHCH_2CH{=}CH_2} \quad (3\text{-}16)$$

Some substituted allyltins, like crotyl (766c) and 3-phenylallyl compounds (707a), are now known.

Radical addition to the C=C bond is not known with certainty. Other radical reactions, such as polymerizations, are actually inhibited by allyltin compounds. Apparently, the weak Sn–allyl bond (see Table 2-4) acts as radical scavenger. Attempts at hydrogenation have so far been unsuccessful. On the other hand, there is evidence of polar addition reactions, e.g. of organotin hydrides with organoaluminum catalysts (Chapter 9), and also of polymerizations by polar mechanisms (Chapter 21).

3-3-3 Alkynyltins

Alkynyltins $R'C{\equiv}CSnR_3$ or $R_3SnC{\equiv}CSnR_3$ are accessible by several routes. The ready hydrolytic cleavage of the C≡C–Sn link (Chapter 4) is reversible (561, 792, 793), enabling the synthesis to be carried out under mild conditions, as in reaction 3-17. The water formed must of course be removed continuously from the reaction mixture, e.g. by azeotropic distillation with benzene, but very good yields are then obtained (561).

$$2\,p\text{-}MeOC_6H_4C{\equiv}CH + Et_3SnOSnEt_3 \rightarrow 2\,p\text{-}MeOC_6H_4C{\equiv}CSnEt_3 + H_2O \quad (3\text{-}17)$$

Alkynols react first at their hydroxy group, and this is often the only reaction (561):

$$2\,HC{\equiv}CCH_2OH + Et_3SnOSnEt_3 \rightarrow 2\,HC{\equiv}CCH_2OSnEt_3 + H_2O \qquad (3\text{-}18)$$

Only then can an excess of stannoxane condense with the acetylenic hydrogen (792).

Condensation of free acetylenes with stannylamines proceeds very smoothly (333):

$$RC{\equiv}CH + Me_2NSnMe_3 \rightarrow RC{\equiv}CSnMe_3 + HNMe_2 \qquad (3\text{-}19)$$

Quite a large number of organotin acetylides, including some with functional groups, has thus become accessible in recent years.

The general methods of alkylation of organotin chemistry are of course also applicable to the acetylides, and some examples are shown by reactions 3-20. The choice of functional groups is naturally very restricted.

$$Bu_3SnCl + LiC{\equiv}CPh \rightarrow Bu_3SnC{\equiv}CPh + LiCl \qquad \text{(ref. 286)}$$

$$Ph_3SnCl + BrMgC{\equiv}CCH{=}CH_2 \rightarrow Ph_3SnC{\equiv}CCH{=}CH_2 + MgBrCl \qquad \text{(ref. 680)}$$

$$(3\text{-}20)$$

$$Ph_3SnNa + BrC{\equiv}CC(R){=}CHR' \xrightarrow[NH_3]{liq.} Et_3SnC{\equiv}CC(R){=}CHR' + NaBr \qquad \text{(ref. 908)}$$

$$2\,Ph_3SnCl + NaC{\equiv}CC{\equiv}CNa \xrightarrow[NH_3]{liq.} Ph_3SnC{\equiv}CC{\equiv}CSnPh_3 + 2\,NaCl \qquad \text{(refs. 284, 908)}$$

Preparation of dialkynyl- (285), trialkynyl-, and even tetraalkynyl-tin derivatives is therefore possible, e.g. by reaction 3-21 (283).

$$4\,p\text{-}XC_6H_4C{\equiv}CLi + SnCl_4 \rightarrow (p\text{-}XC_6H_4C{\equiv}C)_4Sn + 4\,LiCl \qquad (3\text{-}21)$$

$$(X = \text{cyclohexyl, Cl, Br})$$

Some special syntheses present interesting problems with respect to reaction path and mechanism. Amongst these might be mentioned (a) decarboxylation of the stannyl esters of acetylenedicarboxylic acids (494) (reaction 3-22), and (b) addition of sodium acetylides to poly-diethyltin (654) (reaction 3-23).

$$Bu_3SnOCOC{\equiv}CCOOSnBu_3 \rightarrow 2\,CO_2 + Bu_3SnC{\equiv}CSnBu_3 \qquad (3\text{-}22)$$

$$n\,RC{\equiv}CNa + (Et_2Sn)_n \rightarrow n\,RC{\equiv}CSn(Et_2)Na \xrightarrow{n\,EtBr} n\,RC{\equiv}CSnEt_3 + n\,NaBr \qquad (3\text{-}23)$$

$$[R = Me,\ i\text{-}Pr,\ CH{=}CH_2,\ C(Me){=}CH_2]$$

Two of the most important reactions of the alkynyltins are ready cleavage by iodine [a method of quantitative determination (560)] and protolytic cleavage (see Chapter 4). The cause of this high reactivity could lie in the appreciable partial double-bond character of the C–Sn link in the alkynyltins, as shown by valence-bond calculations (387a) and IR studies (560). The high reactivity of Sn–alkynyl bonds also permits preparation of other Group IV and Group V organometallic derivatives of acetylene, phenylacetylene and 1,3-butadiyne. The corresponding alkynyltins are used as starting materials (283a).

Quite different reactions can take place in aprotic solvents, such as the Diels–Alder reaction 3-24 (771).

$$\text{2,3,4,5-}R_4\text{-cyclopentadienone} + R'_3Sn{-}C{\equiv}C{-}SnR'_3 \longrightarrow \text{1,2-}(SnR'_3)_2\text{-3,4,5,6-}R_4C_6 + CO \qquad (3\text{-}24)$$

$$(R = Ph,\ R' = Me)$$

Hexachlorocyclopentadiene adds to alkynyltins in a similar way (766b).

Addition of diazomethane takes place without breaking the C≡C–Sn group (reaction 3-25), forming a pyrazoline derivative which can rearrange to a substituted cyclopropane with elimination of nitrogen (910).

$$R_3Sn{-}C{\equiv}C{-}CH{=}CH_2 \xrightarrow{CH_2N_2} R_3Sn{-}C{\equiv}C{-}\underset{\text{1-pyrazoline ring: }CH{-}CH_2{-}N{=}N}{CH{-}CH_2} \xrightarrow{\text{(spontaneously)}} R_3Sn{-}C{\equiv}C{-}\underset{\text{2-pyrazoline ring: }C{=}N{-}NH{-}CH_2}{C{-}CH_2} \xrightarrow{120\text{—}140^\circ} R_3Sn{-}C{\equiv}C{-}\underset{\text{cyclopropyl}}{CH{-}CH_2{-}CH_2} + N_2 \qquad (3\text{-}25)$$

Table 3-2
Some Reactions of Tetraorganotins Containing Functional Groups

Reaction	REF.
$Ph_3Sn\text{-}C_6H_2(t\text{-}Bu)_2\text{-}OH \xrightarrow[-\frac{1}{2}H_2O]{+\frac{1}{4}O_2} Ph_3Sn\text{-}C_6H_2(t\text{-}Bu)_2\text{-}O\cdot$	824
$p\text{-}Ph_3SnC_6H_4CH_2OH \xrightarrow{KMnO_4} p\text{-}Ph_3SnC_6H_4COOH$	238
$Ph_3Sn\text{-}C_5H_5 + O_2 \xrightarrow{h\nu,\ sens.} Ph_3Sn\text{-}C_5H_5O_2 \xrightarrow{NaBH_4} Ph_3Sn\text{-}C_5H_7(OH)_2$	731
$Ph_3SnCH_2CH{=}CMe_2 + O_2 \xrightarrow{h\nu,\ sens.} Ph_3SnCH_2CH(OOH)C(Me){=}CH_2 + Ph_3SnCH{=}CHC(Me_2)OOH$	731
$(Me_3SnCH_2)_3B \xrightarrow[-H_2O,\ -NaBO_2]{+2H_2O_2,\ +NaOH} 3\ Me_3SnCH_2OH$	762
$R_3Sn(CH_2)_nCN \xrightarrow{LiAlH_4} R_3Sn(CH_2)_nCH_2NH_2 \quad (n > 1)$	364, 580[a]
$R_3Sn(CH_2)_nCOOMe \xrightarrow{LiAlH_4} R_3Sn(CH_2)_nCH_2OH \quad (n > 1)$	364[a]
$R_3Sn(CH_2)_nCOOR' \xrightarrow{KOH} R_3Sn(CH_2)_nCOO^-K^+ \quad (n > 1)$	706[a]
$R_3Sn(CH_2)_nCN + MeMgI \xrightarrow{Hydrol.} R_3Sn(CH_2)_nCOMe \quad (n > 1)$	364
$p\text{-}Ph_3SnC_6H_4NMe_2 + ArN_2{}^+Cl^- \xrightarrow[-HCl]{} Ph_3Sn\text{-}C_6H_3(NMe_2)(N{=}NAr)$	238, 249
$p\text{-}Ph_3SnC_6H_4NMe_2 + MeI \rightarrow p\text{-}Ph_3SnC_6H_4NMe_3{}^+I^-$	255

Reaction	Ref.
$Sn(CH_2Br)_4 + 4NaI \rightarrow Sn(CH_2I)_4 + 4NaBr$	305
$Me_3SnPh + Cr(CO)_6 \rightarrow Me_3Sn(C_6H_5)Cr(CO)_3 + 3\,CO$	765
$Et_3SnC{\equiv}CCH{=}CH_2 + Ph_3CCPh_3 \rightarrow Et_3SnC{\equiv}CCH(CPh_3)CH_2CPh_3$	909
$Et_3SnC{\equiv}CC(Me){=}CH_2 + Ph_3CCPh_3 \rightarrow Et_3SnC(CPh_3){=}C{=}C(Me)CH_2CPh_3$	909
$R_3SnCH_2C{\equiv}CH \overset{Bases}{\rightleftharpoons} R_3SnCH{=}C{=}CH_2 \overset{Bases}{\rightleftharpoons} R_3SnC{\equiv}CMe$	676
$Sn\left(p\text{-}C_6H_4\langle\overset{O}{\underset{O}{}}\rangle\right)_4 \xrightarrow{H^+} Sn(C_6H_4CHO\text{-}p)_4 \xrightarrow{Ph_3PCH_2} Sn(C_6H_4CH{=}CH_2\text{-}p)_4$	180
$Et_3SnC{\equiv}CCH_2OH + H_2C{=}CHOBu \rightarrow Et_3SnC{\equiv}CCH_2CH(Me)OBu$	794
$SnCl_4 \xrightarrow[+CH_2N_2\ +\ Cu]{Ether,\ -20°} Sn(CH_2Cl)_4$ (71 %)	402b

[a] Cleavage of the C–Sn bond occurs if $n = 1$ (see Chapter 4-3).

3-4 TETRAORGANOTINS WITH FUNCTIONAL GROUPS

From the earliest days it has been one of the aims of organotin chemists to provide one or more of the organic groups in tetraorganotins with a functional group, so as to extend their organic chemical reactions. Substitution by the normal methods of organic chemistry (e.g. halogenation or nitration) in alkyl or aryl residues already attached to tin is not usually possible, since cleavage of the C–Sn bonds occurs (see Chapter 4). The functional groups must therefore normally be in position in the alkyl or aryl residue before this is attached to the tin. For many years this imposed severe limitations. Nevertheless, success was achieved in a few cases by Grignard reactions (315, 416), e.g. in linking the tin to vinyl and allyl groups (see Chapter 3-3), propargyl and allenyl (679), *p*-vinylphenyl (613, 655), and pentafluorophenyl (302) groups. Cyanoalkyl moieties could also be attached electrolytically, and a few other functionally substituted alkyl groups by special techniques (Chapter 3-1).

Synthetic routes of more general applicability have been discovered only recently. It has now been found that Sn–H, Sn–N, and Sn–O groups in organotin compounds are sufficiently reactive to add to numerous unsaturated systems, e.g. C=C, C≡C, C=N, C=O, N=C=O, N=C=S, or N=C=N, if necessary with the aid of catalysts. Introduction of stannyl residues into complicated organic molecules is now generally possible, and the number of known organotin compounds with functional groups is considerable and increasing rapidly. These compounds really belong in this chapter, but are discussed more conveniently in the context of their method of preparation (Chapters 9 and 19).

As far as changes of the functional groups themselves and syntheses with the aid of these groups are concerned, an upper limit for possible reaction conditions is set by the reactivity of the C–Sn bond (Chapter 4). Nevertheless, a large number of such reactions is known. Table 3-2 lists a selection, and further examples are discussed in Chapter 3-3. Only the tetraorganotins are included, since these are the most sensitive compounds. With substituted derivatives of type R_3SnX and R_2SnX_2 the number of possible reactions is much greater.

4 Reactions of the carbon–tin bond

For several decades there has been perhaps more experimental work in this sector of organotin chemistry than in any other. Yet our knowledge of the various relationships and underlying causes is still largely incomplete. A review of this field is therefore at present an unrewarding task.

Hundreds of alkyltins, especially those with several different ligands, have been prepared in order to study the fission of particular C–Sn bonds. Amongst the many older investigations one might mention here merely those carried out by the groups of Grüttner and E. Krause (416), of C. A. Kraus, and of H. Gilman. Reference should also be made to several detailed reviews (315, 416, 491).

Attempts were made for a long time to relate the chemistry of the C–Sn bond to a few simple and constant factors, such as the electronegativity of the tin (and thus the polarity of the C–Sn bond) or steric factors. Consequently, results had restricted validity and their interpretation was often unsatisfactory. Luijten and van der Kerk's comment on the situation (1955) (491, p. 76) is somewhat resigned but to the point: "This ... must primarily be considered as of practical importance." "Care must be taken in interpreting the experimental results."

Today we take a more flexible view of the above factors (see Chapter 2). The C–Sn bond is rather long (2.17 Å) and not particularly polar, but somewhat sensitive to the influence of substituents, other reaction products, and solvents. The bond is stable below 200° and quite resistant to oxidation (see Chapter 4-5). However, it is easily polarized, and may certainly become polarized in either direction. The number of possible polar and radical reactions is therefore large. The empty 5*d*- and (possibly) 4*f*-orbitals of the covalently bonded tin offer numerous possibilities to attacking reagents, including formation of transition states with penta-

or hexa-coordinate tin atoms. Concrete ideas on reaction mechanisms began to take shape rather late. There was also delay in applying the knowledge of modern theoretical organic chemistry to these reactions, and reluctance to regard them from the viewpoint of the carbon, i.e. as substitutions at a saturated or unsaturated alkyl group or aromatic system in which a stannyl residue is replaced by another substituent [see the fundamental exposition by C. K. Ingold (315a)].

This change of viewpoint, accompanied by appropriate kinetic studies, has proved fruitful, so that today we have at least a clear, theoretically well founded picture of some parts of the problem. (The above quotation: "Care must be taken ...," applies as much as ever.) To give but a few names, the groups of C. Eaborn, H. G. Kuivila, J. Nasielski, G. Razuvaev, and D. Seyferth have distinguished themselves especially in this field. Some of the most important points will be discussed below. For the remainder the reader must be referred to the reviews quoted already and to some annual literature surveys (768).

4-1 HYDROLYSIS: ATTACK BY ACIDS AND BASES

Water and aliphatic alcohols generally have little effect on the symmetrical, saturated compounds. (Anyway, these are practically insoluble in water.) On the other hand, cleavage occurs already with phenols and mercaptans, better with thiophenols, and easily with carboxylic acids (463, 514, 720). The rate depends on the dissociation of the acid. A second alkyl group is often attacked before the first has been completely removed. The main reaction is usually similar to the following:

$$Et_4Sn + 2\,MeCOOH \rightarrow Et_2Sn(OCOMe)_2 + 2\,C_2H_6$$

Mono- and tri-esters can be isolated as by-products (514). Different groups attached to the same tin atom are generally split off in the same sequence as in cleavage by halogens (see Chapter 4-2).

The action of NH_2^- (and likewise, presumably, other strong bases) towards Sn–R bonds depends to a remarkable extent on the nature of R. In liquid ammonia, compounds containing methyl–tin bonds react slowly at 0°C, those with ethyl–tin bonds only at 100°C, and cyclohexyl compounds not at all. On the other hand, phenyl–, vinyl–, or benzyl–tin bonds are cleaved easily by NH_2^-. Nucleophilic attack at the tin atom can be assumed (see Chapter 2-2). Apparently, polarizability of the corresponding C–Sn bond and stability of the expelled carbanion are important here (261b).

The C–Sn bond in unsaturated organotins is generally more reactive, especially if these are unsymmetrical; e.g. all four C–Sn bonds in tetravinyltin are cleaved ultimately by carboxylic acids (290) [review up to 1961 by D. Seyferth (763)]. Also, pentafluorophenyl-alkyltins (alkyl = Me or Ph) while stable against EtOH–H_2O produce pentafluorobenzene when catalyzed by as much as a trace of halide or cyanide ions (130). 2-Stannylpyridine is cleaved smoothly by water and alcohols, whereas the 3- and 4-stannyl derivatives are stable (25):

$$(\alpha\text{-Pyridyl})SnMe_3 + ROH[H_2O] \rightarrow (\alpha\text{-Pyridyl})H + ROSnMe_3[HOSnMe_3]$$

The mechanisms and the factors controlling them are generally more complicated than might appear at first sight; e.g. when compounds of formula (**1**) are hydrolyzed in dilute aqueous-alcoholic perchloric acid, the rates are as (M =) Si 1 : Ge 36 : Sn 3.5×10^5 : Pb 2×10^8. These values cannot be explained by any set of electronegativities, but might be interpreted in terms of decreasing strength of the C–M bond and increasing availability of empty *d*-orbitals (186). The importance of the intermediate state is then more apparent. The size of the leaving group (alkyl or aryl) must also be relevant, as in the action of aqueous-methanolic alkali:

MeO–C_6H_4–MEt_3

(**1**)

the C–Sn bond becomes more labile in the order benzyl < diphenylmethyl < triphenylmethyl < fluorenyl (89). In the cleavage of aryl–tin bonds in a series of *m*-$XC_6H_4SnMe_3$ compounds, the rate-determining step is probably the separation of the carbanion *m*-$XC_6H_4^-$. The effect of different *meta*-substituents X shows good correlation with the respective Hammett σ-constants ($\rho = 2.18$) (186a). Methanolic HCl is 10^4 times less active than iodine (see Chapter 4-2) against a series of trialkylmonoaryltins, yet the electronic effects must be broadly the same. Cleavage must therefore be regarded as an electrophilic aromatic substitution (108).

Compounds with an allyl or similar group attached to tin, e.g. (**2**)**a**—**c**, are cleaved particularly easily: (**2a**) and (**2c**) by nucleophilic reagents, (**2b**) instantaneously even by water (706). The nature of the action of methanolic HCl on (**2a**) has been analyzed in considerable detail, as has

$R'CH{=}CHCH_2SnR_3$ (**2a**) $R'OCOCH_2SnR_3$ (**2b**) $N{\equiv}CCH_2SnR_3$ (**2c**)

the influence of steric factors. The reaction is an S_N2' protolysis in which a proton and a molecule of methanol participate in transition states (**3a**) or (**3b**) (440).

$$\left[\begin{array}{l} R_3Sn\text{---}CH_2{=}C{=}CHR \\ \quad\quad\quad\quad\quad | \quad\;\; | \\ \quad\quad\quad\quad\;\; R' \;\; H\text{---}O\langle^{CH_3}_{H} \end{array}\right]^+ \qquad \left[\begin{array}{c} \quad\quad\quad R' \\ CH_2\text{—}C \\ R_3Sn \quad\quad CHR \\ O\text{---}H \\ CH_3 \;\; H \end{array}\right]^+$$

(**3a**) (**3b**)

The base-catalyzed cleavage of allyl–tin bonds has also been investigated with 3-phenylallyl-triethyltin (707a). In aqueous-ethanolic alkali, both solvent isotope and primary salt effects are consistent with a rate-determining S_N2 attack on the tin with expulsion of a carbanion.

C≡C–Sn groups show variable sensitivity towards water which depends on the acidity of the respective free alkyne. For example, propargyl cyanide, which shows considerable C–H acidity, condenses exothermically with stannoxanes with separation of water, but phenylacetylene reaches a 50% equilibrium within 24 hours when stoichiometric amounts of the reactants are brought together at 60° (560):

$$2\,PhC{\equiv}CSnEt_3 + H_2O \overset{THF}{\rightleftharpoons} 2\,PhC{\equiv}CH + Et_3SnOSnEt_3$$

In this case, and in many analogous reactions, the equilibrium can be displaced by removal of the water to give a quantitative yield of the alkynyltin (see Chapter 3-3-3).

4-2 CLEAVAGE BY HALOGENS

This must be the most frequently studied reaction of the C–Sn bond. One or more organic moieties are split off, depending on reaction conditions (see Chapter 5-1-4), e.g.

$$R_4Sn + Br_2 \rightarrow R_3SnBr + RBr$$

$$R_4Sn + 2\,Br_2 \rightarrow R_2SnBr_2 + 2\,RBr$$

$$R_3SnCl + Cl_2 \rightarrow R_2SnCl_2 + RCl$$

To illustrate the effect of reaction conditions: for the last reaction, with R = Et, a solution of the reactants in methylene chloride is stable for several hours at −70°, but reacts slowly at −40° and rapidly at +20°. Of particular interest is the dependence of cleavage rate on the nature of the organic moiety, when the other reaction conditions are kept constant

(315, 416, 710). The rate of cleavage of R–Sn increases in the sequence R = Bu < Pr < Et < Me < CH=CH_2 < Ph < CH_2Ph* ≈ CH_2CH=CH_2 ≈ CH_2CN ≈ CH_2COOR.†

In competitive reactions a group further to the right in the sequence is generally replaced before one further to the left, but there are overlaps and exceptions. There are no systematic quantitative investigations covering the entire range of these groups. *iso*-Alkyls appear to react at the same rate as the corresponding *n*-alkyls, but a group bonded to tin through a secondary carbon atom is detached more easily than one bonded through a primary carbon.

Chlorine, bromine, and iodine mostly give analogous results, but show decreasing reactivity. For example, whereas Br_2 displaces phenyl residues from Ph_3SnBr in toluene slowly even at −70°, the same compound in the same solvent remains unaffected by I_2 at +20°. This is important in the analysis of polytins by stepwise degradation (Chapter 22-3).

It has often been supposed that the position of a group in the above sequence is determined entirely by its $+I$ or $-I$ inductive effect. The sequence would then represent an "electronegativity series" of organic groups, analogous to that proposed by M. S. Kharasch (370) for organomercury compounds. But there are discrepancies at too many points, and the idea appears too simple to account for all the factors. Furthermore, the reaction mechanism is unlikely to be the same in all cases.

Kinetics and the influence of various factors have recently been studied in greater detail. Cleavage of a number of tetraorganotins R_4Sn by iodine is always first-order in either reactant, but is strongly dependent on the solvent. It has been concluded from this that the reaction is of the same type as electrophilic substitution at a saturated C atom, but that the transition state differs from case to case (237). The relationship between Hammett σ-constants and reaction rates suggests that the displacement of substituted aromatic residues from tin by iodine is an electrophilic aromatic substitution (107, 491). The alternative interpretation as nucleophilic attack by iodide ions on the tin can be excluded here since an excess of iodide ions does not accelerate the reaction. The relationships to the σ-constants do not hold for very bulky aromatic residues. Steric factors appear predominant here, and reaction is impeded (107).

Nucleophilic attack by iodide ion on tin has, however, been suggested for the case of the strongly electron-withdrawing group $-CF_3$. In the presence of an olefin, the trifluoromethyl anion (assumed as intermediate)

* An earlier observation, that phenyl is displaced by bromine more rapidly than benzyl (416), could not be confirmed in several examples, and certainly not with iodine.

† The last four groups mentioned, including their substituted derivatives, all react very rapidly; relative rates are still unknown.

forms *gem*-difluorocyclopropane derivatives and a fluoride anion, as found by D. Seyferth et al. (766a). A difluorocarbene mechanism is suggested.

$$Me_3SnCF_3 + NaI + \text{cyclohexene} \xrightarrow{\text{Reflux}} \text{7,7-difluoronorcarane} \; (89\%) + Me_3SnI + NaF$$

Besides cyclohexene, twenty-one other cyclic and acyclic (terminal and internal) olefins have been used successfully in this reaction.

4-3 CLEAVAGE BY HYDROGENATION

In symmetrical tetraorganotins, e.g. Et_4Sn or Ph_4Sn, the C–Sn bond can be cleaved by hydrogen under somewhat critical conditions (surveyed in ref. 315), e.g. as in

$$Ph_4Sn + 2\,H_2 \xrightarrow[200°]{60\text{ atm.}} 4\,PhH + Sn$$

This type of cleavage takes place more readily with unsymmetrical organotins. For groups having a multiple bond at the β-carbon the C–Sn bond is cleaved even by $LiAlH_4$ (462, 491), but this does not appear to have been tried with the allyl group:

$$N{\equiv}CCH_2SnR_3 \xrightarrow{LiAlH_4} N{\equiv}CMe + HSnR_3 \qquad (R = Bu, Ph)$$

$$EtOOCCH_2SnR_3 \xrightarrow{LiAlH_4} HOCH_2Me + HSnR_3 \qquad (R = Bu)$$

If the multiple bond is more remote, i.e. at least one further CH_2 group is interposed, the substances are stable against cleavage by $LiAlH_4$ and the functional group is hydrogenated instead (364) (see Table 3-2). Alkynyl groups are cleaved by $LiAlH_4$ (908):

$$H_2C{=}CHC{\equiv}CSnEt_3 \xrightarrow{LiAlH_4} H_2C{=}CHC{\equiv}CH + HSnEt_3$$

A similar reaction occurs with $MeC{\equiv}CSnMe_3$, but in the analogous silicon compound the alkynyl group is hydrogenated to the alkenyl without cleavage occurring (764).

4-4 CLEAVAGE BY FREE RADICALS

For many years polar reactions were the only ones known for the C–Sn group; now there are numerous examples of radical reactions. The

C–Sn bond is in fact a fairly good radical trap, depending on the nature of the organic group, but not nearly as good as the Sn–H bond (see Table 2-4 and Chapter 8-4-1). In this connection should be mentioned the sensitivity of C–Sn bonds in some organotins towards air (oxygen), especially in the presence of UV light. When the C–Sn bonds are strained as in stannacyclohexanes (see Chapter 20-1), this sensitivity is enhanced to the extent of quite rapid autoxidation.

Cleavage of C–Sn bonds depends not only on the organic group attached to the tin atom, but sometimes also on the nature of the attacking radical. (In general, detailed knowledge of radical mechanisms is still fragmentary.)

The photoxidation of triethyl-benzyltin has been investigated more closely (-20 to $+30$°C) and the reaction mechanism elucidated. Besides other products, triethyl- and diethyl-tin benzoates, stannoxanes, and benzaldehyde are formed (7a).

The action of decomposing benzoyl peroxide on trimethyl-phenyltin results mainly in cleavage of the phenyl–tin bond; the trimethylstannyl ester of benzoic acid is formed (besides other products) (692). Some other organotins are cleaved not only by acyloxy but also by alkoxy radicals (865). For example, triorganotin hydrides and benzoyl peroxide yield the benzoate (**4**) and mostly, depending on the mole ratio, also the dibenzoate (**5**) (585a). This is especially so when R = Ph.

$$\underset{(\mathbf{4})}{PhCOOSnR_3} \qquad\qquad \underset{(\mathbf{5})}{(PhCOO)_2SnR_2}$$

Action of di-*t*-butyl peroxide on hexaethylditin results in cleavage of Sn–Sn and also of C–Sn bonds; the ethyl radicals form mainly ethane and ethylene, and a little butane (870). A similar decomposition of hexaethylditin can be produced purely thermally at 260°; the products are Et_4Sn, elemental tin, and approx. 2 ethyl radicals per molecule, which again appear as an ethane–ethylene–butane mixture (690). *t*-Butoxy radicals also act on hexabutylditin to produce mainly tin–tin scission as well as some cleavage of organic groups from tin (585a). Cleavage of the C–Sn bond can occur in principle also by way of a stoichiometric radical-chain reaction. Tetraethyltin reacts with propyl bromide at 35° under UV irradiation or at 85° with a little benzoyl peroxide to give triethyltin bromide (689).

The action of ozone on C–Sn bonds should be mentioned here. Tetraethyltin, on treatment with ozone in *n*-nonane, yields acetaldehyde, diethyltin oxide, and an organotin peroxide of unknown structure (8a, 8b).

4-5 CLEAVAGE BY STRONGLY POLAR ALKYL HALIDES AND SOME OTHER COMPOUNDS

Trifluoroiodomethane and similar compounds effect cleavage of tetraorganotins, but the products decompose further under these reaction conditions (342):

$$Me_4Sn + ICF_3 \xrightarrow{-MeI} Me_3SnCF_3 \longrightarrow Me_3SnF + 1/n(CF_2)_n$$

They can be obtained under milder conditions from ditins (142, 324); the Sn–Sn bond again reacts more easily than the C–Sn bond.

Cleavage by alkyl halides (e.g. isopropyl chloride or bromide) is also known with catalysis by $AlCl_3$ (691). In this case, however, transmetalation may play a part as it does in the so-called "redistribution reactions" (see Chapter 4-6).

In hexamethylphosphoramide as solvent, allyl bromide and benzyl bromide will effect cleavage, if the stannyl residue is attached to an activated carbon, α to e.g. carbonyl or cyanide (649b).

$$\left.\begin{array}{l} \text{RCH(SnBu}_3\text{)COMe} \\ \text{RCH(SnBu}_3\text{)COOR} \\ \text{RCH(SnBu}_3\text{)CN} \end{array}\right\} + R'Br \xrightarrow{120^\circ} \left\{\begin{array}{l} \text{RCH(R')COMe} \\ \text{RCH(R')COOR} \\ \text{RCH(R')COOR} \end{array}\right\} + Bu_3SnBr$$

$$(R' = CH_2CH{=}CH_2,\ CH_2Ph)$$

Nothing definite is known about the reaction of the C–Sn bond with oxygen* (see above, Chapter 4-4), and little more about its reaction with other oxidizing agents.† Reactions with sulfur and selenium (Chapter 17) and tellurium (Chapter 4-6) are, however, well defined. Dinitrogen tetroxide converts Me_4Sn into dimethyltin dinitrate (Chapter 7-2), but no mass balance has been established (3).

4-6 ALKYLATION BY C–Sn BONDS (TRANSMETALATION)

A mixture of tetraethyltin and tetrabutyltin can be heated to 150° for several hours without any trace of mixed organotins appearing on a gas

* Certainly, many organotins in air slowly deposit white precipitates which contain Sn–O links.

† Hot concentrated nitric acid oxidizes up to SnO_2 (see Chapter 22-1); oxidation to intermediate stages seems to be unknown.

chromatogram. Exchange of alkyl groups, such as occurs e.g. with aluminum alkyls because of their electron deficiency and consequent association, cannot therefore take place. This also explains the stability of mixed organotins R_3SnR' or $R_2SnR'_2$. The picture can be changed entirely by introduction of catalysts: e.g. tetramethyltin and tetraethyltin with 2.5 mole % $AlCl_3$ in pentane at 50° exchange all their alkyl groups within 5 hours to attain a statistical equilibrium between Me_4Sn, Me_3SnEt, Me_2SnEt_2, $MeSnEt_3$, and $SnEt_4$. The reaction was discovered by G. Calingaert and his co-workers and called a "redistribution reaction" (124).*

The catalytic process obviously involves transmetalation which in some cases is stoichiometric, e.g.

$$Et_4Sn + AlCl_3 \rightarrow Et_3SnCl + EtAlCl_2$$

Even triethyltin chloride can act as a weak alkylating agent when heated (587):

$$Et_3SnCl + AlCl_3 \rightarrow Et_2SnCl_2 + EtAlCl_2$$

Exchange of groups also takes place between tetraphenyltin and triphenylaluminum, as shown by labeling with carbon-14 (540). Boron halides (X = Cl, Br) are alkylated to the mono-alkyl stage:

$$R_4Sn + 2\,BX_3 \rightarrow R_2SnX_2 + 2\,RBX_2 \qquad (R = \text{e.g. Bu}, C_8H_{17})$$

Usually some by-products are formed ($SnCl_2$, alkanes, olefins) (234). The earliest known alkylations by C–Sn bonds appear to be those of thallium (Goddard, 1922) (259, 416):

$$Et_4Sn + TlCl_3 \rightarrow Et_2SnCl_2 + Et_2TlCl$$

Aryl derivatives, and more recently also vinyl and substituted vinyl derivatives, are obtained in the same way. The organotin compound often reacts rapidly in ether or chloroform even at room temperature, and sometimes transfers all four organic groups to the thallium (86).

It is interesting that the configuration at the C=C link is retained, so that e.g. the following system can be realized (544):

* This concept should be distinguished clearly from "comproportionation" (Chapter 5-1-3) which produces a unique compound and cannot therefore proceed statistically. Qualitative and quantitative studies of redistribution equilibria, and the principles underlying the mathematical treatment thereof, have been summarized by Moedritzer (1968) (534d).

CH3 H H CH3
C=C C=C
H Sn H
Br Br

SnBr2 HgBr2 TlCl3 SnBr2

CH3 H H CH3
C=C C=C
H Hg H

TlCl3 →

CH3 H H CH3
C=C C=C
H Tl H
Cl

The particular mechanism of these transmetalations is not known (see ref. 764). It would be especially important to decide whether the primary attack occurs at the carbon or the tin* of the affected C–Sn link. An S_E reaction at the carbon (315a) would be a possibility.

Reactions of this type have recently aroused much interest. They have been called redistribution reactions, ligand exchange, transmetalation, or syn- or com-proportionation reactions (see Chapter 5-1-3). Experimental facts and theoretical interpretations were brought together in a symposium organized by the New York Academy of Sciences in June 1967, when the present author reported on organotin compounds (554a).

Transmetalations have proved particularly useful in recent years for the preparation of relatively inaccessible organolithium compounds from more easily accessible ones [D. Seyferth et al. (781)].†

$$(H_2C{=}CH)_4Sn + 4\,PhLi \rightarrow 4\,H_2C{=}CHLi + Ph_4Sn \qquad \text{(ref. 780)}$$

$$(MeCH{=}CH)_4Sn + 4\,BuLi \rightarrow 4\,MeCH{=}CHLi + Bu_4Sn$$

Both *cis*- and *trans*–alkenyl groups retain their configuration, as proved by NMR spectroscopy. These reactions reach an equilibrium which is much influenced by the geometric configuration of the alkenyl group. The

* Complex derivatives of tetraorganotins have not been either isolated or detected by any method. Their existence is sometimes assumed (see Chapters 2-2 and 4-1) to explain a reaction mechanism (e.g. in refs. 261b and 706), and could also provide an explanation of primary attack in transmetalations.

† For a review of vinyl derivatives see ref. 763; for propenyls see ref. 764.

position of the double bond is also of importance: whereas vinyltins and allyltins react readily, the corresponding 3-butenyl and 4-pentenyl derivatives do not react with butyllithium in ether (780). Methallyllithium is formed from phenyllithium and methallyltriphenyltin (779). Again, benzyllithium (253) is now easily accessible (773), especially since the tin compound can be made directly from benzyl chloride and elemental tin.

$$(PhCH_2)_3SnCl + 4\,MeLi \xrightarrow{\text{Ether}} 3\,PhCH_2Li + Me_4Sn + LiCl$$

Both *cis*- and *trans*-crotyl-trimethyltin give, on consecutive treatment with butyllithium and Me_3SiCl, a 2:3 mixture of *cis*- and *trans*-crotyl-trimethylsilanes (766c).

Displacement of tin from vinyl or propenyl groups can also be effected by metallic lithium:

$$(H_2C{=}CH)_4Sn + 4\,Li \xrightarrow{\text{Ether}} 4\,H_2C{=}CHLi + Sn$$

In this case reaction is, however, accompanied by isomerization of the propenyl derivative to equilibrium. There are good grounds for assuming that this proceeds via a short-lived radical-anion produced by the lithium, e.g. (774)

$$\underset{\text{H}\qquad\text{H}}{\overset{(CH_3)_3Sn\qquad CH_3}{C{=}C}} \underset{}{\overset{\text{Li}}{\rightleftharpoons}} \underset{\text{H}\qquad\text{H}}{\overset{\overset{\text{Li}^{\oplus}}{(CH_3)_3Sn\qquad CH_3}}{\overset{\ominus}{C}{-}\overset{\bullet}{C}}} \underset{}{\overset{\text{Li}}{\rightleftharpoons}} \underset{\text{H}\qquad CH_3}{\overset{(CH_3)_3Sn\qquad\text{H}}{C{=}C}}$$

Alkynyl residues can be transferred from tin to lithium in the same way (908), as well as to other Group IV and Group V elements (283a).

The reactions of triethyl-phenyltin and diethyl-diphenyltin with silver nitrate are also to be viewed as transmetalations (109, 362, 416). Phenyl-silver can be isolated as a complex with $AgNO_3$, and the reaction is presumably an electrophilic substitution of the aromatic nucleus (Sn by Ag).

Diphenyltellurium can be prepared easily from tetraphenyltin and tellurium at 240° (745).

A number of salts, e.g. $SnCl_2$, $SnBr_2$, $CuBr_2$, $HgCl_2$, $TiCl_4$, $VOCl_3$, and $PdCl_2$, convert tetraethyltin into the respective compound Et_3SnHal (29).

4-7 GRIGNARD-LIKE REACTIONS

Some recently discovered Grignard-like additions of allyltriethyltin to aldehydes provide clear indication of the strong polarizability of the

C–Sn link (562). The reaction rate depends on the polar character of the C=O group: caprylaldehyde and benzaldehyde require 2—4 hours at 150°, *p*-nitro- and *p*-chloro-benzaldehyde react at the same rate at 100°, but chloral shows an exothermic reaction analogous to 4-1 at as low a temperature as 20°. Cinnamaldehyde affords 1-phenyl-1,5-hexadien-3-ol in better yield than via the Grignard reagent:

$$PhCH{=}CHCHO + Et_3SnCH_2CH{=}CH_2 \rightarrow$$

$$\underset{\displaystyle OSnEt_3}{PhCH{=}CHCHCH_2CH{=}CH_2} \xrightarrow[-RCOOSnEt_3]{RCOOH}$$

$$PhCH{=}CHCH(OH)CH_2CH{=}CH_2 \qquad (4\text{-}1)$$

Work-up by hydrolysis is unsuccessful, as the liberated carbinol re-condenses with the stannoxane when fractional distillation is attempted. Benzoic acid is effective, but malonic acid is still better because its stannyl ester precipitates quantitatively from petroleum ether and can be removed simply by filtration.

The polar nature of the C–Sn link in this addition reaction also follows from the effectiveness of dry zinc chloride as a catalyst (although by-products are then formed). It allows temperatures to be lowered by approx. 50° for the same reaction rate.

The C≡C–Sn group can add to chloral, too, but only at much higher temperatures (534a):

$$Cl_3CCHO + PhC{\equiv}CSnEt_3 \xrightarrow[\text{pressure tube}]{150\text{–}200°}$$

$$\underset{\displaystyle \underset{70\,\%}{OSnEt_3}}{PhC{\equiv}CCHCCl_3} \xrightarrow{HCl/Et_2O} PhC{\equiv}CCH(OH)CCl_3 + Et_3SnCl$$

In a special case, phenyl–tin bonds have been added to sulfur dioxide to give a well crystallized phenylsulfinic acid derivative (190a):

$$[\pi\text{-}C_5H_5Fe(CO)_2]_2SnPh_2 + 2\,SO_2 \rightarrow [C_5H_5Fe(CO)_2]Sn(OSOPh)_2$$

5 Organotin halides

All the organic groups of tetraorganotins can be replaced in stages by halogen. One arrives thus at three series R_3SnX, R_2SnX_2, and $RSnX_3$ (where R can be aliphatic or aromatic, and X = F, Cl, Br, or I) (see Table 5-1). This route has occasional uses in the preparation of organotin halides (see Chapter 5-14).

Organotin halides, with the exception of some fluorides (see below), dissolve readily in the usual organic solvents. Conductivity in non-polar media is very slight (see Chapter 2-1) and even in dimethylformamide (e.g.) triphenyltin chloride conducts forty times worse than triphenylmethyl chloride. The tin–halogen bond is covalent but polarized according to $Sn^{\delta+}$–$X^{\delta-}$, and this is also evident from dipole measurements (see Chapter 2-1). The extent of polarization can be greatly influenced by the nature of the medium and the other reactants, so that both radical and polar reactions are known. Dissociation is possible as a limiting case, e.g.

$$R_3Sn^{\delta+}\text{–}Cl^{\delta-} \rightleftharpoons R_3Sn^+ + Cl^-$$

This would then explain why organotin halides with small alkyl groups are distinctly soluble in water and aqueous acids, and why hydrolysis is rapid and reversible (equations 5-1). Hydrolysis is particularly marked in the trihalides, which are stable only in concentrated aqueous acid.

$$\begin{aligned} 2\,R_3SnX + H_2O &\rightleftharpoons R_3SnOSnR_3 + 2\,HX \\ n\,R_2SnX_2 + n\,H_2O &\rightleftharpoons (R_2SnO)_n + 2n\,HX \qquad (5\text{-}1) \\ n\,RSnX_3 + 1.5n\,H_2O &\rightleftharpoons (RSnO_{1.5})_n + 3n\,HX \end{aligned}$$

Hydrolysis of the di- and tri-halides naturally proceeds through intermediate stages (equations 5-1). The intermediate and final products are

Table 5-1
Organotin Chlorides

FORMULA	B.P. (°C/mm.)	M.P. (°C)	n_D^{20}	METHOD OF PREPARATION
$MeSnCl_3$	—	45—46 42—43	—	669, analogous to 264, 556
Me_2SnCl_2	185—190	107—108	—	804, analogous to 264, 404
Me_3SnCl	152—154	41—42	—	analogous to 264, 404
$EtSnCl_3$	196—198 38/1	−10	1.5408°	549, 556
Et_2SnCl_2	277 102—107/12	84—85	—	404, 512, 545, 549, 879
Et_3SnCl	210 97/12	15.5	1.5078	404, 549
$PrSnCl_3$	98—99/12	—	—	analogous to 556
Pr_2SnCl_2	118—121/10	82.5—83	—	404
Pr_3SnCl	123/13	−23.5	1.496	analogous to 404
$BuSnCl_3$	93/10 44—45/0.1	—	1.523	367, 527, 556
Bu_2SnCl_2	143/12 91—94/0.1	43	1.499 (50°)	327, 337
Bu_3SnCl	152/10 98/0.45	—	1.4909	337, 357
i-$BuSnCl_3$	34—35/0.25	—	1.521	527, 556
i-Bu_2SnCl_2	129—130/11 69—71/0.25	9	1.5090	analogous to 404, 907
i-Bu_3SnCl	142/13 78—80/0.25	30.2	1.4836 (30°)	556, unobtainable analogous to 404
t-Bu_2SnCl_2	117/14	42	—	677

t-Bu_3SnCl	132/12	4	—	677
$C_8H_{17}SnCl_3$	97—100/0.005	—	1.507	analogous to 556
$(C_8H_{17})_2SnCl_2$	164—165/0.15	47.5—48.5	—	analogous to 404
$(C_8H_{17})_3SnCl$	163—166/0.008	—	1.4835	analogous to 404
(cyclo-$C_6H_{11})_2SnCl_2$	—	88—89	—	417
(cyclo-$C_6H_{11})_3SnCl$	—	129—130	—	417
$(PhCH_2)_2SnCl_2$	—	163—164	—	800
$(PhCH_2)_3SnCl$	—	143—145	—	800
$CH_2{=}CHSnCl_3$	64—65/15	—	1.5361 (25°)	711, 772
$(CH_2{=}CH)_2SnCl_2$	54—56/3 46/1	74.5—75.5	—	711, 772
$(CH_2{=}CH)_3SnCl$	90—96/26 59—60/6	—	1.5237 (25°)	711, 772
$PhSnCl_3$	109—110/10 96—97/1.4	−31	1.5844	243, 403, 891
Ph_2SnCl_2	333—337	42	—	243, 403, 542
Ph_3SnCl	240/13.5	105—107	—	53, 405, 545

discussed in Chapter 17. Reactions according to equations 5-1, with acid or alkali added as appropriate for completion, are of importance for the preparation and purification of these materials.

Aliphatic organotin halides with X = Cl, Br, or I are colorless, monomeric in solution, mostly liquid, and less often crystalline (see Table 5-1). The lower members have a strong and unpleasant odor and are more or less toxic (see Chapter 24); their vapors irritate the mucous membranes and cause headaches. This is particularly true for the trialkyltin chlorides with R = Me, Et, and Pr. Work with them should be carried out carefully and tidily in an efficient fume hood, preferably in an enclosed apparatus. This is desirable also for all other volatile organotin compounds.

Aromatic organotin halides, apart from some trihalides, are mostly crystalline at room temperature but are otherwise closely similar to the aliphatic ones. Minor differences arise from the easier scission of the aryl–tin bond.

Table 5-2
Ethyltin Halides

$EtSnF_3$	Et_2SnF_2	Et_3SnF
Unknown	M.P. 287—290° (partial decomp.)	M.P. 293—294.5°
	(414)	(414)
$EtSnCl_3$	Et_2SnCl_2	Et_3SnCl
M.P. −10°	M.P. 84—85°	M.P. 15.5°
B.P. 196—198°	B.P. 277°	B.P. 210°
38°/1 mm.	102—107°/12 mm.	97°/12 mm.
n_D^{20} 1.5408		n_D^{20} 1.5078
(549, 556)	(404, 549)	(404, 549)
$EtSnBr_3$	Et_2SnBr_2	Et_3SnBr
M.P. <20°[a]	M.P. 63—64°	M.P. −13.5 to −14°
B.P. 46°/0.1 mm.	B.P. 233°	B.P. 223—225°
	70—71°/0.1 mm.	96—97°/12 mm.
		n_D^{20} 1.5424
(556)	(404, 907)	(404, 549)
$EtSnI_3$	Et_2SnI_2	Et_3SnI
	M.P. 45.5—47°	M.P. −34.5°
B.P. 181—184.5°/19 mm.	B.P. 245—246°	B.P. 234°
115—117°/3 mm.	105—107°/3 mm.	120—122°/15 mm.
		n_D^{20} 1.568
(anal. to 669)	(657)	(879, anal. to 371)

[a] An earlier, much quoted, report (315) that $EtSnBr_3$ is solid and decomposes at 310° must be in error.

Organotin monofluorides have some special structural characteristics (see Chapter 2-2) and do not fit the remainder of the series as regards their poor solubility, great ease of crystallization, and high melting points (mostly above 200° with decomposition). The dialkyltin difluorides also have high melting points and poor solubility, as far as can be judged from the few available examples (15). They form soluble complexes with potassium fluoride. The poor solubility of organotin fluorides is often used for separation of organotin halides from tetraorganotins by means of potassium fluoride (see Chapter 3-1).

The great reactivity of halogen attached to tin makes the organotin halides highly versatile starting materials for almost all other organotin compounds. Therein lies their importance and the considerable interest in their technical preparation (see Chapter 25-1). Their chemical reactions embrace replacement of halogen by other negative, positive, or purely covalent groups. It would therefore be inappropriate to deal here with all these transformations; they are discussed instead under the preparation of the relevant reaction product.

5-1 PREPARATION OF ORGANOTIN HALIDES

5-1-1 Starting from tin metal

The best route to organotin halides would, in principle, be by a single-stage synthesis from elemental tin. As a matter of fact this was accomplished by E. Frankland as early as 1849 using ethyl iodide at 160—180° (reaction 5-2) (212) in the earliest known synthesis of an organotin com-

$$2\,RX + Sn \rightarrow R_2SnX_2 \qquad (X = Cl, Br, I) \qquad (5\text{-}2)$$

pound. Frankland also reported that the reaction went more smoothly and at no more than 20—50° when the mixture was exposed to sunlight by means of a parabolic mirror (214). This may well be one of the earliest examples of a photochemical reaction, and a very fine one at that. It is remarkable that the method pioneered by Frankland had to wait several decades for further investigation (195).

This synthesis has received further major attention only in recent years since the observation of the reaction, albeit at a slow rate, of methyl chloride with tin metal in the presence of copper (804). Improvements have included the use of solvating media such as butanol, esters, and especially glycol ethers, and the use of complex salts and finely divided metals as catalysts. Table 5-3 gives a summary of some of this work, which was initiated mainly by Japanese and Czech workers. Other examples may be found in the considerable patent literature. The method

Table 5-3
Preparation of Organotin Halides from Alkyl Halides and Tin Metal

ALKYL HALIDE	METAL	TEMP. (°C)	OTHER CONDITIONS	PRODUCT (IN SOME CASES AFTER HYDROLYSIS)	YIELD (%)	REF.
MeCl	Sn foil	—	pressure vessel/trace Mg, MeI, THF	Me_2SnCl_2	90—93	518
MeI[a]	Sn foil	—	pressure vessel/trace Mg, THF (or butanol)	mixture[b]	[b]	516[c]
EtBr	Sn powder	160	autoclave, vigorous stirring, 10 mole % Et_3N	Et_2SnBr_2 (mainly) + Et_3SnBr	69	798b
EtI	Sn foil	160—180	pressure vessel	Et_2SnI_2	'good'	212, 344[d]
EtI	Sn foil	20—50	sunlight	Et_2SnI_2	'v. good'	214
BuBr	Sn powder	160	autoclave, vigorous stirring, 10 mole % Et_3N	Bu_2SnBr_2	65	798b
BuI	Sn foil	130—135	trace Mg, THF (or butanol)	Bu_2SnI_2	'good'	519, 520, see also 515, 517
BuI	Sn foil	120—160	trace of metal, e.g. Li and alcohol	Bu_2SnI_2	60—93	625
s-BuI	Sn foil	130—135	trace Mg, THF (or butanol)	*s*-Bu_2SnI_2	'good'	519[e]
$C_5H_{11}I$	Sn + 5% Cu (alloy)	—	reflux in $O(CH_2CH_2OMe)_2$	$(C_5H_{11})_2SnO$	51	905[f]
$C_8H_{17}Br$	Sn + 5% Cu (alloy)	—	reflux in $O(CH_2CH_2OMe)_2$ with trace $C_8H_{17}I$	$(C_8H_{17})_2SnO$	47	905[g]
$CH_2{=}CHCH_2Br$	Sn powder	110	in toluene, trace $HgCl_2$ + pyridine	$(H_2C{=}CHCH_2)_2SnBr_2$	82	799[h]
p-$YC_6H_4CH_2X$[i]	Sn powder	5—150	vibratory ball milling	(*p*-$YC_6H_4CH_2)_2SnX_2$[i]	50—60	268

Y-$C_6H_4CH_2Cl$[j]	Sn powder	110	in toluene, trace H_2O	$(Y\text{-}C_6H_4CH_2)_2SnCl_2$[j]	80—90	169, 623, 800
Y-$C_6H_4CH_2Cl$[j]	Sn powder	100	in H_2O	$(Y\text{-}C_6H_4CH_2)_3SnCl$[j]	50—85	169, 800
$EtOCOCH_2I$	Sn foil	60—70	some iodine	$(EtOCOCH_2)_2SnI_2$	84	195[k]

[a] See also references 119 and 344.
[b] 25—30% Me_3SnI, 40—46% Me_2SnI_2, and 20—27% $MeSnI_3$.
[c] The same method with EtI yields a mixture of 14—18% Et_3SnI, 63—67% Et_2SnI_2, and 3—6% $EtSnI_3$.
[d] Similar methods in ref. 344 for Me_2SnI_2, Pr_2SnI_2, $i\text{-}Pr_2SnI_2$, $i\text{-}Bu_2SnI_2$, and $(i\text{-}C_5H_{11})_2SnI_2$.
[e] Analogous method for Pr_2SnI_2, $i\text{-}Pr_2SnI_2$, $i\text{-}Bu_2SnI_2$, $(cyclo\text{-}C_6H_{11})_2SnI_2$, $(C_8H_{17})_2SnI_2$, $(C_{12}H_{25})_2SnI_2$, and others.
[f] Analogous method gives Bu_2SnO in 63% yield.
[g] Analogous method gives Bu_2SnO in 51% and $(C_6H_{13})_2SnO$ in 47% yield.
[h] Analogous reaction for $MeCH{=}CHCH_2Br$, but some $[CH_2{=}CHCH(Me)]_2SnBr_2$ is formed as by-product in consequence of allyl rearrangement. As expected, $CH_2{=}C(Me)CH_2Br$ affords mainly $(MeCH{=}CHCH_2)_2SnBr_2$.
[i] Y = Cl, Br; X = Cl, Br, I.
[j] Y = e.g. *o*-, *m*-, and *p*-halogen, Me, OMe, and $CHMe_2$.
[k] Analogous reaction with the ethyl ester of *o*-iodobenzoic acid. Similar reactions are also described in ref. 244.

usually yields the dialkyltin dihalide, as in reaction 5-2. The halogen can be I, or more rarely Br or Cl. Mixtures are often formed, but in some cases the trialkyltin halide can be obtained in good yield by reaction 5-3.

$$3\,PhCH_2Cl + 2\,Sn \rightarrow (PhCH_2)_3SnCl + SnCl_2 \qquad (5\text{-}3)$$

Dibenzyltin dichloride is formed first and then converted into tribenzyltin chloride, especially in polar solvents. In water, bis(dibenzylchlorotin) oxide is an intermediate in the conversion (798a). Analogous alkylation of the primary products, dialkyltin dihalides, to trialkyltin halides and even to tetraalkyltins seems possible in several cases (797a) (Sisido et al.).

Early work by Emmert et al. (195) (see Table 5-3) on reactions of tin metal with halo-carboxylic esters has recently been extended to halo-succinates, -malonates, β-halo-isobutyrates, and β-halo-propionamides by Matsuda et al. (288b, 621a, 639a).

The mechanism of these syntheses is not known for certain, but a radical one is assumed for at least some of the examples (268, 800).

5-1-2 By partial alkylation of inorganic tin compounds

Alkylation of tin tetrahalides to a particular stage, e.g. as in reaction 5-4 or 5-5, is accompanied by side-reactions giving more or less alkylated derivatives. The resulting mixtures have to be separated, usually by addition of alkali or aqueous ammonia. The dihalide then forms the insoluble dialkyltin oxide according to reaction 5-1. In this way Bu_2SnCl_2 has been obtained by a special technique (785) in 80% yield, and i-Bu_2SnCl_2 in 70% yield isolated as i-Bu_2SnO (904).

$$2\,RMgX + SnCl_4 \rightarrow R_2SnCl_2 + 2\,MgXCl \qquad (5\text{-}4)$$

$$2\,R_3Al + 3\,SnCl_4 \rightarrow 3\,R_2SnCl_2 + 2\,AlCl_3 \qquad (5\text{-}5)$$

The method is more successful when the aluminum chloride in reaction 5-5 is complexed by ether or alkali halide and removed. Alkylation is then quantitative and the reaction product more uniform (324, 549). When trialkyltin halides are prepared by reaction 5-6, the crude mixture is best heated, after removal of the $AlCl_3$ complex, to 100—150°, when comproportionation (Chapter 5-1-3) raises the yield to over 90% (549).

$$R_3Al + SnCl_4 + R'_2O \rightarrow R_3SnCl + R'_2O{\cdot}AlCl_3 \qquad (5\text{-}6)$$

Partial alkylation of $SnCl_4$ is also possible with diazomethane. Mixtures of $ClCH_2SnCl_3$, $(ClCH_2)_2SnCl_2$, and $(ClCH_2)_3SnCl$ are obtained in benzene at 3—6°C. The last of these can be obtained in 95% yield in ether at −20°C with Cu as catalyst (402b).

For many years alkyltin trihalides were rather inaccessible materials. They have been prepared from alkali stannite and alkyl halide via the alkylstannonic acid (530) (reaction 5-7) and more recently also by a successful partial alkylation of $SnCl_4$ (549) (reaction 5-8).

$$RSn(O)OH + 3\,HCl \rightarrow RSnCl_3 + 2\,H_2O \qquad (5\text{-}7)$$

$$Et_2AlOEt + 2\,SnCl_4 \xrightarrow{R_2O} 2\,EtSnCl_3 + EtOAlCl_2 \qquad (5\text{-}8)$$

In most cases the organotin trihalides are prepared more conveniently by the following method.

5-1-3 By comproportionation

A rather surprising preparative method for many organotin mono-, di-, and tri-halides is the comproportionation of organotins with tin halides discovered by K. A. Kocheshkov (403—405). Heating of the starting materials in the required stoichiometric proportions affords the desired organotin halide as sole end-product (reactions 5-9—5-11).

$$3\,R_4Sn + SnCl_4 \rightarrow 4\,R_3SnCl \qquad (5\text{-}9)$$

$$R_4Sn + SnCl_4 \rightarrow 2\,R_2SnCl_2 \qquad (5\text{-}10)$$

$$R_4Sn + 3\,SnCl_4 \rightarrow 4\,RSnCl_3 \qquad (5\text{-}11)$$

As far as is known, the reactions are all exothermic in the gas phase ($\Delta H = -5.3$ to -9.6 kcal./mole of end-product) (802). The same is true for methyltin chlorides in the liquid phase ($\Delta H = -7$ to -9 kcal./mole) (264).

For the ethyltin compounds these overall reactions have been resolved into their various stages by means of gas chromatography (556). The first stage (reaction 5-12) is usually distinctly exothermic. Thorough cooling

$$R_4Sn + SnCl_4 \rightarrow R_3SnCl + RSnCl_3 \qquad (5\text{-}12)$$

stops the reaction here, when the products can be isolated (527, 556). Further reaction 5-13 is spontaneous for methyltin- (556) and cyclopropyltin-compounds (766), but heating is required in the case of other organic groups. The overall reaction 5-12 plus 5-13 is that for the preparation of R_2SnX_2 (reaction 5-10).

$$R_3SnCl + RSnCl_3 \rightarrow 2\,R_2SnCl_2 \qquad (5\text{-}13)$$

R_3SnCl acts as a good alkylating agent for $SnCl_4$ (reaction 5-14), and combination of this reaction with reaction 5-12 provides a simple route to the organotin trihalides by reaction 5-15. [Reactions of bromides and iodides are similar (556).] This is often an excellent method, since complete comproportionation by reaction 5-11 is successful only for vinyltins (711, 772) and a few aryltin compounds [e.g. for R = Ph (405, 930)]. The

$$R_3SnCl + SnCl_4 \rightarrow R_2SnCl_2 + RSnCl_3 \quad (5\text{-}14)$$

$$R_4Sn + 2\,SnCl_4 \rightarrow R_2SnCl_2 + 2\,RSnCl_3 \quad (5\text{-}15)$$

reason is that the essential step 5-16 fails in other cases (e.g. for R = Et or Bu). This step, and therefore reaction 5-11 also, can however be carried out successfully in polar media, especially in a mixture of $POCl_3$ and P_2O_5 (556).

$$R_2SnCl_2 + SnCl_4 \rightarrow 2\,RSnCl_3 \quad (5\text{-}16)$$

In this context, the behavior of methyltins is generally the same as that of the corresponding ethyltins, as shown by NMR studies (71a, 264). The rate of comproportionation to products Me_nSnX_{4-n} decreases in the sequence X = Cl > Br > I (71a).

For compounds of type R_3SnX and R_2SnX_2 the overall reactions 5-9 and 5-10 proceed smoothly when R = straight-chain aliphatic groups or phenyl groups (403—405, 556, 766). For R = branched chain groups, e.g. neopentyl (929), *s*-butyl, or benzyl (568), and for some aromatic groups, e.g. *p*-tolyl, *p*-ethoxyphenyl, or *p*-biphenylyl, these comproportionations often give rather poor results and sometimes fail altogether (565).

Using the different reactivities of individual organotins towards comproportionation, organotin dichlorides with two different alkyl or aryl groups can be prepared, starting from organotin trichlorides and tetraalkyltins or trialkyltin chlorides. Some of the chlorides $RR'SnCl_2$ have been converted into chlorides $RR'R''SnCl$ by alkylation with R''_4Sn (439b).

Comproportionations are not restricted to alkyl and halogen groups. Alkoxy groups exchange with a considerable number of other negative groups (175a):

$$R_2Sn(OMe)_2 + R_2SnX_2 \rightarrow 2\,R_2Sn(OMe)X$$

[X = F, Cl, Br, I, SCN, OCOR (R = Me, $C_{11}H_{23}$), camphorsulfonyl]

(For tin alkoxides and the structure of reaction products, see Chapter 17-4.)

It is remarkable that comproportionation reactions should yield a single end-product, and no satisfactory explanation has been advanced. One might have expected a statistical distribution of all the substituents among the available tin atoms such as occurs in the $AlCl_3$-catalyzed redistribution reaction (Chapter 4) or in a reaction mixture of R_3Al and $AlCl_3$: the latter immediately rearranges to a statistical mixture of $R_3Al + R_2AlCl + RAlCl_2 + AlCl_3$ (923). Instead, the system reaches a final stable state as shown in Figure 5-1 (556).

Because of this pronounced "antistatistical" character, the concept of comproportionation (sometimes called synproportionation) should be distinguished clearly from that of the randomized redistribution reactions.

For the three ethyltin chlorides it has been shown that reactions 5-9—5-11 are in fact equilibrium reactions, even if the equilibrium lies almost entirely on the right-hand side. Slight disproportionation (reaction 5-17) sets in at 170—200° and is most noticeable with the monochloride (556).

$$2\,Et_3SnCl \xrightarrow[2\text{ hr.}]{180°} Et_2SnCl_2 + Et_4Sn \qquad (\text{each} \sim 5\%) \qquad (5\text{-}17)$$

The following dissociation constants have been found for methyltin and ethyltin chlorides:

$$R = Me^* \qquad\qquad R = Et^\dagger$$

$$\frac{[R_4Sn][R_2SnCl_2]}{[R_3SnCl]^2} = K_1 = 3 \times 10^{-3}/175° \qquad \leqslant 2 \times 10^{-6}/70°$$

$$\frac{[R_3SnCl][RSnCl_3]}{[R_2SnCl_2]^2} = K_2 = 1 \times 10^{-4}/175° \qquad \leqslant 2 \times 10^{-7}/70°$$

$$\frac{[R_2SnCl_2][SnCl_4]}{[RSnCl_3]^2} = K_3 = 7 \times 10^{-2}/175° \qquad \leqslant 4 \times 10^{-6}/70°$$

$AlCl_3$ will, in principle, speed attainment of equilibrium, but it also produces some decomposition of the organotin halides especially the trihalides according to reaction 5-18. Without $AlCl_3$ this occurs only above 170°, and for the higher alkyl groups one isolates HCl + olefin, or a secondary product of the olefin, instead of the normal RCl (e.g. EtCl) (556).

$$RSnCl_3 \xrightarrow{AlCl_3,\ 100°} SnCl_2 + RCl \qquad (5\text{-}18)$$

$AlCl_3$ and other Lewis acids have, on the other hand, been used successfully in similar antistatistical comproportionations of organogermanium compounds (427, 707). This complicates matters even further since $AlCl_3$ is the typical catalyst for the statistical redistribution reactions, even in mixtures of different organolead and organosilicon compounds (125).

* Solvent-free system; comproportionation reactions; NMR measurements (264).
† 0.5M solution in benzene; pure compounds; GLC data (573a).

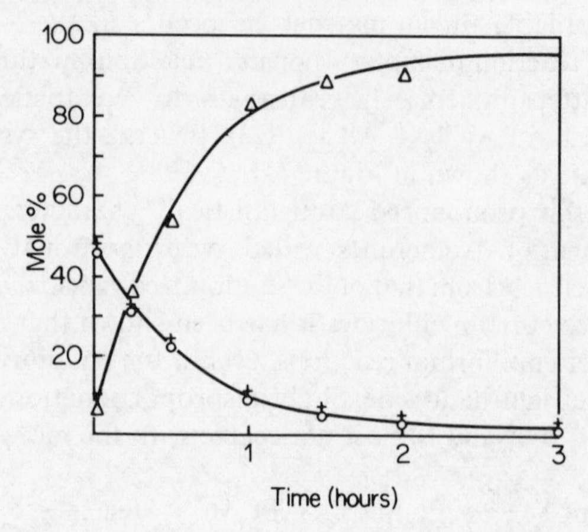

FIG. 5-1 Comproportionation of Et_4Sn and $SnCl_4$ according to reaction 5-10, at 200°C, O Et_3SnCl; △ Et_2SnCl_2; + $EtSnCl_3$

In the systems Bu_4Sn–$GeCl_4$ and Bu_4Ge–$SnCl_4$ ligand exchange is observed only above 200°. It is remarkable that Lewis acids have no catalytic effect here.

Theories concerning the mechanism of comproportionation are so far entirely conjectural. Polar solvents accelerate the reaction; at least some of the steps should therefore have a polar mechanism. One could be dealing with a series of electrophilic substitutions at the saturated carbon of the C–Sn link (see Chapter 4). The ability of the central tin atom to form adducts with donor compounds (Chapter 2) could also be playing a part, especially with some of the reaction products acting as donors. This has also been concluded from the complicated kinetics of comproportionation 5-12 for R = Me (264).

A single example may demonstrate that the mechanisms involved are more complicated than might be assumed. When equimolar amounts of Et_3SnCl and $SnCl_4$ are mixed in benzene at 50° and the reaction (5-14) is followed by GLC, a surprisingly low order of reaction is found, namely 1.27 in the first half of the reaction and 1.30 subsequently (573a). Perhaps a transition state is involved which depends only on the concentration of one reaction partner.

For many years little effort was concentrated on mechanisms of comproportionation. Only in the last few years have modern views, theories, and methods been brought to bear in this field. However, a quickly growing interest can now be observed. Van Wazer (875a) and Moedritzer (534c) have reviewed a considerable number of the known facts and have developed reasonable mathematical methods for understanding and calculating some of the known comproportionation (and statistical redistribution) reactions. (For a recent review by the present author, see ref. 554a.)

5-1-4 By cleavage of C–Sn or Sn–Sn bonds

Tetraorganotins are cleaved by halogens, with chlorine acting more vigorously than bromine, and bromine more vigorously than iodine (see Chapter 4-2). The halogens are best used in solution, e.g. in methylene chloride. The first two organic groups can be replaced by halogen in distinct stages as in reactions 5-19. Heating with an excess of halogen leads in the end to simultaneous replacement of the last two organic groups, so that compounds $RSnX_3$ cannot be obtained in this way.

$$\begin{aligned} R_4Sn + X_2 &\rightarrow R_3SnX + RX \\ R_3SnX + X_2 &\rightarrow R_2SnX_2 + RX \end{aligned} \qquad (5\text{-}19)$$

The nature of R has a considerable influence on the reaction rate; e.g. Et_3SnBr in toluene at $-70°$ is stable against further attack by bromine, but Ph_3SnBr under these conditions slowly reacts further by reaction 5-19. Treatment with halogen is therefore best carried out at as low a temperature as possible. These cleavage reactions were investigated in considerable detail for the purpose of studying the reactivity of the C–Sn bond for different organic groups (see Chapter 4). Nowadays they are of little preparative importance except for isolated cases such as unsymmetrical organotin halides [e.g. as in reaction 5-20 (365)].

$$Bu_3SnCH_2CH_2COOMe + Br_2 \rightarrow Bu_2Sn(Br)CH_2CH_2COOMe \qquad (5\text{-}20)$$

Stepwise cleavage of organic groups from tetraorganotins can also be achieved with hydrogen halides. These are either passed as gases through the hot reaction mixture or used as their boiling aqueous solutions.

The Sn–Sn link is cleaved by halogen even more easily than the C–Sn bond. With iodine the reaction is rapid and quantitative (reactions 5-21 and 5-22) and therefore of analytical interest (see Chapter 22-3), but

$$R_3SnSnR_3 + I_2 \rightarrow 2\,R_3SnI \qquad (5\text{-}21)$$

$$(\text{-}SnR_2\text{-}SnR_2\text{-})_n + 2n\,I_2 \rightarrow 2n\,R_2SnI_2 \qquad (5\text{-}22)$$

it is used very rarely for preparative purposes, as the required starting materials (hexaorganoditins or polymeric diorganotins) are themselves prepared from organotin halides as discussed in Chapter 14. Further cleavage reactions are given in the same chapter.

5-1-5 Preparation of individual organotin halides from other organotin halides

Direct access to some of the compounds in these series can be difficult or even impossible. Transformation of one organotin halide into another is therefore of preparative importance. Some generally useful methods, which are also adopted for purification of organotin halides, proceed via stannoxanes or diorganotin oxides respectively as in examples 5-23 and 5-24.

$$2\,R_3SnCl \xrightarrow[-H_2O,\ -2\,NaCl]{+2\,NaOH} R_3SnOSnR_3 \xrightarrow[-H_2O]{+2\,HBr} 2\,R_3SnBr \qquad (5\text{-}23)$$

$$R_2SnBr_2 \xrightarrow[-H_2O,\ -2\,NaBr]{+2\,NaOH} R_2SnO \xrightarrow[-H_2O]{+2\,HCl} R_2SnCl_2 \qquad (5\text{-}24)$$

Hydriodic and hydrofluoric acids act exactly as HCl and HBr. The reactions are reversible (see equations 5-1).

The fluorides (examples in Table 5-2) are generally prepared by shaking more easily accessible organotin halides in alcohol with aqueous KF solution. Aliphatic compounds R_3SnF are then precipitated quantitatively as colorless, odorless crystals; compounds R_2SnF_2 are soluble in an excess of KF solution and can thus be separated. Isolation of R_2SnF_2 for the same reason requires precisely calculated amounts of KF.

The poor solubility of the organotin fluorides is also used for separation of other compounds R_3SnX and R_2SnX_2 from mixtures by treatment with KF.

6 Organotin pseudohalides

Just as a halogen bonded to tin can be changed for another (see Chapter 5-1-5) so it can be replaced by a pseudohalogen, e.g.

$$R_3SnCl + NaN_3 \rightarrow R_3SnN_3 + NaCl \qquad (6\text{-}1)$$

Another versatile method of preparation is as follows (479a):

$$R_{4-n}Sn(NR'_2)_n \xrightarrow[-n\,HNR'_2]{n\,HN_3} R_{4-n}Sn(N_3)_n \qquad (n = 1, 2; R' = Me, Et)$$

Several organotin cyanides, isocyanates, isothiocyanates, fulminates, and azides of type R_3SnX, and a few of type R_2SnX_2, are known (69, 265, 315, 479, 490, 494, 819, 848). A recent review deals with azide derivatives (847a).

Slight dissociation may occur in polar media and especially in water, but it would be more in line with the general character of these compounds if one assumed a covalent but polarizable bond between tin and pseudohalogen. This is also indicated by their good solubility in organic media and by the ease of sublimation, e.g. of Et_3SnCN which can be seen to sublime well below the melting point and even in solvent vapor. The fact that Bu_3SnN_3, a liquid at room temperature, melts without decomposition and can be distilled (490), must also be interpreted in the same way. That this compound is analogous to the alkyl azides can also be seen from the addition to acetylenic compounds, when a triazole ring is formed (494):

```
                                                   N—C—COOR
                                                   ‖  ‖
(C4H9)3Sn—N3 + ROOC—C≡C—COOR ——→ N  C—COOR
                                                    \/
                                                    N—Sn(C4H9)3
```

Higher organotin azides react with trialkylphosphines as follows (479a, 741a):

$$R_3SnN_3 + Me_3P \rightarrow R_3SnN{=}PMe_3 + N_2$$

whereas the methyl compound disproportionates:

$$2\ Me_3SnN_3 \xrightarrow{Me_3P} Me_2Sn(N_3)_2 + Me_4Sn$$

The diazide reacts with the phosphine (R = Me, Et, Ph), but utilizes only one azido group:

$$Me_2Sn(N_3)_2 + R_3P \rightarrow Me_2Sn(N_3){-}N{=}PR_3 \quad (741a)$$

The structure of the organotin cyanides poses special problems. Trimethyl-silicon and -germanium cyanides exist in an equilibrium (equation 6-2) in which measurable amounts of isocyanide are present even at room temperature ($\nu_{N=C}$ at 2105 cm.$^{-1}$ as well as $\nu_{C\equiv N}$ at 2198 cm.$^{-1}$) (767). Equilibrium is established rapidly so that treatment with $Fe(CO)_5$ gives the typical reaction of organic isocyanides, displacement of CO molecules. The IR spectrum of Me_3SnCN lacks NC bands (767), but

$$Me_3GeC{\equiv}N \rightleftharpoons Me_3GeNC \qquad (6\text{-}2)$$

treatment with $Fe(CO)_5$ yields some $Me_3SnN{=}C{-}Fe(CO)_4$ which can originate only from the isocyanide. It therefore seems probable that the organotin cyanides are also in equilibrium with the isocyanide, but that its share in the equilibrium mixture is much smaller, and attainment of equilibrium much slower, than in the analogous Si and Ge derivatives.

There is the further point that treatment of tributyltin cyanide with sulfur in CS_2 at 46° (and by other methods) yields a product which from IR spectroscopic and other indications must be regarded as an isothiocyanate $Bu_3SnN{=}C{=}S$ (166, 265). One would conclude that this must have been formed from the isocyanide Bu_3SnNC, unless there were again an equilibrium thiocyanate $\rightleftharpoons$ isothiocyanate. [All the SCN-compounds, prepared by different methods, were identical (166).]

7 Stannyl esters

These compounds are treated here separately from the remaining compounds containing the Sn–O group (Chapter 17) because their properties and modes of reaction bear a close relationship to those of the organotin halides and pseudohalides just discussed.

7-1 ESTERS OF CARBOXYLIC ACIDS

These compounds can be allocated to three sub-divisions:

$$RCOOSnR'_3 \qquad (RCOO)_2SnR'_2 \qquad (RCOO)_3SnR'$$

Table 7-1 contains some examples. The Sn–O bonds in stannyl esters are covalent but mainly undergo polar reactions, depending on the solvent and the attacking reagent. The situation is similar in this respect to that of tin–halogen and tin–pseudohalogen bonds (Chapters 5 and 6). It is thus clear why even stannyl esters with small organic groups are usually more soluble in alcohol or ether than in water (416). Many of them have low melting points and sublime easily, e.g. triethyltin acetate at 230°. Classification of this type of compound as acid esters has received further support from recent studies. Admittedly there are features additional to those associated with simple alkyl esters, but these can be attributed to the high reactivity of the Sn–O bond, and to the size and the vacant orbitals of the tin atom. Thus the Sn–O bond in these esters will undergo scission, e.g. with diazomethane (382), whilst on the other hand the esters themselves can be obtained by radical cleavage of C–Sn and Sn–Sn bonds with diacyl peroxides (584, 869). (Preparative references are in Table 7-1.)

Bonding in the stannyl esters of acetic acid and higher fatty acids is essentially similar to that in the covalent alkyl esters, as evidenced by

Table 7-1
Stannyl Esters of Carboxylic Acids

FORMULA	B.P. (°C/mm.)	M.P. (°C)	METHOD OF PREPARATION
$Me_3SnOCOMe$	—	196.5—197.5	355
$Et_3SnOCOPh$	132—134/1 114/0.5	71; 74.5	26, 355, 719
$Ph_3SnOCOPh$	—	82.5—84	241
$Et_3SnOCOCH_2Br$	—	99.5	28
$Pr_3SnOCOCH_2CN$	—	91—92	716
$Pr_3SnOCOCF_3$	88—90/1	80; 94—95	716, 721
$Et_3SnOCOC(Me){=}CH_2$	—	75.5	28, 396
$BuSn(Et_2)OCOMe$	—	71—73	356
$NC(CH_2)_3Sn(Pr_2)OCOMe$	—	71—72	365
$(Et_3SnOCO)_2CH_2$	—	175—175.5 (decomp.)	491
$Pr_2\overline{SnCH_2CH_2OCO}$	—	158—160	365
$Bu_2Sn(OCOMe)_2$	142—145/10	8.5—10	188, 728
$Et_2Sn(OCOPh)_2$	—	121	595, 685
i-$Pr_2Sn(OCOCHCl_2)_2$	—	69—71	720
$Me_2SnOCOC(Me){=}CH_2$	—	140—144	716
$[(Bu_2SnOCO)_2 \cdot H_2O]_n$ (oxalate)	—	195 (decomp.)	34
$Et_2Sn(Cl)OCOMe$	—	94	872
$BuSn(OCOMe)_3$	117—119/1	46	31
$PhSn(OCOPr)_3$	171—173/1	50.5	31
$EtSn(OCOPh)_3$	—	185—188 (decomp.)	694
$Bu_2Sn(Br)OCOMe$	—	67—68.5	15, 726, 728

IR spectra and dipole moments (e.g. 2.2 D for tributyltin acetate in benzene as against 1.9 for alkyl acetates). With the exception of the formic acid esters, the compounds are monomolecular in dilute solution. There are no indications of ionic structures (166, 318, 628). Analogous plumbyl esters behave similarly both in the solid state and in solution. The corresponding germyl compounds must also be regarded as esters, but these do not form polymers (318) (see below). The same is true for the silyl esters (636).

Interpretations of IR spectra of solid organotin carboxylates were originally in favor of salt-like structures such as $R_3Sn^+ + RCOO^-$ (636). However, more recent and comprehensive studies have shown that the solids contain covalent coordination polymers with penta-coordination around the tin. Similar units exist in the melt immediately above the melting point, and to some extent even in concentrated solutions (318, 628). Stannyl esters of formic acid retain this structure even in dilute solutions, depending on the concentration (631, 639). (For details see Chapter 2.)

Stannyl esters can generally be prepared by a straightforward reaction from the appropriate acid and the oxide, e.g. in accordance with reactions 7-1 (15, 29, 492, 761, 914).

$$MeCOOH + \tfrac{1}{2}Et_3SnOSnEt_3 \rightarrow MeCOOSnEt_3 + H_2O$$

$$2\,(\beta\text{-Pyridyl})COOH + Pr_2SnO \xrightarrow{83\,\%} (\beta\text{-Pyridyl-COO})_2SnPr_2 + H_2O \quad (29)$$

$$3\,MeCOOH + BuSnO_{1.5} \rightarrow (MeCOO)_3SnBu + 1.5\,H_2O \quad (31)$$

(7-1)

Triphenyltin acetate, which is of technical importance (see Chapter 25), is prepared in this way, and so are numerous other aliphatic and aromatic stannyl esters as shown in Table 7-1. These processes, or variations thereof, are also described in a number of patents. Other syntheses utilize the reactivity of the tin–halogen bond. From the many examples a few are reactions 7-2—7-4.

$$2\,RCOOH + Bu_2SnCl_2 \xrightarrow[-2\,HCl]{Et_3N} Bu_2Sn(OCOR)_2 \quad (15) \qquad (7\text{-}2)$$

$$3\,i\text{-}PrCOOAg + BuSnCl_3 \xrightarrow{CCl_4} BuSn(OCO\,i\text{-}Pr)_3 + AgCl \quad (31)$$

(7-3)

$$(MeCOO)_2Pb + 2\,R_3SnBr \rightarrow 2\,R_3SnOCOMe + PbBr_2 \quad (914)$$

(7-4)

Di-esters (and probably also tri-esters) undergo comproportionation with diorganotin dihalides. The reaction, e.g. 7-5, can be spontaneous in pentane (15).

$$(MeCOO)_2SnBu_2 + Bu_2SnBr_2 \rightarrow 2\,MeCOOSn(Br)Bu_2 \qquad (7\text{-}5)$$

Comproportionation with diorganotin dihydrides (Chapter 8) is similar. Stannyl esters are also formed in a ready reaction from organotin hydrides and aliphatic carboxylic acids (see Chapter 8), from ditins and polymeric diorganotins with the aid of diacyl peroxides (869) (see Chapter 14), or by cleavage of tetraorganotins with carboxylic acids:

$$2\,RCOOH + R'_4Sn \rightarrow (RCOO)_2SnR'_2 + 2\,RH \qquad (7\text{-}6)$$

Depending on the strength of the acid (phenols also give this reaction), the nature of the alkyl group R′, and the temperature, the alkyl groups are displaced in successive stages as alkane (29, 226, 392, 572, 716, 720, 722, 761, 914). (See also Chapter 4.) However, several of these reactions often proceed simultaneously. Not all the products actually formed have been isolated in every instance, and this may account for some of the discrepancies in the literature (572). The analogous reaction of hexaphenylditin with glacial acetic acid yields hexaacetoxyditin (882). Stannyl esters are also obtained in the reaction of organotin oxides with alkyl esters, but the original interpretation of this reaction (26) has not been confirmed. The real course of the reaction (574) is

$$PhCOOEt + Et_3SnOSnEt_3 \longrightarrow PhCOOSnEt_3 + Et_3SnOEt \qquad (7\text{-}7)$$

Some other syntheses are important only in special cases, and these have been summarized (315).

Stannyl esters have recently found preparative uses; e.g. some esters of unsaturated carboxylic acids have been used for polymerizations (Chapter 21), and esters of amino-acids for protection of carboxy groups in peptide synthesis (211).

7-2 ESTERS OF INORGANIC ACIDS

Stannyl esters of almost any desired acid (Table 7-2) can be prepared by methods similar to the syntheses described in Chapter 7-1, and these are covered by numerous patents. Many of the esters have been known for a long time,* some for over 100 years (315, 416), but new examples are being added all the time (819), e.g. esters of sulfonic acids (30) (reaction 7-8) and sulfinic acids (642):

$$2\,EtSO_2OH + Et_2SnO \longrightarrow (EtSO_2O)_2SnEt_2 + H_2O \qquad (7\text{-}8)$$

$$R_3SnOSnR_3 + 2\,RSO_3H \longrightarrow 2\,R_3SnOSO_2R + H_2O \qquad (7\text{-}9)$$

Triorganotin esters of strong acids require mild conditions for preparation by reaction 7-9. When heated, the acid cleaves organic groups from the tin and forms diorganotin diesters (30).

Like the analogous carboxylic esters, these compounds were orignally regarded as salts although some of them had been known for some time to be readily soluble in alcohol (416). Trimethyltin perchlorate and nitrate and some other compounds of this class can be sublimed *in vacuo* without decomposition (139, 628, 898). IR studies show that, at least in the solid, there are no ions but only covalent links. In addition, some association with penta-coordination around the tin has been observed for the perchlorate and some related materials just as for the analogous carboxylic esters (see Chapter 2). These are therefore closely related to the esters.

* It should be pointed out, however, that many of these compounds are still mentioned in the literature without proper characterization, proof of purity or certain identification.

Table 7-2
Stannyl Esters of Inorganic Acids

FORMULA	M.P. (°C)	METHOD OF PREPARATION
Me_3SnONO_2	140 (decomp.)[a]; sealed tube	898
Ph_3SnONO_2	182—185	783, 856
$(Et_3SnO)_2CO$	137—140 (decomp.); absence of air	295
$Ph_3SnOS(O)Ph$	228.5—230.5	642
$Me_3SnOClO_3$	125—127, 128[a]; sealed tube	139
$Me_2Sn(ONO_2)_2$	Explodes when heated	3, 261a
$Et_2Sn(ONO_2)OH$	214 (decomp.)	898
Me_2SnAsO_4H	Turns brown at 350; no M.P.	709
Me_2SnPO_4H	>345	772
$Me_2Sn(OPH_2O)_2$	213 (decomp.)	772
$Et_2Sn(OSO_2Me)_2$	234	30
$Pr_2Sn(OSO_2Et)_2$	298	30
$Me_2Sn[OP(O)HPh]_2$	159 (decomp.)	772
$(H_2C{=}CH)_2SnO_3PPh$	>355	772

[a] Sublimes *in vacuo.*

Triphenyltin nitrate can be obtained from the chloride (783) (with advantage in acetonitrile) or from hexaphenylditin (856) by the action of silver nitrate; the cleavage of the Sn–Sn bond in the latter case should be noted (Chapter 14). It is stable at 20°, but when heated it decomposes in various ways depending on the conditions (856). Even the colorless dimethyltin dinitrate is stable at 20°, but it explodes when heated. The compound dissociates in polar solvents, the Me_2Sn^{2+} ion being heavily solvated. The arrangement in the solid is covalent, but the Sn–O bonds are polarized. The NMR spectrum shows that the Sn is electron-deficient compared with that in Me_4Sn. Surprisingly, the NO_3 groups do not act as bridging groups; they are purely monodentate (3).

Some obviously complex hydroxy nitrates $R_2Sn(ONO_2)OH$ can be obtained from diorganotin dioxides with dilute nitric acid. They are soluble in water and methanol (898).

8 Organotin hydrides

8-1 PREPARATION OF ALKYLTIN HYDRIDES R_3SnH, R_2SnH_2, and $RSnH_3$

Although it is the least stable tin–hydrogen compound, tin tetrahydride SnH_4 was prepared as early as 1919 by Paneth and Fürth (641), while the far more stable organotin hydrides remained unexplored for a considerable time. Granted that Kraus and Greer had prepared trimethyltin hydride by 1922 (409), the preparation in quantity of representatives of all three series R_3SnH, R_2SnH_2, and $RSnH_3$ was made possible only by the introduction of lithium aluminum hydride as a reductant for organotin halides. The properties and reactions of these hydrides were then open to investigation.

The original preparation by Finholt, Bond, Wilzbach, and Schlesinger (206) (1947) has been modified by several authors, and is now the most commonly employed laboratory method (367, 465, 578).

$$R_3SnCl + \tfrac{1}{4}LiAlH_4 \rightarrow R_3SnH + \tfrac{1}{4}LiAlCl_4$$

$$R_2SnCl_2 + \tfrac{1}{2}LiAlH_4 \rightarrow R_2SnH_2 + \tfrac{1}{2}LiAlCl_4$$

$$RSnCl_3 + \tfrac{3}{4}LiAlH_4 \rightarrow RSnH_3 + \tfrac{3}{4}LiAlCl_4$$

It is remarkable that not all of the hydrogen of the $LiAlH_4$ is available for reduction. An approximately 100% excess of reductant is required because of this.

More recently, dialkylaluminum hydrides R_2AlH have also been used as reductants. These are particularly useful for the readily volatile di- and tri-hydrides, which can then be distilled off directly from the reduction mixture (321, 575): e.g.

$$R_2SnCl_2 + 2\,R'_2AlH \rightarrow R_2SnH_2 + 2\,R'_2AlCl$$

$$RSnCl_3 + 3\,R'_2AlH \rightarrow RSnH_3 + 3\,R'_2AlCl$$

It is usual to add to these homogeneous reaction mixtures only as much ether as is required for complex formation with the R'_2AlCl (e.g. for $R' = Ph$ or *i*-Bu)..The product R'_2AlCl can then be used in its turn for alkylation of $SnCl_4$ (see Chapter 3-1-2). Table 8-1 shows a selection of the many known organotin hydrides. Organotin deuterides (Table 8-1) are also well prepared by this method (594), since all the deuterium of the R'_2AlD is available for reduction.

Alkoxides $R_nSn(OR')_{4-n}$ (575), too, may be reduced to organotin hydrides with either $LiAlH_4$ or R_2AlH. Reaction of organotin sodium derivatives with NH_4Br in liquid ammonia was formerly the commonest preparative method (315, 409). Alkylation of sodium tin hydrides is also a possible route (193), which is now unimportant except possibly for some special preparations. Organotin sodium compounds have, however, come back into the picture recently. They give good yields of organotin deuterides with strongly complexing solvents (e.g. tetrahydrofuran or diglyme) and stoichiometric amounts of D_2O (427a):

$$R_3SnNa + D_2O \rightarrow R_3SnD + NaOD$$

Reaction of the appropriate organotin methoxide with diborane or $NaBH_4$ has been recommended for preparation of rather sensitive compounds, e.g. phenyltin trihydride, in a pure state (21, 22). Reduction of the respective organotin chlorides with $NaBH_4$ in glyme or diglyme at $-10°C$ gives poor yields in the case of butyltin tri- and di-hydrides, but good or excellent yields for Me_2SnH_2 and a number of monohydrides (72a).

The rapid hydrogen–halogen exchange, which is the basis of the preparation of organohalotin hydrides (Chapter 8-2), also gives mild preparative access to readily volatile organotin hydrides (580, 648, 725a). A tin hydride of high boiling point, e.g. Ph_3SnH, is mixed with the organotin halide to be reduced, e.g. Me_2SnCl_2 or $(H_2C{=}CHCH_2)_2SnBr_2$, and the new organotin hydride is removed continuously from the equilibrium by vacuum-distillation.

$$2\,R_3SnH + R'_2SnX_2 \rightleftharpoons 2\,R_3SnX + R'_2SnH_2$$

$$3\,R_3SnH + R'SnX_3 \rightleftharpoons 3\,R_3SnX + R'SnH_3$$

A similar hydrogen exchange can be effected with stannoxanes (591, 724, 816); this time the equilibrium lies well over to the right-hand

Table 8-1
Organotin Hydrides and Deuterides[a]

FORMULA	B.P. (°C/mm.) (M.P.) (°C)	ν(Sn–H)[b] (cm.$^{-1}$)	n_D^{20}	METHOD OF PREPARATION
$MeSnH_3$	0—1.4	1870	—	206
Me_2SnH_2	35	1845	1.4480°	206
Me_3SnH	59	1833	1.4461	206
Me_3SnD		(1327)	—	427a
$EtSnH_3$	25	1869	1.4491	177, 575
Et_2SnH_2	96—98	1822	1.4679	177, 575
Et_3SnH	148—150 39/12	1813	1.4709	177, 575, 367
Et_3SnD	37/11	(1299)	1.4702	427a, 594
Pr_3SnH	76—78/12	1809	1.4715	288a, 367
Pr_3SnD		(1301)	—	427a
i-Pr_3SnH	69—70/12	1787 (neat)	—	346, 443
Bu_3SnH	76—81/0.7	1812	1.4726	288a, 367
Bu_3SnD		(1301)	—	427a
i-Bu_3SnH	101—104/12	1810	1.4697	578
i-Bu_3SnD	104—106/11	(1302)	1.4697	594
s-Bu_2SnH_2	60/12	1825	1.4745	750
s-Bu_3SnH	65/0.4	1790	1.486	591
s-Bu_3SnD		(1284)	—	427a
t-Bu_2SnH_2	38/11	1812	1.4634	582
t-Bu_3SnH	92—97/11	1777		750
$C_8H_{17}SnH_3$	29/0.3 (−52)	1863	1.4680	750
$(C_8H_{17})_2SnH_2$	110/10^{-3} (−15)	1836	1.4738	750
$(C_8H_{17})_3SnH$	164—166/10^{-3}	1805	1.4742	443, 750
$PhSnH_3$	57—64/105	1880	—	21, 575
Ph_2SnH_2	89—93/0.3 (−17)	1857	1.5950	564, 575
Ph_3SnH	168—172/0.5 (28)	1843	1.632 (supercooled melt)	367
Ph_3SnD	152—156/0.002(28)	(1323)	1.6318$^{28°}$	427a, 594
(cyclo-$C_6H_{11})_2SnH_2$	94—96/10^{-3}	1820	1.5295	582
(cyclo-$C_6H_{11})_3SnH$	147—150/10^{-3}	1782	1.5411	591
(*p*-$ClC_6H_4)_3SnH$	(224—226)	—	—	826[c]
(*p*-$MeC_6H_4)_3SnH$	(78—79)	1835 (neat)	—	443
$(PhCH_2)_2SnH_2$	120/10^{-3} (decomp.)	1843	1.6215	566
(*p*-$EtOC_6H_4)_2SnH_2$	(53—54)	1850	—	565[d]
(α-Naphth)$_2SnH_2$	(102)	1870, 1888	—	565
(*p*-Biphenylyl)$_2SnH_2$	(141)	1848, 1862	—	565
$H_2C{=}CHSnH_3$	—	1892, 1910 (gas)		94
$Bu_2SnHSnHBu_2$	122—124/10^{-3}	1795	—	730, 816
i-$Bu_2SnHSnH$*i*-Bu_2	109—112/10^{-3}	1793	1.518	816
$Et_2SnH(CH_2)_6SnHEt_2$	115/10^{-3}	1810	1.5091	580, 648
$Et_2SnH(CH_2)_2CN$	101/12	—	1.4994	580, 648
i-$Bu_2SnH(CH_2)_3NH_2$	91/0.7	—	—	648
$Et_2SnH(CH_2)_4CH{=}CH_2$	101/13	1805 (neat)	1.4817	750
$Me_2SnHCF_2CHF_2$	78	1877 (gas)	—	134

[a] For other organotin hydrides see H. G. Kuivila (429).
[b] In cyclohexane (10—40 %) unless stated otherwise; ν(Sn–D) in parentheses.
[c] Here also: tri-*o*-tolyl-, tri-*m*-tolyl-, and tri-*o*-biphenylyl-tin hydride.
[d] Here also: di-*β*-naphthyl- and di-*p*-tolyl-tin dihydride.

side:

$$R_3SnOSnR_3 + R'_2SnH_2 \xrightarrow{20°} 2\,R_3SnH + \tfrac{1}{n}(R'_2SnO)_n$$

The driving force for the reaction is provided by the formation of insoluble $(R'_2SnO)_n$. Another possible preparative method springs from the observation of R. Okawara and his co-workers that stannyl esters of formic acid eliminate CO_2 when heated (626):

$$R_3SnOCHO \xrightarrow{160-180°} R_3SnH + CO_2 \qquad (R = Pr, Bu)$$

This is the reverse of the hydrostannation of CO_2 (see Chapter 9).

A new and promising method for the preparation of organotin hydrides involves reduction of Sn–O bonds in stannoxanes or alkyltin alkoxides by polymeric organosilicon hydrides (available as industrial by-products) (262a, 288a, 315b), e.g.

$$R_3SnOSnR_3 + 2/n\,(-\overset{\overset{\large Me}{|}}{\underset{\underset{\large O-}{|}}{Si}}-H)_n \rightarrow 2\,R_3SnH + 2/n\,(-\overset{\overset{\large Me}{|}}{\underset{\underset{\large O-}{|}}{Si}}-O-)_n$$

Yield: R = Pr 66%; Bu 79%; Ph 32%

Tributyltin ethoxide gives 86% of hydride, and dibutyltin diethoxide 66% of dihydride. The driving force for these reactions appears to reside in the insolubility of the branched silicone. Monomeric silicon hydrides of different types give low yields of organotin hydride, or none at all.

Impure samples of organotin hydrides decompose slowly even at 20°C. Cleavage occurs not only at the Sn–H, but also at the Sn–C bond, so that tetraalkyltins and tin metal are sometimes formed in addition to ditins and polytins (443, 575). Decomposition is generally affected by catalysts, especially bases and heavy metals. It is therefore understandable that widely divergent statements have been made concerning the stability of organotin hydrides, depending on their purity and method of preparation.

UV irradiation leads to decomposition, but only short-wave UV is effective since there is little or no absorption of longer wavelengths (see Chapter 23-1); e.g. irradiation of dimethyltin dihydride at 130°C rapidly yields Me_4Sn, Me_3SnH, H_2, and tin metal (134). UV irradiation of dibutyltin dihydride yields butane, butene, and H_2, as well as a complicated mixture of polytins (864). Radical reactions of organotin hydrides can be catalyzed by UV radiation, however (see relevant chapters).

Pure organotin hydrides (see Table 8-1) are generally stable in the absence of air, and can be distilled without decomposition, if necessary *in vacuo*. When pure, they are far more stable than they were assumed to be in the past. However, aliphatic hydrides are more stable than aromatic ones, and *i*- or *n*-butyl compounds in their turn are more stable than ethyl compounds. Very large differences are found in any series $RSnH_3$, R_2SnH_2, R_3SnH: the monohydrides are about as stable at 150°C as the trihydrides at 0°C.

Ditin dihydrides of the type $H(R_2)Sn–R'–Sn(R_2)H$ have gained some importance as intermediates (see Table 8-1). R′ can be $–C_6H_4–$ (474).

Organotin hydrides of the aliphatic series are colorless liquids of not too unpleasant odor (certainly not as bad as that of organotin halides). Their vapor must not be inhaled (see Chapter 24). Some of the aromatic hydrides are crystalline at room temperature (see Table 8-1). Behavior towards organic solvents resembles that of the paraffin hydrocarbons. Since organotin hydrides are attacked rapidly by oxygen, they must necessarily be handled under an inert gas (usually nitrogen or argon).

Organotin hydrides are not associated, at least in dilute solution or in the vapor. In this respect they resemble the hydrides of the other elements of Group IV and contrast with those of Group III in which the pronounced association, especially of organo-boron and -aluminum hydrides (928), has been thoroughly investigated. Nevertheless, some reactions of organotin hydrides are known in which complexing of the Sn–H group must be assumed at least in a short-lived transition stage (see below).

8-2 PREPARATION OF ORGANOTIN HYDRIDES CONTAINING NEGATIVE SUBSTITUENTS

As described in Chapter 8-4-3, hydrogen bonded to tin is exchanged spontaneously for halogen in an equilibrium reaction. From a mixture of dialkyltin dihalide and dihydride at 20°C the dialkylhalotin hydride is formed almost quantitatively (580, 648, 725, 725a, 727).

$$R_2SnH_2 + R_2SnCl_2 \rightleftharpoons 2\,R_2Sn(Cl)H$$

However, if the mixture is heated *in vacuo* the dihydride distils off first (for R = Bu) and is thus removed continuously from the equilibrium. Fluoro-, bromo-, and iodo-hydrides are obtained similarly (724a).

These mixtures often decompose as low as 50–60°C with production of hydrocarbons, hydrogen, and tin metal. Organohalotin hydrides are thus rather labile. However, their reactivity is greater than that of the simple hydrides, so they react smoothly and with excellent yields (see Chapter 9). In this respect, mixtures of dihydrides and dihalides behave like pure halo-hydrides (580, 725). This is the case also in the violent, exothermic elimination of H_2 with pyridine or other amines, in which

1,2-dihalo-ditins (Chapter 14-2) are formed quantitatively in accordance with reaction 8-1 (580, 725).

$$2\,R_2Sn(Cl)H \xrightarrow{NR_3} ClSn(R_2)Sn(R_2)Cl + H_2 \qquad (8\text{-}1)$$

Comproportionation reactions of this kind may also be used to form equilibrium mixtures of halo-hydrides of types $RSnX_2H$ and $RSnXH_2$, but these have found little preparative use. Cleavage of organotin hydrides by HBr (224) at low temperatures provides an alternative route; phenyl–tin bonds are attacked easily, ethyl–tin bonds are not.

$$EtSnH_3 + HBr \xrightarrow{-78^\circ} EtSn(Br)H_2 + H_2$$

$$Ph_3SnH + 2\,HBr \xrightarrow{-78^\circ} PhSn(Br_2)H + 2\,PhH$$

Acyl groups attached to tin behave similarly to halogen. If dibutyltin dihydride is mixed with the diacetate, a new absorption band at 1875 cm.$^{-1}$ appears in addition to the Sn–H band of the dihydride at 1835 cm.$^{-1}$. The new band must clearly be attributed to dibutylacetoxytin hydride present in a temperature-dependent equilibrium (726, 728, 729):

$$Bu_2Sn(OCOMe)_2 + Bu_2SnH_2 \rightleftarrows 2\,Bu_2Sn(OCOMe)H$$

The synthetic utility of this compound remains unexplored. A similar mixture is formed from the dihydride and an equimolar quantity of acetic acid, with evolution of one mole of H_2. If allowed to stand for longer periods at 20°C, it slowly evolves the remaining hydride hydrogen and again forms a 1,2-disubstituted ditin (Chapter 14-2).

The analogous reaction of dialkyltin dihydrides with dialkyltin dialkoxides fails because of immediate condensation (582, 588).

$$R_2Sn(OR')_2 + R_2SnH_2 \rightarrow [2\,R_2Sn(OR')H] \xrightarrow{-2\,R'OH} 2/n\,(-R_2Sn-)_n$$

8-3 TETRAALKYLDITIN DIHYDRIDES

Hydrogenation of tetraalkylditin dihalides, easily accessible by reaction 8-1, yields ditin dihydrides of the type $H(R_2)SnSn(R_2)H$ (580, 730, 816) (see Table 8-1). If R is one of the lower aliphatic groups, the compounds may be distilled *in vacuo* without decomposition. The Sn–H band (about 1795 cm.$^{-1}$) lies somewhat low for monohydrides; its position is similar to that in trialkyltin hydrides with secondary alkyl groups (see Table 8-1).

Tetraalkylditin dihydrides are important for the preparation of tetratins (816) and cyclic polytins (582). They are closely related to other 1,2-substituted ditins which are discussed in Chapter 14-2.

8-4 REACTIONS OF ORGANOTIN HYDRIDES

As recently as 1960 the authors of a comprehensive literature survey (315) observed: "Because of their instability, the organotin hydrides have received little attention." The situation is now changed entirely. Organotin hydrides have their uses in the most diverse syntheses, and are amongst the most interesting intermediates in the entire organic chemistry of tin. Originally, it was assumed that the Sn–H bond would, in consequence of the electronegativity difference, be polarized as $Sn^{\delta+}$–$H^{\delta-}$ and that polar reactions with transfer of a hydride ion to the reaction partner would therefore be predominant. Similar predictions were actually confirmed for the analogous organosilicon hydrides, which would at first sight appear similar in character.

However, interpretation of the electronegativity concept is now more elastic (see Chapter 2). Furthermore, it is found that the bond-length between hydrogen and elements of Group IV increases rapidly down the Group (Table 8-2). On both grounds it becomes more reasonable to

Table 8-2

Lengths of Bonds between Hydrogen and Group IV Elements

C–H	Si–H	Ge–H	Sn–H	Pb–H	
1.05	1.46	1.50	1.68	1.80	Å

assume that the Sn–H bond is fairly non-polar, but that it can become markedly polarized under the influence of the reaction partner and the reaction medium. This does mean, however, that the Sn–H bond must be able to display the entire range of reactions possible (and known) for the C–H bond. Hydrogen should be removable as a hydride ion (i.e. together with both bonding electrons), as a free radical, or as a proton. This has now been confirmed, and examples of all three reaction types have recently been studied in detail. However, the mechanism of many of these reactions is unknown. No reliable values are available for the dissociation energy of the Sn–H bond (see below); this is a serious handicap in the present context. Some detail features of this bond have also received only partial clarification (for this see Chapter 23).

The reactions of the organotin hydrides can be divided into the following groups.

(a) Reactions with radical sources and short-lived radicals (Chapter 8-4-1).

(b) Hydrostannation of carbon–carbon multiple bonds (Chapter 9).

(c) Hydrostannation of polar double bonds (Chapter 9).

(d) Condensations with proton-donor compounds (Chapter 8-4-2).

(e) Ligand exchange with other stannyl compounds (Chapter 8-4-3).
(f) Condensation reactions leading to Sn–Sn bonds (Chapter 8-4-4).
(g) Organotin hydrides as reductants (Chapter 10).
(h) Various reactions, e.g. elimination of H_2, and reactions with diazomethane and carbenes (Chapter 8-4-5).

Reactions belonging to two groups may often take place concurrently or consecutively; e.g. in the reduction of carbonyl compounds to carbinols, addition of type (c) may be followed by condensation of type (f). Similarly, hydrostannation of –C=C–, i.e. type (b), must be preceded by a radical reaction (a) yielding stannyl radicals.

Hydrostannation (b) and (c) and utilization of organotin hydrides as reductants (g) are topics in their own right. Moreover, both have been used so extensively in preparative work that their treatment here would exceed the scope of this chapter. Consequently separate treatment in Chapters 9 and 10 seems justified.

8-4-1 Reactions with radical sources and short-lived radicals

The Sn–H bond is easily cleaved by free radicals, indeed far more easily than C–H, Si–H, or Ge–H. Radical cleavage of the Pb–H bond seems to occur even more readily, as evidenced by the decreased stability and higher reactivity of organolead hydrides (571). No quantitative comparison is possible for the time being, as this would require an exact value for the dissociation energy of the Sn–H bond. In SnH_4 this has been estimated as 60.4—73.3 kcal./mole (154, 553, 828), but experience with organotin hydrides indicates that in these the value must be much lower, most likely below 50 kcal./mole (438) (see also Chapter 2-1). An average value of about 35 kcal./mole has been used in the author's laboratory and has proved quite satisfactory (see Table 2-4).

Free radicals attack the Sn–H bond with liberation of stannyl radicals:

$$R\cdot + R_3SnH \rightarrow RH + R_3Sn\cdot$$

The radical R· must however be capable of exceeding a certain energy threshold; e.g. triethyltin hydride is stable against the phenoxy radical (**1**) (galvinoxyl) up to 70°C (597),* whereas most other phenoxy radicals smoothly abstract the hydride hydrogen to form the phenol; the stannyl radicals dimerize to a ditin.

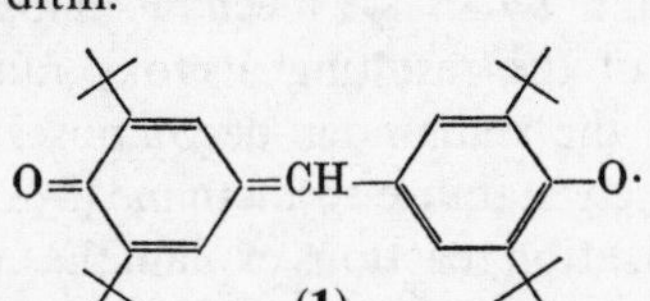

O=⟨ ⟩=CH–⟨ ⟩–O·

(**1**)

* For a short time. In concentrated mixture there is slow reduction to hydrogalvinoxyl and then to the symmetrical diphenol even at 25° within 16 hr. (T. N. Mitchell, unpublished results from the author's laboratory.)

Cleavage can also be effected by short-wave UV radiation. Using this technique, triorganostannyl radicals can be isolated at −180°C (746):

$$2\,R_3SnH \xrightarrow{h\nu} H_2 + 2\,R_3Sn\cdot$$

The triphenylstannyl radical is light yellow; color and ESR signal disappear if the temperature is allowed to rise to −110°C and the ditin is formed. The tri-*i*-butylstannyl radical is yellow; its ESR signal is rather broad and anisotropic, indicating stronger localization of the unpaired electron on the tin.

It can now be seen why many reactions of organotin hydrides are accelerated or indeed initiated by radical sources or UV radiation (see also Chapters 9 and 10).

It is also apparent why organotin hydrides are such excellent radical scavengers, as shown in the author's laboratories. They are amongst the most effective compounds of this class (599), capable even of trapping and preserving rather short-lived radical species.

All of this may be demonstrated with benzoyl peroxide. Its decomposition is induced more effectively by triethyltin hydride than by any other material tested; e.g. at 50°C decomposition depends only on the rate of addition of the organotin. A stannyl radical attacks one of the peroxide oxygens, cleaves the O–O bond, and at the same time becomes attached to the respective benzoyloxy moiety by an S_R2 mechanism as shown for peroxide labeled with ^{18}O (584, 585, 585a). The other half of the peroxide molecule is left as benzoyloxy radicals which are then trapped almost quantitatively by an excess of trialkyltin hydride. Diethyltin dihydride reacts in the same way. Only a minute proportion of the benzoyloxy radicals survives to decompose in the usual way to phenyl radicals and CO_2(585a). By use of unsymmetrical peroxides $MeCOOOCOC_6H_4X$-*p*, and by varying the inductive effect of X, it has been shown that electron deficiency at the peroxide bridge is important for attack by stannyl radicals (713a). The latter are thus shown to be strongly nucleophilic.

The process is similar for acetyl peroxide. Its decomposition is again induced strongly by organotin hydrides, and at 40°C depends solely on the rate of addition. A stannyl radical cleaves the peroxy bridge and forms the ester (stannyl acetate) by an S_R2 mechanism (553, 585a) (see reactions 8-2). The main part of the resulting acetoxy radicals is trapped (8-2, right-hand side), and the remainder decomposes to CO_2 and methyl radicals which ultimately stabilize to methane (8-2, left-hand side). These extremely rapid stabilization reactions explain the failure of acyl peroxides as hydrostannation catalysts (551) (see Chapter 9). Organotin hydrides also accelerate the decomposition of other acyl peroxides and *t*-butyl peresters (for products and mechanisms see refs. 555, 585, and 585a), but not of *t*-butyl peroxide. In this last case they still act as powerful scavengers.

$$MeCOOC(O)Me \rightarrow MeCOO\cdot ;\quad MeCOO\cdot + R_3Sn\cdot \rightarrow MeCOOSnR_3 ;\quad MeCOO\cdot \rightarrow CO_2 + Me\cdot ;\quad Me\cdot + R_3SnH \rightarrow R_3Sn\cdot + CH_4$$

$$MeCOO\cdot + R_3SnH \rightarrow R_3Sn\cdot + MeCOOH ;\quad R_3SnH + MeCOOH \rightarrow MeCOOSnR_3 + H_2 \qquad (8\text{-}2)$$

The *t*-butoxy radicals thus have no opportunity to decompose in the normal way to acetone and methyl radicals (585a):

$$t\text{-BuOO}t\text{-Bu} \rightarrow 2\,t\text{-BuO}\cdot \xrightarrow{2\,R_3SnH} 2\,t\text{-BuOH} + R_3SnSnR_3$$
$$2\,t\text{-BuO}\cdot \not\rightarrow Me\cdot + MeCOMe$$

Rates of decomposition of azobisisobutyronitrile and azobiscyanocyclohexane (**2**) are unaffected by the presence of trialkyltin hydrides (599). However, the additive suppresses the formation of the normal stable end-products [in the latter case 1,1′-dicyanobicyclohexyl (924) (**3**)] which now appear in small amounts only. The principal products, arising from trapping of primary radicals, are now the monocyclic nitrile (**4**) and an unstable adduct which eliminates trialkyltin cyanide (599). This is one part of the reaction. Another part of the azo-compound undergoes radical hydrostannation (see Chapter 9). The expected product (**5**) cannot however be isolated, since it rapidly loses R_3SnCN and then, more slowly, HCN to form the cyclohexanone azine (**6**). Azobisisobutyronitrile, the most frequently employed hydrostannation catalyst (Chapter 9), behaves similarly to the azocyclohexane (**2**) (599).

On the other hand, decomposition of substituted azo-compounds like $PhN{=}NSO_2Ph$, $PhN{=}NSR$, and $RON{=}NOR$ ($R = t$-Bu or CH_2Ph) is induced strongly by organotin hydrides. It appears that stannyl radicals attack the $-SO_2-$, $-SR$, or $-OR$ group respectively by an S_R2 mechanism

(571a). The stannyl radical is again shown to be nucleophilic.

R_3SnCN + Resin

+ R_3SnH

H CN (4) + N_2 + [R_3Sn—N=C=]

NC CN (2) —N_2

H CN NC SnR_3 (5) — R_3SnCN H CN

NC CN (3)

HCN + =N—N= (6)

8-4-2 Reactions with proton donors

Tin hydrides react with the protons of numerous acids with evolution of H_2. Generally, the stronger the acid and the more pronounced the hydride character of the organotin hydrogen, the faster the reaction (see also Chapter 23-2). Consequently, trialkyltin hydrides react faster than dihydrides or triaryltin hydrides, and these in their turn react faster than trihydrides. Kuivila et al. (435) have suggested that the initial step is an electrophilic attack of an undissociated acid molecule on the hydride hydrogen. Reaction 8-3 of trialkyltin hydrides with carboxylic acids usually proceeds quite smoothly, yielding the respective stannyl esters (315, 728, 876). This reaction has been used for quantitative estimation of

$$R'COOH + HSnR_3 \rightarrow R'COOSnR_3 + H_2 \qquad (8\text{-}3)$$

the Sn–H content of reaction mixtures; the best procedure is to mix the sample with an excess of dichloroacetic acid and to measure the volume of hydrogen produced (584, 585a). However, the effectiveness of the acid does not depend solely on its acid strength; e.g. while the reaction rate against triethyltin hydride increases understandably in the order:

$$MeCOOH < ClCH_2COOH < Cl_2CHCOOH < Cl_3CCOOH,$$

benzoic acid reacts more slowly than the weaker acetic acid. When R = Ph (reaction 8-3), complications arise because of the ready cleavage of the aryl–tin bond (728, 876) (see Chapter 4).

Dihydrides rarely undergo the simple condensation

$$2\,R'COOH + H_2SnR_2 \rightarrow (R'COO)_2SnR_2 + 2\,H_2$$

but rather ligand exchange occurs first to form the dialkyl-acyloxytin hydride (see Chapter 8-2). This then reacts further to give a substituted ditin with evolution of H_2, or alternatively $(R'COO)_2SnR_2$ by reaction with more acid (728). The latter predominates when R is aliphatic.

Inorganic acids, e.g. HBr, also evolve H_2 with aliphatic organotin hydrides. With aromatic ones, cleavage of the aryl–tin bond is again the principal reaction (cf. reaction with carboxylic acids above). The effect has already been discussed in Chapter 8-2.

Most phenols, alcohols, and mercaptans have insufficient acidity to effect condensations analogous to 8-3. However, their more strongly acidic complexes, e.g. with $ZnCl_2$ (609, 811), react smoothly; catalytic amounts of the complexing agent are sufficient. As a matter of interest, base catalysis is also feasible (435). An intermediate state involving penta-coordinate tin seems likely (see Chapter 2), e.g.

$$(C_4H_9)_3Sn{-}H + CH_3O^- \rightleftharpoons \left[(C_4H_9)_3Sn\begin{smallmatrix}H\\ \\ OCH_3\end{smallmatrix}\right]^-$$

$$\xrightarrow{CH_3OH} (C_4H_9)_3Sn{-}OCH_3 + H_2 + CH_3O^-$$

8-4-3 Exchange reactions with other stannyl compounds

Among the reactions in which the organotin hydride hydrogen shows unambiguous electronegative character are the exchange reactions 8-4 with nucleophilic groups X. They are not confined to monofunctional

$$R_3SnH + R'_3SnX \rightleftharpoons R_3SnX + R'_3SnX \qquad (8\text{-}4)$$

In general:

$$sn{-}H + sn'{-}X \rightleftharpoons sn\begin{smallmatrix}H\\ \\ X\end{smallmatrix}sn' \rightleftharpoons sn{-}X + sn'{-}H$$

groups but occur in many complicated reactions of organotin hydrides. Examples of such exchanges are therefore discussed also in other chapters (e.g. 8-1, 8-2, and 8-4). This type of migration from one tin atom to another is observed only for negative (nucleophilic) substituents. A four-centre intermediate state with penta-coordinate tin seems likely, as indicated in equation 8-4 (594). Formation of dibutylacetoxytin hydride from dibutyltin dihydride and dibutyltin diacetate constitutes the first recorded example (726); this was diagnosed at once as an equilibrium reaction (see Chapter 8-4-2). An equimolar mixture of dibutyltin dihydride and dichloride (725, 727) behaves chemically as pure $Bu_2Sn(Cl)H$ (580), but is really yet another equilibrium system (see Chapter 8-2). Any attempt to distil the chloro-hydride produces a distillate of the dihydride.

As this is removed continuously, pure dichloride remains. Other exchange reactions have been discovered subsequently, many of them again yielding equilibrium mixtures.

These ligand exchanges provide further proof that covalent bonds around tin are far more mobile than they were assumed to be in the past. There are indeed certain similarities to the organic compounds of Group III. In these, association and ligand exchange [investigated thoroughly (928)] are ascribed to electron deficiency. In the case of tin, there must be significant participation by unfilled *d*-orbitals, as discussed in Chapter 2. The halogen–hydrogen exchange reaction may be used to prepare sensitive organotin hydrides (580, 648) (see Chapter 8-1).

Exchange between hydrogen and deuterium is also rapid (594); e.g. triethyltin deuteride and tri-*i*-butyltin hydride at 40°C equilibrate completely in 25 minutes (594):

$$Et_3SnD + i\text{-}Bu_3SnH \overset{40°}{\rightleftharpoons} Et_3SnH + i\text{-}Bu_3SnD$$

Interpretation in terms of ionic dissociation is ruled out, just as it is for halogen–hydrogen exchange.

Similar exchanges of nucleophilic substituents on tin have been observed for trialkyltin oxides (548, 588, 724) and sulfides (591, 816), organotin methoxides (591, 724) and cyanides, stannylphosphines (591, 816), and stannylamines (591, 816). Exchange of hydrogen for nucleophilic groups of other organotin molecules is more rapid for triaryltin hydrides than for trialkyltin hydrides. Also, organotin dihydrides exchange faster than the corresponding monohydrides. The rate of exchange thus increases with increasing electrophilicity of the tin (591, 816). Little can be said concerning the effect of the substituent X (in equation 8-4), since comparative studies are incomplete and exchange is frequently accompanied by condensation (Chapter 8-4-4). When $X = NEt_2$ or OMe, exchange cannot be demonstrated directly. On the evidence of the end-products (591, 816) it must be accompanied by a very rapid condensation reaction (Chapter 8-4-4). As far as can be seen at the moment, the rate of exchange increases with increasing nucleophilicity of the substituent X; e.g. $i\text{-}Bu_2SnCl_2$ reacts far more slowly with $(Bu_3Sn)_2S$ than with $(Bu_3Sn)_2O$. These are again mostly equilibrium reactions. However, if one component is removed from the equilibrium, e.g. the precipitated insoluble R_2SnO, then the exchange becomes quantitative and can be used e.g. to prepare R_3SnH (588) from $(R_3Sn)_2O$ (see Chapter 8-1).

8-4-4 Condensations leading to Sn–Sn bonds

Until recently there were just a few indications that organotin hydrides might be able to interact with stannyl compounds carrying a negative substituent to form Sn–Sn bonds with elimination of HX.

$$R_3SnH + X^{\delta-}\text{–}Sn^{\delta+}R'_3 \rightarrow R_3SnSnR'_3 + HX \qquad (8\text{-}5)$$

For example, ditins were often obtained as by-products in the hydrostannation of aldehydes and ketones (Chapter 9-5) with trialkyltin hydrides. We now know that the primary adduct, a stannyl alkoxide, condenses with an excess of tin hydride according to reaction 8-5. (X is here $OCHR_2$ or OCH_2R.) With triaryltin hydrides and organotin dihydrides reaction 8-5 is so fast that the primary adduct cannot be isolated. The reaction has been turned into a smooth reduction technique for carbonyl compounds (see Chapter 10-1).

The long-known elimination of the trifluorovinyl group as trifluoroethylene (769) (reaction 8-5 with X = $CF{=}CF_2$) remained an isolated observation, as did the cleavage of the Sn–N bond in the adducts of organotin hydrides with azodicarboxylic esters (158) and azobenzene (295).

Recent work by the teams in Giessen and Utrecht has proved beyond doubt that these cleavage reactions constitute a very general and versatile type.

For the same organotin hydride, the rate of reaction 8-5 is seen to increase markedly with increasing nucleophilicity of the group X (or rather of the atom by which group X is attached to the tin). The following order has been found for X (588, 591):

$$\text{–}SeSnR_3 < \text{–}SSnR_3 < \text{–}OR \approx \text{–}PR_2 < \text{–}OSnR_3 < \text{–}N(R)CHO < \text{–}NR_2$$

For example, bistributyltin selenide and sulfide show no measurable reaction with tributyltin hydride at 160°C, while tributyltin methoxide at 110°C reacts within 180 hours, the dibutylphosphide within 130 hours, the stannoxane within 24 hours, and the diethylamide instantaneously and exothermically at 20°C (591).

Any $+I$ or $-I$ effect of the substituent R of course exerts additional influence; e.g. compounds $R'_3Sn^{\delta+}\text{–}X^{\delta-}$ react faster with X = $OSnBu_3$ than with X = $OSnPh_3$. Similar effects are observed in stannylamines (159) with X = NR_2. The less electron-withdrawal by R, the faster the reaction. As expected, the fastest reactions are shown by the most strongly nucleophilic dimethylamino- and diethylamino-derivatives (591, 815). Since these are easily prepared (see Chapter 15), they can be strongly recommended for condensation reactions such as 8-5 (591, 815). However, stannoxanes and alkoxides are also very useful (159a, 591, 724).

On the other hand, the reaction rate also increases with increasing electrophilicity of the hydrogen of the Sn–H group. For constant X (in reaction 8-5), one observes the order *s*-Bu_3SnH < *i*-Bu_3SnH < *n*-Bu_3SnH <

Ph_3SnH (591). This sequence is also reflected in the positions of the Sn–H absorption band, which moves in the same order from 1790 to 1808, 1810, and finally 1841 cm.$^{-1}$ (see also Chapter 23-2). As expected, triaryltin hydrides with electron-donor substituents on the benzene ring (e.g. –OMe) react more slowly than the corresponding unsubstituted tin hydrides (159, 159a).

It follows from the above, and also from dependence of reaction on solvent, that a polar mechanism is operating. All observations are compatible with an initial approach of the nucleophilic group X towards the hydride hydrogen, which is thus rendered positive and ultimately detached as a proton (159, 159a, 591). The suggested mechanism is further supported by the rates of condensation of tin hydrides with various *para*-substituted triethyltin phenoxides (159a). Here, electron-donor substituents on the phenoxy group increase the electron density on the oxygen in accordance with their $+I$ effect and their Hammett σ-constants. Reaction is thus enhanced, whereas it is retarded by electron-withdrawing substituents.

The mechanism might be formulated as follows:

$$\begin{array}{c} R_3Sn{-}H \\ \\ R'_3\overset{\delta+}{Sn}{-}\overset{\delta-}{X} \end{array} \rightleftharpoons \begin{array}{ccc} R_3\overset{\delta-}{Sn} & \cdots\cdots & \overset{\delta+}{H} \\ \vdots & & \vdots \\ R'_3\overset{\delta+}{Sn} & \cdots\cdots & \overset{\delta-}{X} \end{array} \longrightarrow \begin{array}{c} R_3Sn \\ | \\ R'_3Sn \end{array} + \begin{array}{c} H \\ | \\ X \end{array}$$

It shows once again that the Sn–H bond is easily polarized by an attacking reagent. The relevant molecular form $R_3Sn^{-}{-}H^{+}$ would have been considered as unacceptable only a few years ago.

$$\begin{array}{ccc} R_3\overset{\delta+}{Sn} & \cdots\cdots & \overset{\delta-}{H} \\ \vdots & & \vdots \\ \overset{\delta-}{X} & \cdots\cdots & \overset{\delta+}{Sn}R'_3 \end{array}$$

Depending on the nature of X, there is of course another possible intermediate state (above) identical with that in equation 8-4. Condensation reactions, especially those which are slow, are therefore complicated by ligand exchange (see Chapter 8-4-3). It follows that attempts to produce unsymmetrical substituted ditins $R_3SnSnR'_3$ yield the two symmetrical ditins as by-products. (For full discussion see refs. 591, 816.)

Provided that a suitable partner is chosen, best of all one with $X = NEt_2$, these condensations offer a smooth synthesis of hexaorganoditins, especially unsymmetrical ones (see Chapter 14-1).

Condensations are not confined to monofunctional stannyl compounds. Organotin di- and tri-hydrides react similarly to monohydrides. Again, the other organotin may carry not one, but two or three negative substituents X. Appropriate combinations lead to linear or branched tri-, tetra-, or poly-tins (see Chapter 14-3). Cyclic polytins are particularly easily prepared by this technique starting from R_2SnH_2 and R_2SnX_2 (see Chapter 14-4).

Whereas synthesis of cyclic polytins is unaffected by H–X exchange, the remaining condensations are even more susceptible to interference from ligand exchange reactions than the preparation of ditins, e.g. condensation of 2 moles of trialkyltin diethylamide with 1 mole of organotin dihydride gives an excellent yield of the tritin. But, the reverse combination—2 moles of trialkyltin hydride with 1 mole of dialkyltin tetraalkyldiamide—gives an almost inseparable mixture of homologous polymers $R_3Sn{-}(R'_2Sn)_n{-}SnR_3$ with $n = 0, 1, 2, 3, 4, 5, \ldots$ (816), especially with the lower alkyls. (For full discussion see ref. 816.)

Halide, cyanide, and carboxylate can be added to the list of substituents X suitable for condensation reactions 8-5, if an amine is added as acceptor for the HX liberated. Even then the reaction succeeds only with the more reactive aryltin hydrides (588, 591).

8-4-5 Reactions with metal alkyls

Organotin hydrides attack the metal–carbon bond of other metal alkyls. The exchange reaction with trialkylaluminums (reaction 8-6) has been the most intensively studied (274, 750). This also plays a part in hydrostannation with the aid of organoaluminum catalysts (576) (see Chapter 9-1-2).

$$R_3SnH + R'_3Al \rightleftharpoons \left[R_3Sn \begin{matrix} R' \\ \\ H \end{matrix} AlR'_2 \right] \longrightarrow R_3SnR' + R'_2AlH \tag{8-6}$$

The reverse reaction does not seem to occur. Exchange is therefore quantitative and may be complete in as little as 20 minutes, e.g. when stoichiometric amounts of the ethyl compounds are mixed at 80°C. *n*-Butyl and *n*-octyl groups are transferred almost as quickly, *i*-butyl much more slowly (see Table 8-3).

Most probably, the reaction proceeds via a four-centre intermediate state (576, 750),* just as in the case of the deuteride–hydride equilibrium

* Such a state is well known and stable in the trialkylaluminum dimers. Association is again less pronounced for *i*-butyls than for *n*-alkyls (928).

Table 8-3
Half-times of the Exchange Reaction 8-6 *at* 80°C *as a Function of R and R'* (750)

$R' = n$-Bu, R =	Et	Pr	n-Bu	C_8H_{17}	i-Bu	s-Bu	t-Bu
$t_{\frac{1}{2}}$ (min.)	6.2	14.8	14.0	13.0	120	420	>10,000
$R = n$-Bu, R' =	Et		n-Bu	C_8H_{17}	i-Bu		
$t_{\frac{1}{2}}$ (min.)	13.2		14	23	131		

(Chapter 8-4-3). The effect of volume of the group R on reaction rate also indicates a four-centre state: the rate increases markedly in the order R = t-Bu ≪ s-Bu < i-Bu ≪ n-Bu. For less sterically hindered organotin hydrides, the rate of exchange increases with increasing negativity of the hydride hydrogen: $Me_3SnH < C_8H_{17}SnH_3 < (C_8H_{17})_2SnH_2 < t\text{-}Bu_2SnH_2$. Compounds which can saturate the coordination sphere of the aluminum by complexing, e.g. ethers or amines, inhibit the exchange reaction 8-6 altogether.

The reaction of phenyllithium with triphenyltin hydride is formally analogous: the products are tetraphenyltin and lithium hydride (247). [There are contradictory reports concerning the reaction of triphenyltin hydride with methyllithium (251, 895).] The hydrogen behaves here like other electronegative substituents, e.g. electronegative organic groups, which are readily displaced by alkyllithiums [see the work of D. Seyferth et al. (765a)].

Labile C–Pb bonds like those in lead allyl compounds give an equilibrium

$$\geqq SnH + \geqq PbCH_2CH{=}CH_2 \rightleftarrows \geqq SnCH_2CH{=}CH_2 + \geqq PbH$$

Only a little lead hydride is formed, but sufficient for some smooth reactions of this highly active species. Lead–oxygen bonds behave similarly (554, 554b), e.g. in organolead acetates which have been found especially useful (157a):

$$Ph_3SnH + Bu_3PbOCOMe \xrightarrow{HC\equiv CR} Bu_3PbCH{=}CHR + Ph_3SnOCOMe$$

C–Hg bonds, too, are cleaved readily by organotin hydride with production of various condensates. Sn–Hg bonds must be present in some intermediates. The reactions are discussed in Chapter 18.

8-4-6 Other reactions

Organotin hydrides react with aliphatic diazo-compounds (462) to yield ultimately the tetraorganotin. Many of the reactions are catalyzed by copper.

$$R_3SnH + CH_2N_2 \xrightarrow{0^\circ} R_3SnMe + N_2$$

$$R_2SnH_2 + 2\,CH_2N_2 \xrightarrow{0^\circ} R_2SnMe_2 + 2\,N_2$$

Reaction with diazo-esters and diazo-ketones gives functionally substituted organotins (see also Chapters 3-4 and 9).

$$R'CHN_2 + HSnPr_3 \xrightarrow{70^\circ,\ Cu} R'CH_2SnPr_3 + N_2$$

$$(R' = EtOCO,\ MeCO,\ PhCO,\ NC\text{–})$$

These reactions deserve further attention because of a possible connection with carbene chemistry. Carbene intermediates are claimed for the following insertion reaction (and others) starting with sodium trichloroacetate (131a):*

$$Et_3SnH \xrightarrow{:CCl_2} Et_3SnCHCl_2$$

Carbene-producing species with C–Br bonds [e.g. $PhHgC(Cl_2)Br$, $PhHgC(Cl)Br_2$ and $PhHgCBr_3$], which are successful in insertion reactions into Si–H and Ge–H bonds, yield only R_3SnBr in reaction with R_3SnH (765b).

Some organotin hydrides (especially slightly impure samples) decompose on prolonged storage, with evolution of hydrogen. With the aid of suitable catalysts, the reaction can be used to prepare hexaorganoditins and aliphatic or aromatic cyclic polytins (see Chapter 14-4):

$$R_3SnH + HSnR_3 \xrightarrow{Cat.} R_3SnSnR_3 + H_2$$

$$n\,R_2SnH_2 \xrightarrow{Cat.} (\text{–}SnR_2\text{–})_n + n\,H_2$$

Trihydrides similarly give red oils containing branched tin chains, but these have not been studied in detail.

Diverse materials are catalytically active towards this reaction: heavy metals [especially Pd, Ni, Pt (609)], bases† (e.g. pyridine, triethylamine), formamide (439, 548, 825), sodium methoxide, B_2H_6 (110), organotin halides, and others. The mechanism is unlikely to be the same for all catalysts; only for sodium methoxide has it been elucidated completely (435). Catalytic elimination of hydrogen from dialkyltin dihalide plus

* Other workers were not able to confirm these results, and found only $R_3SnOCOCHCl_2$ (175c), which may decarboxylate on heating.

† H_2 elimination with the aid of primary amines was at one time interpreted erroneously as deamination with evolution of ammonia (366, 617). See also Chapter 10.

amine may involve a dialkylhalotin hydride (Chapter 8-2) as an intermediate stage which readily eliminates hydrogen to give the tetraalkylditin dihalide (see Chapter 14-2). The latter then forms the corresponding cyclic polytin, perhaps by repeated hydrogen–chlorine exchange and elimination of H_2 (see Chapter 14-4). Details of this reaction await elucidation.

Various isolated observations, too numerous to mention here, complete the picture of the versatility of organotin hydrides. Reference should be made to review articles (143, 315, 429, 551, 616, 828) and to the literature surveys published annually since 1964 (768).

9 Hydrostannation of unsaturated compounds

For a long time, alkyl and aryl groups could be grafted on to tin only with the aid of highly reactive organometallic reagents. In consequence, the number of organotin compounds with functional groups in the alkyl or aryl portion was small, and the choice of functional groups was restricted (see Chapter 3-4). This picture changed completely with the discovery that organotin hydrides would add easily to unsaturated linkages, sometimes with and sometimes without a catalyst. This reaction, nowadays called hydrostannation, in a very short time became one of the most fruitful synthetic pathways in the entire field of organotin chemistry. It has now been supplemented by the equally versatile addition of Sn–N- and Sn–O-compounds to unsaturated systems (see Chapter 19).

Additions of this kind had been known for some time for organoaluminum hydrides (928) and organoboron hydrides (97), but in these cases the olefins have to be devoid of other functional groups because of high reactivity of the Al–H and B–H bonds. In Group IV, organosilicon (656) and organogermanium hydrides (256) had already been added successfully to olefins. Similar reactions have succeeded recently with organolead hydrides (571).

9-1 OLEFINS

In 1956 van der Kerk and his associates reported that organotin hydrides could add to α-olefins $R'CH{=}CH_2$, where R' could contain functional groups. Experimental data multiplied rapidly (361, 363, 366, 553, 616).

These uncatalyzed hydrostannations sometimes take place instantaneously, more often within a few hours at 60—100°C. Aliphatic organotin hydrides react more slowly than aromatic ones; with some olefins, e.g. *n*-octene or allyl alcohol, they fail entirely unless a catalyst is added (see below). Electron-withdrawing substituents on the $H_2C{=}CH-$ group, such as Ph or CN, increase the receptivity of the olefin substrate to a considerable extent. The stannyl group may add either at the terminal carbon (reaction 9-1) or at the inner carbon (reaction 9-2) of the olefinic double bond.

$$R'CH{=}CH_2 + HSnR_3 \longrightarrow R'CH_2CH_2SnR_3 \quad (9\text{-}1)$$

$$R'CH{=}CH_2 + HSnR_3 \longrightarrow R'CH(SnR_3)Me \quad (9\text{-}2)$$

Early experiments with radical sources and radical scavengers were unsuccessful (227, 361, 363, 366, 616); the reaction was therefore assumed originally to be heterolytic and to proceed as 9-1 via a four-centre mechanism (472, 553, 620). Degradation with Br_2 (see Chapter 4) led to the conclusion that the straight-chain adduct was formed as the major product in all cases examined (reaction 9-1) (361, 363, 366, 616); e.g. the adduct of tributyltin hydride and acrylonitrile was thought to be a pure product with the structure $NCCH_2CH_2SnBu_3$ (616).

More recently, gas-chromatography has shown the adducts to be far from uniform, at least in the few cases that have been studied carefully, e.g. acrylonitrile and styrene (471, 554, 814). Addition of aliphatic organotin hydrides to the strongly polar acrylonitrile gives up to 80% of branched product (reaction 9-2), depending on the nature of R and R′. Polar solvents, such as isobutyronitrile, increase this percentage. The mechanism is polar in this case, but does not involve the four-centre intermediate assumed originally (620). Details, such as steric effects and the influence of inductive effects on the Sn–H and C=C groups, still require clarification.

9-1-1 Use of radical catalysts

Besides the mechanism just discussed, there is also a radical mechanism leading to the straight-chain product of reaction 9-1; with triphenyltin hydride this is indeed the only mechanism. This radical mechanism was discovered by the Giessen group in 1961 (577) and has since been explored by them in detail (553, 578, 597). In the examples investigated, scavengers such as hydroquinone or, far better, a particular phenoxy radical called Galvinoxyl* [formula (**1**) in Chapter 8] inhibit the uncatalyzed hydro-

* This was recognized as an excellent scavenger by Bartlett and Funahashi (60), and has proved particularly valuable in organotin hydride chemistry (597) (see also Chapter 8-1).

stannation appreciably or halt it entirely. Radical initiators like azobisisobutyronitrile promote the reaction strongly. This hydrostannation is therefore unambiguously a radical process. In the absence of catalysts, it appears the predominant one for α-olefins. The aforementioned polar mechanism appears to operate only for strongly electron-withdrawing groups R′ or in consequence of particular steric effects (see ref. 471). Certainly, free-radical sources ensure a uniformly radical mechanism and consequently a uniform product. Figure 9-1 illustrates these relation-

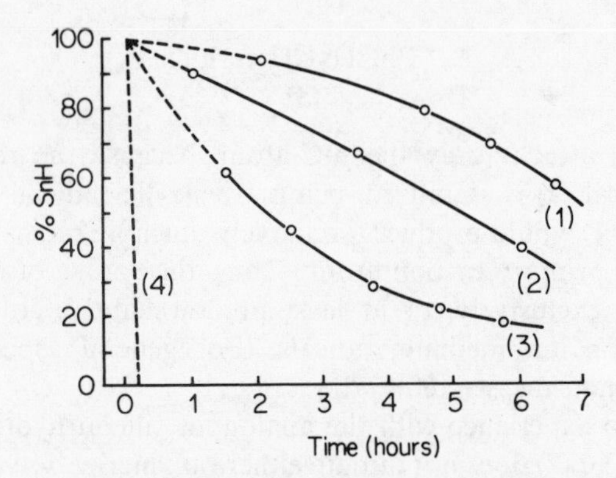

FIG. 9-1 Styrene stabilized with scavengers (commercial monomer) undergoes hydrostannation by triethyltin hydride in about 7 hours at 100°C (curve 2). Slow at first, the reaction gathers speed as the scavenger is used up. Unstabilized styrene reacts faster from the very beginning (curve 3). Small amounts of azobisisobutyronitrile enhance the rate greatly (curve 4); scavengers, e.g. galvinoxyl [Chapter 8, formula (**1**)] retard strongly (curve 1) (553, 597)

ships, using styrene as an example. Radical initiators cause rapid (Figure 9-1, curve 4) and uniform production of the straight-chain adduct as in reaction 9-1. Uniformity has been attested by gas-chromatography (554, 814), which has a lower detection limit of $\leqslant 0.1\,\%$ for the branched adduct. Entirely uniform reaction has also been observed in other radical-catalyzed hydrostannations (554, 814), e.g. with acrylonitrile (471).

It is remarkable also that styrene can react at 100°C in the presence of radical initiators (Figure 9-1, curve 4) without any polystyrene being formed. An entirely normal radical-chain process takes place, starting with a short-lived radical derived from the initiator (see Chapter 8-4-1). A stannyl radical (**1**) remains and reacts further as in the following scheme

(553, 597):

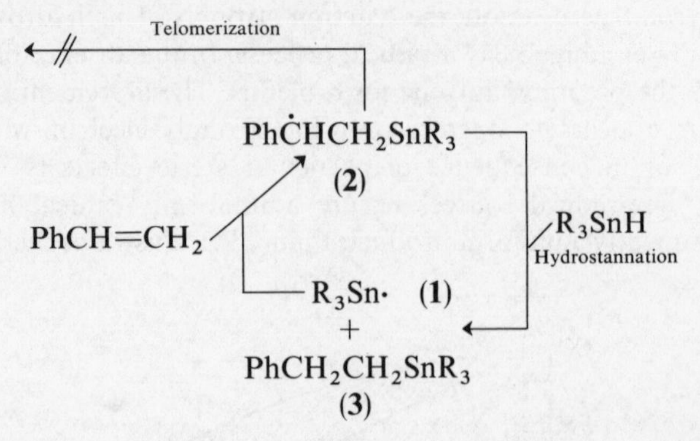

Radical (**1**) attacks only the β-C-atom, because the resulting intermediate radical (**2**) is stabilized, being a benzylic radical. [Attack of an α-C-atom by (**1**) would produce a relatively unstable radical with the odd electron on a primary carbon atom.] Thus, the course of radical hydrostannation is exclusively, or at least predominantly, governed by the stability of the intermediate radicals. (For general aspects of radical reaction mechanisms, see ref. 677b.)

Contrary to experience with the analogous silicon hydrides (221), the resulting radical (**2**) does not initiate either polymerization or telomerization of the olefin, but is trapped exclusively by further organotin hydride to yield the end-product (**3**) and another stannyl radical (**1**). Thus, organotin hydrides are seen once again to be powerful scavengers (see also Chapter 8-4-1). In the uncatalyzed hydrostannations referred to earlier, the part of the catalyst radical seems to be taken by a thermal radical from the olefin. H. G. Kuivila (429) arrived at the same conclusion from different premises.

For preparative purposes, the most successful catalysts have proved to be azobisisobutyronitrile, azoisobutyric esters, azocyanocyclohexane, and bisbenzoxydiimide (553, 597). Catalysts can be used to advantage even with the more reactive aromatic tin hydrides (210).

The first step of the above radical-chain mechanism is reversible (439a, 555, 555a, 812a) (more pronounced with internal than with α-olefins):

$$R_3Sn^{\cdot} + \;>C=C<\; \rightleftharpoons R_3Sn-\overset{|}{\underset{|}{C}}-\dot{C}<$$

Thus, in the presence of stannyl radicals, pure isomers of olefins equilibrate to a *cis–trans* mixture, in some cases quite rapidly. In the case of styrene (above) this has been demonstrated by means of deuterostyrene.

The preparative application of radical catalysis has been pursued particularly by the author's team (553, 578). Catalysis allows hydrostannation of olefins to be effected under milder conditions. This is especially helpful for rather sensitive compounds, quite apart from the advantage of uniform reaction referred to earlier. Some examples of currently accessible materials are given in Table 9-1. A comprehensive list of preparations including many references and full discussion will be found in references 495b and 553. It is now feasible for α-olefins to interact smoothly with all three classes of organotin hydride:

$$R'CH{=}CH_2 + HSnR_3 \rightarrow R'CH_2CH_2SnR_3$$

$$2\,R'CH{=}CH_2 + H_2SnR_2 \rightarrow (R'CH_2CH_2)_2SnR_2$$

$$3\,R'CH{=}CH_2 + H_3SnR \rightarrow (R'CH_2CH_2)_3SnR$$

Dialkyltin dihydrides react in stages; the mono-adduct can be isolated (134, 578):

$$R'CH{=}CH_2 + H_2SnR_2 \rightarrow R'CH_2CH_2SnHR_2$$

So far, only α-olefins have been discussed. Other olefins can also undergo hydrostannation, but the number of examples is as yet small (361, 363, 366, 597, 616, 651). It seems to be a necessary condition that the C=C group must have a neighboring electron-withdrawing group capable also of stabilization of the intermediate carbon radical by mesomerism; e.g. cyclohexene and methyl oleate do not react even in the presence of radical-forming substances (553), whereas crotonic esters react smoothly. Strained internal double bonds are more active; e.g. cyclopentene adds organotin hydrides slowly, with azobisisobutyronitrile as catalyst (555).

The reaction is rather slow at 80°C and can be halted completely by a little galvinoxyl. It is accelerated greatly by 1 mole % of azobisisobutyronitrile to reach completion within 30 minutes at most (597). Identical products are obtained with or without catalysis (597). Degradation with Br_2 (see Chapter 4) gives the structure (616):

$$MeCH{=}CHCOOR' + R_3SnH \rightarrow MeCH(SnR_3)CH_2COOR'$$

Since 1963, accounts of radical mechanisms in α-olefin hydrostannation have come also from the groups of H. C. Clark (134, 137) and J. Valade (859). They have shown that UV light, too, acts as an accelerator, presumably by forming radicals from the olefin. (Organotin hydrides themselves absorb UV irradiation only at rather short wavelengths.)

Catalysis by UV light was developed subsequently into a practical method (206a, 439a, 812a), and it is now a useful hydrostannation technique. It can be recommended especially for such olefins as react rather

Table 9-1
Hydrostannation of Olefins[a]

OLEFIN	ORGANOTIN HYDRIDE	PRODUCT	B.P. (°C/mm.)	M.P. (°C)	n_D^{20}	METHOD OF PREPARATION
Isobutene	Me_3SnH	$Me_3SnCH_2CHMe_2$	58/37	—	1.4552	812a
2,3-Dimethyl-1-butene	Me_3SnH	$Me_3SnCH_2CH(Me)CHMe_2$	80/22	—	1.4649	812a
Cyclopentene	Me_3SnH	Me_3Sn cyclo-C_5H_9	77/25	—	1.4884	812a
Cyclohexene	Me_3SnH	Me_3Sn cyclo-C_6H_{11}	74—76/10	—	1.4937	812a
Acrylonitrile	Et_3SnH	$Et_3SnCH_2CH_2CN$[b]	128—130/12	—	1.4912	578
Styrene	Et_3SnH	$Et_3SnCH_2CH_2Ph$	153—154/10[c] 110—120/4	—	1.5260	578
p-Divinylbenzene	Et_3SnH	$Et_3SnCH_2CH_2C_6H_4CH{=}CH_2$-*p*	107—109/0.001	—	1.5481	578
Triethylvinyltin	Et_3SnH	$Et_3SnCH_2CH_2SnEt_3$	109/0.35	—	1.5156	812
Methyl acrylate	Pr_3SnH	$Pr_3SnCH_2CH_2COOMe$	145—150/12	—	—	363
Cinnamonitrile	Pr_3SnH	$Pr_3SnCH(CN)CH_2Ph$	130—141/0.0003	—	—	363
1,5-Hexadiene[d]	Bu_3SnH	*n*-$Bu_3Sn(CH_2)_4CH{=}CH_2$	104/0.3	—	1.4795	578
Vinyl butyl ether	*i*-Bu_3SnH	*i*-$Bu_3SnCH_2CH_2O$*n*-Bu	108—110/0.6	—	1.4713	578
1-Octene	*i*-Bu_3SnH	*i*-$Bu_3Sn(CH_2)_7Me$	111—114/0.2	—	1.4753	578
Allyl cyanide	Ph_3SnH	$Ph_3SnCH_2CH_2CH_2CN$	—	80—81	—	363
Acrylamide	Ph_3SnH	$Ph_3SnCH_2CH_2CONH_2$	—	123—124	—	363
Acrolein diethyl acetal	Ph_3SnH	$Ph_3SnCH_2CH_2CH(OEt)_2$	—	35.5—37.5	—	363
Allyl alcohol	Ph_3SnH	$Ph_3SnCH_2CH_2CH_2OH$	—	105	—	363
Triphenylallyltin	Ph_3SnH	$Ph_3SnCH_2CH_2CH_2SnPh_3$	—	182—185	—	366
p-Triphenylstannylstyrene	Ph_3SnH	*p*-$Ph_3SnC_6H_4CH_2CH_2SnPh_3$	—	184—186	—	618

Tetrafluoroethylene	Me_2SnH_2	$Me_2Sn(H)CF_2CF_2H$	77.9	—	—	134
Methyl methacrylate	Et_2SnH_2	$Et_2Sn(H)CH_2CH(Me)COOMe$	63—65/0.5	—	1.4778	578
Acrylonitrile	Et_2SnH_2	$Et_2Sn(CH_2CH_2CN)_2$	144—147/2.5	—	1.5088	578
Styrene	Pr_2SnH_2	$Pr_2SnCH_2CH_2Ph$	172/0.0015	—	—	366
1,7-Octadiene[d]	Bu_2SnH_2	$Bu_2Sn[(CH_2)_6CH{=}CH_2]_2$	135/0.001	—	1.4826	576
Methyl acrylate	Ph_2SnH_2	$Ph_2SnCH_2CH_2COOMe$	191—194/0.003	—	—	366
Methyl methacrylate	i-$BuSnH_3$	i-$BuSn[CH_2CH(Me)COOMe]_3$	165—167/0.01	—	1.4845	578
Methyl acrylate	$BuSnH_3$	$BuSn(CH_2CH_2COOMe)_3$	139—141/0.0004	—	—	366
1,4-Pentadiene	Et_2SnClH	$Et_2Sn(Cl)(CH_2)_5Sn(Cl)Et_2$	—	42	—	580, 581
Acrylonitrile	Et_2SnHI	$Et_2Sn(I)CH_2CH_2CN$	102/0.02	—	—	580, 581
Methyl acrylate	Bu_2SnHCl	$Bu_2Sn(Cl)CH_2CH_2COOMe$	122/0.06	—	1.4991	648
Allyl acetate	Bu_2SnHCl	$Bu_2Sn(Cl)CH_2CH_2CH_2OCOMe$	112/0.0001	—	1.4995	736
Allyl alcohol	i-Bu_2SnHCl	i-$Bu_2Sn(Cl)CH_2CH_2CH_2OH$	152/0.4	—	1.5162	580, 581
1-Octene	i-Bu_2SnHCl	i-$Bu_2Sn(Cl)(CH_2)_7Me$	122/0.2	—	1.4871	580, 581

[a] For numerous other examples see refs. 429 and 495b.

[b] The uniform product can only be made with the aid of radical-initiator catalysts (578). The uncatalyzed product, formerly regarded as uniform, has now been shown to be a mixture of isomers (471).

[c] A B.P. of 110—120°C/10 mm. has been given in error (578) instead of 110—120°C/0.1 mm.

[d] Using a large excess of olefin. The di-adduct is formed as well as the mono-adduct.

slowly or not at all with radical initiators as catalysts. Thus, isolated terminal or non-activated internal C=C groups, e.g. in 2-butene, 2-methyl-2-butene, 2,3-dimethyl-1-butene, cyclohexene, or cycloheptene, react satisfactorily under UV irradiation (see also Table 9-1).

9-1-2 Use of organometallic catalysts

In addition to the direct radical hydrostannation of olefins, just discussed, an indirect pathway has been opened up by the work of the Giessen group. Organotin hydrides have been found to react quantitatively with trialkylaluminums by hydride–alkyl exchange as in reaction 9-3 (see also Chapter 8-4). In presence of α-olefins, the resulting dialkylaluminum regenerates the trialkylaluminum as shown by K. Ziegler and his collaborators (926) (reaction 9-4). The overall reaction (9-5) therefore amounts to a hydrostannation of the olefin. 1—5 mole % of the organoaluminum suffice for adequate catalysis of the hydrostannation, as long as air and moisture are carefully excluded.

$$(R'CH_2CH_2)_3Al + HSnR_3 \rightarrow R'CH_2CH_2SnR_3 + (R'CH_2CH_2)_2AlH \quad (9\text{-}3)$$

$$(R'CH_2CH_2)_2AlH + R'CH{=}CH_2 \rightarrow (R'CH_2CH_2)_3Al \quad (9\text{-}4)$$

$$R'CH{=}CH_2 + R_3SnH \rightarrow R_3SnCH_2CH_2R' \quad (9\text{-}5)$$

Small amounts of transition metals, e.g. finely divided nickel prepared from the acetylacetonate, lead to slight improvement (590). Interaction of organotin di- or tri-hydrides and α,ω-dienes yields linear or branched and crosslinked polymers respectively, in which the main polymer chain carries tin atoms at regular intervals (see Chapter 21).

The advantage of this technique lies in the particularly smooth reaction of aliphatic organotin hydrides with unsubstituted α-olefins, vinylidenes, and dienes. This can be quantitative within 3 hours at 70—80°C (Table 9-1). Organometallic catalysis is here often superior to radical catalysis (Chapter 9-1-1). The former also permits ready hydrostannation of allyltins (590), which otherwise gives poor yields without catalysts (363, 366) and shows no improvement with radical initiators (553). Admittedly, the advantages must be offset against considerable limitation of the choice of functional groups, since most of these react irreversibly with alkylaluminums. It is possible, however, to use olefins with a second inner or ring –C=C–, which can be transformed subsequently into some other functional group, e.g. by epoxidation or other suitable olefinic reaction.

9-1-3 Addition of alkylhalotin hydrides

Substances of the type $R'SnR_2X$ or $R'SnRX_2$, carrying functional groups on R′ and one or two halogens or other negative groups on the tin, have proved of interest in some syntheses and technical applications (see Chapter 25). In the past, these were obtained exclusively by cleavage of alkyl moieties from suitable tetraalkyltins (see Chapter 4-2).

A more attractive pathway derives from the observation that dialkyltin dihydrides react with dialkyltin dihalides to give halotin hydrides (see Chapter 8-2). These are far more reactive than the dihydrides themselves, and are eminently suitable for hydrostannations (580, 581), e.g. reaction 9-6. Most olefins so far tried react at room temperature, often with heat evolution (see Table 9-1). Radical sources act as catalysts, but can usually be dispensed with.

$$HOCH_2CH{=}CH_2 + HSn(Cl)i\text{-}Bu_2 \xrightarrow{cat.} HOCH_2CH_2CH_2Sn(Cl)i\text{-}Bu_2 \tag{9-6}$$

Similar additions using derivatives of the type $RSn(Hal)_2H$ (580, 581) have also been successful.

9-2 CONJUGATED DIENES

Hydrostannation of butadiene and its derivatives is accelerated by free-radical sources and retarded by scavengers, i.e. it proceeds by a radical mechanism (596). This may be illustrated by the example of 2,3-dimethylbutadiene (**4**). Attack by a stannyl radical produces a mesomeric radical (**5**), which is then trapped either in form (**5a**) to yield the 1,2-adduct (**6**) or in form (**5b**) to effect 1,4-hydrostannation (**7**). Abstraction of hydrogen from the R_3SnH creates another stannyl radical which then attacks a further molecule of diene (**4**). Trapping of form (**5b**) is favored, since this form carries the odd electron on a primary C-atom (262); the ratio (**6**):(**7**) depends also on the nature of the trapping radical, i.e. on R in R_3SnH (600) (see reaction scheme).

The ratio of 1,2- to 1,4-hydrostannation* is affected further by substituents on the conjugated system (596). [Statements relating to reactions of R_3SnH are also valid for R_2SnH_2 and $R_2Sn(H)X$, and probably for $RSnH_3$ too.] Attack by the stannyl radical takes place almost exclusively at the unsubstituted end of the conjugated system, i.e. at the terminal carbon as with α-olefins (see Chapter 9-1), and accounts for all products isolated.

* Analogous experiments with thiols yield 1,2-, 1,4-, and 4,1-adducts (640).

The resulting radical—analogous to (**5a**) ↔ (**5b**), but with –CHR at the right-hand end—is bound to have its odd electron on a secondary carbon.

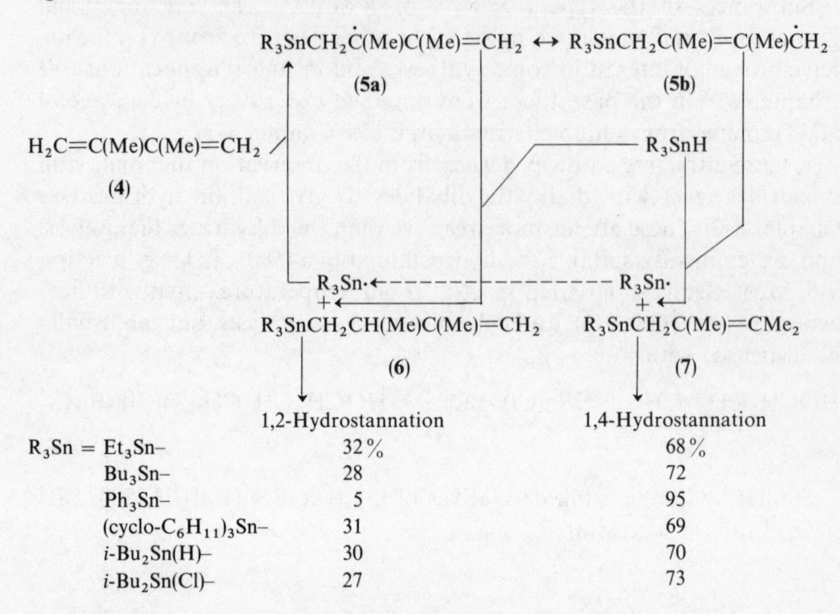

	1,2-Hydrostannation	1,4-Hydrostannation
R_3Sn = Et_3Sn–	32%	68%
Bu_3Sn–	28	72
Ph_3Sn–	5	95
(cyclo-$C_6H_{11})_3Sn$–	31	69
i-$Bu_2Sn(H)$–	30	70
i-$Bu_2Sn(Cl)$–	27	73

The proportion of 1,2-addition then rises to 34%. The 1,4-adduct has predominantly *trans*-configuration (intense absorption at 950—960 cm.$^{-1}$).

The case of 1,3-pentadiene (piperylene) is now understood in some detail (555b). With either radical-initiator catalysts or UV irradiation, six adducts have been isolated and identified unambiguously: the *cis*- and *trans*-forms of the 1,2-, 1,4- (main product), and 4,1-adducts. Results obtained with trimethyl- and triethyl-tin hydride are essentially the same. Using pure *cis*- or *trans*-piperylene a high degree of stereospecificity of the Sn–H addition has been found, at least for the early part of the reaction. Besides, there is isomerization both of olefin (because of reversibility of the primary attack by stannyl radicals; see above) and of addition products. The influence of each of these factors depends greatly on the conditions, especially the mole ratio of compounds in the reaction.

Hydrostannation of 1,3-pentadiene had previously been reported to yield only four isomeric products (153, 664), which were assumed to be the *cis*- and *trans*-isomers of the 1,2- and 2,1-adducts. 1,4-Addition was not then taken into account, but this is in fact the main process (596). Conversely, other authors (137) have reported hydrostannation of butadiene to yield a 1,4-adduct only. The products are stated to show IR absorptions at 902 and 906 cm.$^{-1}$ respectively, but the significance of these bands is not discussed in the paper. In fact, they are C–H deformation frequencies of a vinyl group and thus prove the presence of some 1,2-adduct in the mixture. Agreement with other work (596) seems quite good after all.

9-3 ALLENES

Cumulative systems C=C=C can add organotin hydrides, reaction being subject to radical catalysis. Thorough investigation convinced H. G. Kuivila and his co-workers (438) that polar effects exert only slight, if any, influence on the initial step, attack by the stannyl radical $R_3Sn\cdot$ (R was Me). The nucleophilic stannyl radical behaves here, in principle, like the electrophilic ethyl-thiyl radical. The first product should generally be the mesomeric allyl radical (**8**) (R′ = H or Me):

$$R'_2C{=}C{=}CR'_2 + R_3Sn\cdot \longrightarrow R'_2\dot{C}{-}C(SnR_3){=}CR'_2 \longleftrightarrow R'_2C{=}C(SnR_3){-}\dot{C}R'_2$$

(8)

The nature of R′ and some further factors then determine which canonical form of (**8**) is ultimately stabilized by hydrogen-abstraction from another molecule of R_3SnH. (Structure and constitution of all hydrostannation products from trimethyltin hydride and various allenes were established.) The end-products from the trimethyl and the two dimethyl derivatives carry the stannyl group exclusively on the middle carbon of the original C=C=C system. The monomethyl derivative gives 70% and allene itself 45% attachment to the central carbon, the remaining products having stannyl groups on outer carbons. A possible explanation may lie in intermediate formation of a cyclic radical (**9**) involving pentacoordinate tin:

$$H_2C{=}C{=}CH_2 + R_3Sn\cdot \longrightarrow \underset{(\mathbf{9})}{H_2C{-}C{=}CH_2 \;(\cdots R_3Sn\cdots)} \rightleftharpoons \underset{(\mathbf{8})}{H_2\dot{C}{-}C(SnR_3){=}CH_2}$$

$$(\mathbf{9}) \rightleftharpoons \underset{(\mathbf{10})}{H_2C(SnR_3){-}\dot{C}{=}CH_2}$$

This could then become stabilized by R_3SnH in either of the equilibrium forms (**8**) or (**10**).

These results may have significance also for the better understanding of radical hydrostannation of α-olefins (Chapter 9-1). In particular, they may provide an answer to the question why the stannyl group there becomes attached exclusively to the terminal carbon atom.

9-4 ALKYNES

Hydrostannation of both terminal and internal C≡C groups goes much more smoothly than with the corresponding olefins (361, 363, 366, 553), and can be strongly exothermic. The resulting vinyl stannyl compounds (Table 9-2) can add a second molecule of organotin hydride in a noticeably less vigorous reaction.* Treatment of α,ω-diynes with organotin dihydrides leads to formation of unsaturated polymers (see Chapter 21).

Diacetylene derivatives yield stannyl-en-ynes $R_3SnCH{=}CHC{\equiv}CR'$ (911). Organohalotin hydrides also add easily to alkynes (580, 581).

These hydrostannations were thought originally to proceed by a polar four-center mechanism (363). However, the *cis*-compound (**11**) is the primary product from phenylacetylene, which then rearranges to the *trans*-compound (**12**) (228). These results are incompatible with a four-center mechanism.

$$PhC{\equiv}CH + HSnR_3 \rightarrow PhC(H){=}C(H)SnR_3 \quad \begin{matrix} (\mathbf{11})\ cis \\ (\mathbf{12})\ trans \end{matrix}$$

A recent re-investigation (465c) has shown that stannyl radicals catalyze the rearrangement of the primary *cis*-products of type (**11**) (formed by *trans*-addition of organotin hydride) to *trans*-isomers of type (**12**). Attack of stannyl radicals on the C=C group produces an ethyl radical with two stannyl groups.* Elimination of one of these leads to formation of the isomerized product. The course of isomerization is determined by several factors, including steric ones. The latter are evident from conformational analysis of the intermediate ethyl radical.

Furthermore, free-radical sources accelerate the reaction with acetylene (809) as well as with other alkynes (577, 597) (Table 9-2). An investigation of phenylacetylene and triethyltin hydride at 80°C showed strong inhibition by certain scavengers and good promotion by radical sources (e.g. azobisisobutyronitrile) (597). One possible pathway for hydrostannation of alkynes is therefore a radical one. All cases of radical catalysis investigated to date show exclusive production of the straight-chain adduct $R'CH{=}CHSnR_3$.

This was also thought to be the sole product in the uncatalyzed hydrostannation of alkynes (363, 616), until D. Seyferth and his associates (776)

* Involvement of the 1,2-position of the two stannyl residues had been assumed (363, 616), but reinvestigation showed that both are bound to the same carbon atom in most of the examples (473b).

Table 9-2
Hydrostannation of Alkynes

ALKYNE	ORGANOTIN HYDRIDE	PRODUCT	M.P. (°C)	B.P. (°C/mm.)	n_D^{20}	METHOD OF PREPARATION
Diethyl acetylene-dicarboxylate	Me_3SnH	$Me_3SnC(COOEt){=}CHCOOEt$[a]	—	71—72/0.07	1.4845	468
Ethyl propiolate	Et_3SnH	mixture[c]	—	several fractions	—	469
Phenylacetylene	Et_3SnH	$Et_3SnCH{=}CHPh$[b]	—	86/0.15	1.5579	468
1-Hexyne	Pr_3SnH	$Pr_3SnCH{=}CHBu$[b]	—	75—80/0.1	—	366
Methyl propiolate	Pr_3SnH	$Pr_3SnCH_2CH(COOMe)SnPr_3$	—	120—123/0.0002	—	366
Acetylene	Bu_3SnH	$Bu_3SnCH{=}CH_2$	—	88/0.8	—	809
Propiolonitrile	Bu_3SnH	$Bu_3SnC(CN){=}CH_2$	—	100/0.1	1.4928	468
Diphenylacetylene (Tolan)	$Bu_2Sn(Cl)H$	$Bu_2Sn(Cl)C(Ph){=}CHPh$	—	152/0.0001	1.5913	586
Acetylene	Ph_3SnH	$Ph_3SnCH_2CH_2SnPh_3$	206—207	—	—	366
Phenylacetylene	Ph_3SnH	*trans*-$Ph_3SnCH{=}CHPh$	119—120	—	—	366, 228
Phenylacetylene	$2Ph_3SnH$	$Ph_3SnCH_2CH(Ph)SnPh_3$	139—140	—	—	366
Diphenylacetylene (Tolan)	Ph_3SnH	$Ph_3SnC(Ph){=}CHPh$	105—106	—	—	592

[a] Mainly the fumaric, plus a little of the maleic derivative (ratio 8:1).
[b] *cis*- and *trans*-Compounds; also probably small amounts of $Et_3SnC(Ph){=}CH_2$ or $Pr_3SnC(Bu){=}CH_2$(468).
[c] Apart from *cis*- and *trans*-$Et_3SnCH{=}CHCOOEt$, also $Et_3SnC(COOEt){=}CH_2$, $Et_3SnCH_2CH_2COOEt$, and $Et_3SnC{\equiv}CCOOEt$.

came across small amounts of the branched product (**13**):

$$MeC{\equiv}CH + HSnMe_3 \xrightarrow{100^\circ} \begin{cases} MeCH{=}CHSnMe_3 \; (cis + trans) & 96\% \\ MeC(SnMe_3){=}CH_2 \; (\mathbf{13}) & 4\% \end{cases}$$

Uncatalyzed hydrostannation of propiolic esters was also subsequently found to give non-uniform products (467, 469):

$$R'OCOC{\equiv}CH + HSnR_3 \longrightarrow \begin{cases} R'OCOCH{=}CHSnR_3 \; (cis + trans) \\ R'OCOC(SnR_3){=}CH_2 \end{cases}$$

Among them are 20—45% of theoretical yield of higher-boiling materials and, in addition, smaller amounts of $R'OCOCH_2CH_2SnR_3$ and $R'OCOC{\equiv}CSnR_3$ produced in hitherto unexplained side-reactions. The composition of the product depends largely on the nature of R, and rather less on R′. Propargyl alcohol $HC{\equiv}CCH_2OH$ and ethyl tetrolate $MeC{\equiv}CCOOEt$ react in much the same way, whilst the more polar propiolonitrile yields exclusively and exothermically α-stannylacrylonitrile (in each case in the absence of catalysts).

$$NCC{\equiv}CH + R_3SnH \rightarrow NCC(SnR_3){=}CH_2$$
$$(R = Me, Et, Bu)$$

It can now be seen (465d, 467, 468) that, in uncatalyzed hydrostannation of alkynes, radical and polar reactions proceed side-by-side, and that they yield different products. The polar reaction is favored by polar solvents and by polarizing substituents on the C≡C group; in an extreme case (see above) it may become the only reaction

$$\overset{\delta-}{X}{-}C{\equiv}\overset{\delta+}{C}{-}R' \;\; (\overset{\delta-}{H}{-}\overset{\delta+}{Sn}R_3) \longrightarrow \overset{\ominus}{C}(X){=}C(R')(H) + {}^{\oplus}SnR_3 \longrightarrow (R_3Sn)(X)C{=}C(R')(H)$$

The *trans*-adduct is then formed. This mechanism is also supported by the order of reactivities of the various triorganotin hydrides which is found to be $Ph_3SnH \ll Me_3SnH < Et_3SnH \approx Bu_3SnH$ (467).

Radical catalysts will of course induce a radical mechanism and thus afford a unique product, as mentioned earlier. Choice of conditions (uncatalyzed in a polar medium or radical-catalyzed in a non-polar medium) therefore enables one to steer the hydrostannation reaction in a desired direction. The versatility of organotin hydrides as reagents, already demonstrated in Chapter 8-4, is again obvious.

9-5 ALDEHYDES AND KETONES

Hydrostannation of aldehydes and ketones yields stannyl alkoxides (Chapter 17-4). Polar addition occurs preferentially (553, 557, 558), but radical pathways are possible also (121, 553, 557, 558, 674). Only addition reactions are discussed here; hydrogenation of carbonyl compounds by means of organotin hydrides is discussed in Chapter 10. Hydrostannation with aliphatic trialkyltin hydrides has been studied most thoroughly. Addition (reaction 9-7) takes place spontaneously only with aldehydes, and then only if the carbonyl group is activated by electron-withdrawing substituents, e.g. as in chloral:

$$R'R''C{=}O + HSnR_3 \rightarrow R'R''CHOSnR_3 \qquad (9\text{-}7)$$

Most other aldehydes and all ketones require prolonged heating or catalysts. A recent re-investigation confirms these findings and gives additional examples (465b).

Complexes of aldehydes and ketones with acids (**14**) or Lewis acids (**15**) are attacked appreciably faster than the simple carbonyl compounds (557, 558). The primary reaction consists of transfer of a hydride ion to

$$\overset{\delta+}{>C}\cdots O\cdots H\cdots \overset{\delta-}{A} \qquad\qquad \overset{\delta+}{>C}\cdots O\cdots \overset{\delta-}{L}$$

(14) **(15)**

(A = anion of proton donor, L = Lewis acid)

the now more strongly electrophilic carbonyl carbon; the secondary step depends on the acid. With e.g. H_2SO_4, glacial acetic acid, or phenols, the proton remains on the carbonyl oxygen, i.e. the free alcohol is formed together with trialkyltin sulfate, acetate, or phenoxide.

Methanol, too, behaves like an acid [hydrogen bond as in (**14**)]. With methanol as catalyst, the products are the alcohol and the trialkyltin methoxide. However, the two alkoxy groups exchange to give an equilibrium mixture from which methanol, as the lowest-boiling component, can generally be distilled to leave a pure adduct of carbonyl compound and trialkyltin hydride. Side-reactions are rare, and small amounts of methanol are therefore often the best catalyst for hydrostannation of carbonyl compounds (557, 558). Bases inhibit the catalytic action of methanol.

Complexes of carbonyl compounds with anhydrous zinc chloride react with organotin hydrides to yield trialkyltin alkoxides and free $ZnCl_2$ (557, 558). Zinc chloride is therefore another suitable catalyst, especially for hydrostannation of ketones and aromatic aldehydes.

The products isolated from hydrostannation of hydroxybenzaldehydes are substituted organotin phenoxides (**16**), by rearrangements of the adducts expected from reaction 9-7.

$$HOCH_2-C_6H_4-O-SnR_3$$

(16)

In all these polar hydrostannations, the organotin hydride functions as a nucleophilic reagent. Radical mechanisms, however, are also in evidence. UV light accelerates the addition of R_3SnH to ketones (121, 674). Radical promoters (e.g. azobisisobutyronitrile) and even O_2 accelerate the hydrostannation of aldehydes (557, 558). Hydrostannation is also catalyzed by finely divided platinum (553).

Treatment of α,β-unsaturated ketones with R_3SnH can yield 1,4-adducts in which tin is bonded to oxygen (see Chapter 9-7).

Dialkyltin dihydrides react with aldehydes at room temperature and without need for a catalyst. The mono-adduct is formed first (373), but the di-adduct will result under suitable conditions (557, 558):

$$RCHO + H_2Sn\textit{i}\text{-}Bu_2 \rightarrow RCH_2OSn(H)\textit{i}\text{-}Bu_2 \xrightarrow{RCHO} (RCH_2O)_2Sn\textit{i}\text{-}Bu_2$$

Dropwise addition of dialkyltin dihydrides to a mixture of aldehyde or ketone and dialkyltin dichloride leads smoothly to carbonyl adducts of the highly reactive dialkylhalotin hydrides (Chapter 8-2), e.g. according to (296, 580, 581):

$$C_6H_{10}{=}O + H{-}SnClR_2 \longrightarrow C_6H_{10}(H)(O{-}SnClR_2)$$

9-6 UNSATURATED NITROGEN COMPOUNDS

Various reactions are observed, reflecting the diversity of systems with double-bonded nitrogen. In most cases, the organotin hydride adds to form an N–Sn bond, i.e. to produce a stannylamine. This class of compound is discussed separately in Chapter 15.

9-6-1 Azomethines

Uncatalyzed addition of organotin hydrides to the C=N group has not been achieved. Adequate reaction is however induced by radical initiators or by anhydrous zinc chloride (559), e.g. by reaction 9-8. Here, as with carbonyl compounds (Chapter 9-5), both radical and polar mechanisms are able to operate. Applications are as yet very limited.

$$\begin{matrix} p\text{-MeC}_6\text{H}_4\text{N} \\ \| \\ \text{PhCH} \end{matrix} + \text{HSnR}_3 \rightarrow \begin{matrix} p\text{-MeC}_6\text{H}_4\text{NSnR}_3 \\ | \\ \text{PhCH}_2 \end{matrix} \qquad (9\text{-}8)$$

The resulting stannylamines are cleaved by water quantitatively into the secondary amine and a stannoxane. Even atmospheric moisture produces rapid hydrolysis; the stannoxane or hydroxide then reacts further with CO_2 to form the carbonate (see Chapter 9-7). Alcohols, too, effect cleavage.

9-6-2 Azo-compounds

Addition of organotin hydrides to azo compounds, e.g. as in reaction 9-9, proceeds even without catalysis (295, 553, 611). An excess of hydride produces hydrogenolysis of the N–Sn bond.

$$\begin{matrix} \text{PhN} \\ \| \\ \text{PhN} \end{matrix} + \text{HSnR}_3 \rightarrow \begin{matrix} \text{PhNH} \\ | \\ \text{PhNSnR}_3 \end{matrix} \xrightarrow{\text{HSnR}'_3} \begin{matrix} \text{PhNH} \\ | \\ \text{PhNH} \end{matrix} + \text{R}_3\text{SnSnR}'_3 \qquad (9\text{-}9)$$

The two stages can be performed separately, giving unsymmetrical hexaalkylditins (see Chapter 8-4-4). Apart from azobenzene, work has been done on azodicarboxylic esters (611) and on azo-compounds which were of interest as hydrostannation catalysts (see Chapter 8-4-1). A radical mechanism has been established for these additions. Mention should be made of phenyl azophenyl sulfone (571a) and phenylazoisobutyronitrile (**17**). The latter decomposes above 100°C by a radical process (532). Below 100°C it undergoes hydrostannation, but the primary product (**18**) (detectable at 40°C) decomposes rapidly to acetone phenylhydrazone and trialkyltin cyanide (599). This radical addition of organotin hydrides can be used as a source of initiating radicals for other reactions, e.g. for hydrostannation of olefins (553).

$$\underset{\textbf{(17)}}{\text{PhN=NC(CN)Me}_2} + \text{R}_3\text{SnH} \rightarrow$$

$$\underset{\textbf{(18)}}{\text{PhNHN(SnR}_3\text{)C(CN)Me}_2} \rightarrow \text{PhNHN=CMe}_2 + \text{R}_3\text{SnCN}$$

9-6-3 Nitroso-compounds

Slow, dropwise addition of nitrosobenzene to triethyltin hydride yields the *O*-stannylhydroxylamine (**19**), with evolution of heat. Like the corresponding unsubstituted hydroxylamine, this reacts easily with more nitrosobenzene to give the azoxy-compound (**20**). The latter is obtained also in a single step and in good yield from stoichiometric amounts of the reactants (559). An excess of organotin hydride leads to slow reduction

$$PhNO + HSnR_3 \rightarrow \underset{(\mathbf{19})}{PhNHOSnR_3} \xrightarrow[-R_3SnOH]{+PhNO} \underset{(\mathbf{20})}{PhN(\rightarrow O){=}NPh}$$

to the azo-derivative. Other nitroso-compounds, e.g. *p*-nitroso-dimethylaniline, often afford some hexaalkylditin besides the azoxy-compound. This indicates ready cleavage of the primary adduct [analogous to (**19**)] by organotin hydride which need not be present in excess (559) (see also Chapter 8-4-4).

9-6-4 Isocyanates and isothiocyanates

Phenyl isocyanate, its *p*-chloro and *p*-nitro derivatives, as well as *p*-phenylene diisocyanate (466, 553) undergo strongly exothermic reactions with trialkyltin hydrides, without the aid of catalysts. The adducts (**21**) are crystalline and very hygroscopic, and can be distilled *in vacuo*. Dialkyltin dihydrides react analogously under mild conditions with 2 moles of isocyanate. An excess of organotin hydride leads to cleavage of the adduct into a formanilide and the hexaalkylditin. (Further details of this condensation are in Chapter 8-4-4.) Starting with two equivalents of hydride one obtains the formanilide directly (484).

$$p\text{-}R'C_6H_4NCO + HSnR_3 \rightarrow \underset{(\mathbf{21})}{p\text{-}R'C_6H_4N(SnR_3)CHO}$$

$$(R = Et, Bu, i\text{-}Bu)$$

In principle, hydrostannation of isocyanates can take place either at the C=N or at the C=O group. Experiments with R_3SnD have proved that the hydride hydrogen turns up on the carbonyl carbon (C–D absorption) (553). The strong band at 1630 cm.$^{-1}$ may however be interpreted either as a C=N absorption band or as an unusually far displaced C=O band. On the basis of UV and NMR measurements, the product is assigned the stannylamine structure (**21**) (466). Reaction of aliphatic isocyanates, as represented by the hexyl compound, is noticeably more sluggish than that of aromatic ones. Interpreted initially in another way (466), the adduct is now assigned a structure analogous to (**21**), with n-C_6H_{13} in place of $R'C_6H_4$ (553). Radical sources or scavengers are without effect on the reaction rate; polar solvents enhance it considerably. Reaction is therefore polar. Electron-withdrawing substituents on the N=C=O group accelerate, similar substituents (e.g. Ph) on the Sn–H group retard. It is concluded that the initial step involves a nucleophilic attack by the hydride hydrogen on the electron-deficient carbon of the N=C=O group (470).

9-7 CARBODIIMIDES, CO_2, AND CS_2

CO_2 passed through a solution of triethyltin hydride undergoes hydrostannation to the formate ester at 20°C (295).

$$R_3SnH + CO_2 \rightleftharpoons R_3SnOCHO$$

The reverse reaction may be utilized for preparation of hydrides (see Chapter 8-1). CS_2 seems to undergo a similar hydrostannation with R_3SnH, but secondary products only (bistrialkyltin·sulfides) have been isolated so far (642).

Addition of R_3SnH to carbodiimides can be accelerated by means of azobisisobutyronitrile:

$$(\text{cyclo-}C_6H_{11})N{=}C{=}N(\text{cyclo-}C_6H_{11}) + R_3SnH \rightarrow (\text{cyclo-}C_6H_{11})N(SnR_3)CH{=}N(\text{cyclo-}C_6H_{11}) \quad \mathbf{(22)}$$

Cleavage of the resulting stannyl formamidines (**22**) with HCl in ether affords formamidine hydrochlorides of the type RNHCH=NR·HCl besides R_3SnCl (296).

9-8 POLAR CONJUGATED SYSTEMS

Hydrostannation of conjugated dienes proceeds by a radical mechanism and predominantly in the 1,4-position (see Chapter 9-2). The polar conjugated systems of acrylonitrile $H_2C{=}CHCN$ and cinnamonitrile PhCH=CHCN react with R_3SnH by radical or polar mechanisms, depending on conditions, but always with 1,2-addition to the C=C group (363, 366). The Giessen team has achieved exothermic 1,4-hydrostannation with the malononitrile and malonic ester derivatives (**23**)–(**25**).* These react mostly at 20°C and without catalysts to give the stannyl ketenimines and keten acetals (**26**)—(**28**) (601).

* These are easily accessible by normal procedures from malonic acid derivatives and carbonyl compounds.

$$R'CH{=}C(CN)_2 + HSnR_3 \rightarrow R'CH_2C(CN){=}C{=}NSnR_3$$

(23) **(26)**

(R′ = e.g. Me, EtO, p-$Me_2NC_6H_4$, p-$MeOC_6H_4$, p-ClC_6H_4, Ph, p-$O_2NC_6H_4$, α-furyl)

$$R'CH{=}C(CN)COOEt + HSnR_3 \rightarrow R'CH_2C(CN){=}C(OSnR_3)OEt$$

(24) **(27)**

(R′ = e.g. Me, p-$MeOC_6H_4$, p-ClC_6H_4, Ph)

$$R'CH{=}C(COOEt)_2 + HSnR_3 \rightarrow R'CH_2C(COOEt){=}C(OSnR_3)OEt$$

(25) **(28)**

Compounds (**24**) and (**25**), which are less polar than (**23**), react more slowly. This, and the acceleration produced by electron-withdrawing groups R′, leads to the conclusion that the primary step involves transfer of a hydride ion to the electron-deficient carbon atom. The mesomeric anion (**29**) then becomes stabilized as the stannyl derivative (**30**).

$$>\overset{\delta+}{C}{=}C(CN)_2 \;\cdots\; \overset{\delta-}{H}{-}\overset{\delta+}{Sn}R_3 \longrightarrow \left[>C(H){-}\overset{\ominus}{C}(CN)_2 \longleftrightarrow >C(H){-}C(CN){=}C{=}N^{\ominus} \right] + \overset{\oplus}{Sn}R_3$$

(29)

$$\longrightarrow >C(H){-}C(CN){=}C{=}N{-}SnR_3$$

(30)

The N-stannyl ketenimines (**30**) form coordination polymers containing penta-coordinate tin atoms (**30a**), n being strongly dependent on concentration and solvent. Partial dissociation occurs in polar solvents like DMSO (537b).

$$\cdots Sn \cdots \left[\cdots N{\equiv}C{=}\underset{CH_2R'}{C}{=}C{\equiv}N \cdots Sn \cdots \right]_n \cdots \qquad \textbf{(30a)}$$

O-Stannyl keten acetals (**28**) (812b) are monomeric with intramolecular penta-coordination around the tin atoms. The two ethoxy groups are equivalent, as shown by IR and NMR spectra.

The cyclic acylal $PhCH{=}C(COO)_2CMe_2$ forms crystalline stannyl keten acetals, which are however highly associated in benzene solution ($n \approx 20$). They are stable towards ethanol.

This constitutes the first preparation of stannyl ketenimines and keten acetals, and these compounds are now easily accessible. They are extremely versatile reagents, usually eliminating the stannyl moiety, and are therefore of great synthetic utility (812b, 813). Some reactions of compound (**26**) are given as an example (reaction scheme 9-10), but those of (**27**) and (**28**) are similar.

$$
RCH_2C(CN){=}C{=}NSnR'_3
\begin{cases}
\xrightarrow[-R'_3SnBr]{H_2C=CHCH_2Br} RCH_2C(CN)_2CH_2CH{=}CH_2 \\
\xrightarrow[-R'_3SnCl]{MeCH=CHCH_2Cl} RCH_2C(CN)_2CH_2CH{=}CHMe \\
\xrightarrow[-R'_3SnCl]{PhCH_2Cl} RCH_2C(CN)_2CH_2Ph \\
\xrightarrow[-R'_3SnBr]{Br_2} RCH_2C(CN)_2Br \\
\xrightarrow[-R'_3SnOEt]{EtOH} RCH_2C(CN)_2H \\
\xrightarrow[-R'_3SnCl]{PhCOCl} RCH_2C(CN)_2COPh \\
\xrightarrow[-R'_3SnCl]{MeCOCl} RCH_2C(CN)_2COMe \\
\xrightarrow[-R'_3SnCl]{ClCOOEt} RCH_2C(CN)_2COOEt \\
\xrightarrow[-R'_3SnI]{MeI\ (in\ DMF)} RCH_2C(CN)_2Me \\
\xrightarrow[-R'_3SnBu]{BuLi} RCH_2C(CN)_2Li
\end{cases}
\qquad (9\text{-}10)
$$

Under appropriate conditions (non-polar solvent, radical catalysis), alkylidenemalonic esters (**25**) can also undergo radical hydrostannation.

If R′ is aromatic, this leads to 1,4-addition and formation of compounds of type (**28**). If R′ is aliphatic, β-stannyl alkylmalonic esters are formed by 1,2-addition (812b).

1,4-Hydrostannation of vinyl ketones was observed by M. Pereyre and J. Valade (649a, 650, 651).

$$Me_2C{=}CHCOMe + HSnBu_3 \rightarrow Me_2CHCH{=}C(Me)OSnBu_3 \quad (9\text{-}11)$$

(31)

Apart from mesityl oxide, exclusive 1,4-addition is also observed with phorone, dihydrophorone, and even dimethylacrolein. With less sterically hindered ketones, e.g. methyl propenyl ketone, this is accompanied by 1,2-hydrostannation of the C=C linkage to give the α-stannyl ketone. When triphenyltin hydride is used, the enol ether (**31**) cannot be isolated. Under the prevailing conditions it reacts further with some of the hydride present to yield the saturated ketone (via the enol) and hexaphenylditin (condensation; see Chapter 8-4-4). Reaction between benzylideneacetophenone (chalcone) and triethyltin hydride (812) is also analogous to reaction 9-11. These results have since been confirmed and extended (471a).

The more strongly polar vinyl ketones of type (**32**) undergo heterolytic, and therefore exclusively 1,4-, hydrostannation analogous to (**23**)—(**25**) (601).

$$R'CH{=}C(CN)COPh + HSnR_3 \rightarrow R'CH_2C(CN){=}C(Ph)OSnR_3$$

(32) **(33)**

(R′ = e.g. *p*-MeO–C_6H_4, Ph, *p*-O_2N–C_6H_4, α-furyl)

The resulting enol ethers (**33**) are hydrolytically stable, in contrast to (**31**), and may even be recrystallized from aqueous alcohol. However, *O*-substituted derivatives are formed with e.g. ClCOOEt and PhCOCl.

α,β-Unsaturated nitro-compounds, like α-nitrostilbene or benzylidenenitroethane, undergo polar 1,4-hydrostannation to form *O*-stannyl nitronic esters (537a):

$$PhCH{=}C(R')NO_2 + R_3SnH \rightarrow PhCH_2C(R'){=}\underset{\downarrow \atop O}{N}OSnR_3$$

(R′ = H, Me, Ph)

10 Organotin hydrides as reducing agents

H. G. Kuivila and O. F. Beumel attempted hydrostannation of methyl vinyl ketone with diphenyltin dihydride, but observed only reduction of the ketone to methyl vinyl carbinol instead of the expected addition to the vinyl group (431, 432). (See also footnote to Table 10-1.)

$$H_2C{=}CHCOMe + Ph_2SnH_2 \rightarrow H_2C{=}CHCH(OH)Me + 1/n\,(Ph_2Sn)_n$$

van der Kerk and his associates, in an attempted hydrostannation of allyl bromide, observed instead reduction to propene:

$$H_2C{=}CHCH_2Br + Ph_3SnH \rightarrow H_2C{=}CHMe + Ph_3SnBr$$

They soon reported other selective hydrogenations by means of triphenyltin hydride (366, 617).*

These observations formed the starting point of a systematic investigation into the reducing action of organotin hydrides on various classes of compound. [For a critical review up to mid-1963, see H. G. Kuivila (429).] In homogenous media, reduction will often take place at room temperature, i.e. under mild conditions. Moreover, it is highly selective. Areas of application show little overlap with those of complex metal hydrides ($LiAlH_4$, $NaBH_4$, R_2AlH) or of catalytic hydrogenation with molecular hydrogen. This makes the technique all the more important.

The reactions frequently involve alkyl or aryl radicals (see below) as highly reactive intermediates. This has now provided renewed impetus for the study of their properties and reactions.

* These authors (366, 617) and others (447) originally reported deaminations supposed to have been observed with Ph_3SnH. After some criticism (825) these claims were withdrawn (368), but are still cited occasionally in literature references (531).

Table 10-1
Hydrogenation of Carbonyl Compounds by means of Organotin Hydrides

CARBONYL COMPOUND	HYDRIDE[a]	CAT.[b]	REACTION CONDITIONS			CARBINOL	YIELD (%)	REFERENCES
Methyl vinyl ketone	Ph_2SnH_2	—	15 hr.	moist Et_2O	20°	Methylvinylcarbinol	59	431, 432
Mesityl oxide	Ph_2SnH_2	—		moist Et_2O	20°	Methylisobutenylcarbinol	60	431, 432
Mesityl oxide	Bu_2SnH_2	Pd	2 hr.	no solvent	0°	Methylisobutenylcarbinol	94	373
Crotonaldehyde	Ph_2SnH_2	—		moist Et_2O	20°	Crotyl alcohol	59	431, 432
Crotonaldehyde	Bu_2SnH_2	—		moist Et_2O	20°	Crotyl alcohol	44	432
Crotonaldehyde	Bu_2SnH_2	Pt	1 hr.	no solvent	0°	Crotyl alcohol	83	373
Cinnamaldehyde	Bu_2SnH_2	Pt	$\frac{1}{2}$ hr.	no solvent	0°	Cinnamyl alcohol	84	373
Benzylideneacetophenone	Ph_2SnH_2	—		moist Et_2O	20°	Phenylstyrylcarbinol	75	431, 432
Benzylideneacetophenone	Bu_2SnH_2	Pt	5 hr.	no solvent	60°	1,3-Diphenylpropan-3-one	83	373
Allylacetone	Bu_2SnH_2	Pt	7 hr.	no solvent	50°	1-Hexen-5-ol	80	373
Benzaldehyde	Ph_3SnH	—	15 hr.		100°	Benzyl alcohol	85	432
Benzaldehyde	Bu_3SnH	—	15 hr.		140°	Benzyl alcohol	86	432
Benzaldehyde	Ph_2SnH_2	—		moist Et_2O	20°	Benzyl alcohol	62	431, 432
Veratraldehyde	Bu_2SnH_2	Ni	$3\frac{1}{2}$ hr.	no solvent	60°	Veratryl alcohol	91	373
Benzophenone	Bu_2SnH_2	—	3 hr.	no solvent	40—60°	Benzhydrol	85	432
Phenyl 2-thienyl ketone	Ph_3SnH	—	19 hr.		145°	Phenyl-2-thienylcarbinol	20	876

Cyclohexanone	Ph_2SnH_2	—		moist Et_2O	20°	Cyclohexanol	82	431, 432
Cyclohexanone	Bu_2SnH_2	Ni	4 hr.	no solvent	50°	Cyclohexanol	82	373
Benzil	Bu_2SnH_2	Pt	2 hr.	boiling Et_2O		Benzoin	71	373
Benzil	Bu_2SnH_2	Pt	1 hr.	Et_2O		*meso*-Hydrobenzoin	87	373
Benzil	Ph_2SnH_2	—		moist Et_2O	20°	*meso*-Hydrobenzoin	84, 88	431, 432
Benzil	Bu_2SnH_2	—		moist Et_2O	20°	*meso*-Hydrobenzoin	93	432
Biacetyl	Bu_2SnH_2	Ni		cyclohexane	0°	Acetoin (dimer)	92	373
Furoin	Bu_2SnH_2	Ni	6 hr.	cyclohexane	60°	*meso*-Hydrofuroin	53	373
Benzoquinone	Bu_2SnH_2	—		moist Et_2O	20°	Hydroquinone	66	432
Benzoquinone	Bu_2SnH_2	Pt		Et_2O, reflux		Hydroquinone	83	373
Butyl pyruvate	Bu_2SnH_2	Ni	$\frac{1}{2}$ hr.		50°	Butyl *dl*-lactate	86	373
Ethyl levulinate	Bu_2SnH_2	Ni	$1\frac{1}{2}$ hr.	exothermic		γ-Valerolactone	94	373

[a] As far as investigated (427a), the corresponding deuterides give analogous results and lead to *C*- or *C,O*-deuterated carbinols.

[b] Where no catalyst is shown, the authors appear to have used crude, undistilled organotin hydrides. It has since become apparent that reductions with pure, redistilled materials are much slower, or require higher temperatures (373). α,β-Unsaturated ketones then form the saturated ketone through 1,4-hydrostannation (see Chapter 9-8) (471a).

Most reductions by organotin hydrides can also be effected by the corresponding deuterides R_3SnD or R_2SnD_2. This provides a convenient route to a large number of C-deuterated and *O*-deuterated compounds (427a).

10-1 ALDEHYDES AND KETONES

One advantage of organotin hydrides as reductants for carbonyl compounds lies in the direct production of the carbinol. When metal hydrides (above) are used, the first product is the metal alkoxide. The carbinol has then to be liberated by hydrolysis, which may (for sensitive compounds) lead to side-reactions; e.g. reduction of veratraldehyde with dibutyltin dihydrides gives an excellent yield of the alcohol (see Table 10-1), while $LiAlH_4$ produces resinification (455).

Several reductions of aldehydes and ketones have been shown to be two-stage processes: initial formation of an organotin alkoxide by addition of the hydride (see Chapter 9):

$$R_3SnH + O{=}C\langle \rightarrow R_3SnOCH\langle$$

is followed by a condensation of the type described in Chapter 8-4-4:

$$R_3SnOCH\langle + HSnR_3 \rightarrow R_3SnSnR_3 + HOCH\langle$$

With dialkyltin dihydrides, the first stage leads to a dialkylalkoxytin hydride (373, 560):

$$R_2SnH_2 + O{=}C\langle \rightarrow R_2Sn(H)OCH\langle$$

For example, when equimolar amounts of dibutyltin dihydride and benzophenone are mixed with a trace of $PtCl_4$ (for its function, see below), the carbonyl band diminishes rapidly and disappears. But the adduct $R'_2CHOSn(H)R_2$ is present in only small amounts, because its IR absorption expected at about 1865 $cm.^{-1}$ does not appear. It seems to disproportionate immediately. OH absorption, which is an indication of the second stage (cleavage), is correspondingly weak. However, it soon increases (at 40°C) while the remaining Sn–H absorption disappears.

$$[R_2Sn(H)OCH\langle] \rightleftharpoons \tfrac{1}{2}R_2SnH_2 + \tfrac{1}{2}R_2Sn(OCH\langle)_2 \rightarrow 1/n\,(R_2Sn)_n + HOCH\langle$$

Cleavage of the dialkylalkoxytin hydride into the carbinol and a polymeric dialkyltin must be an intermolecular process. This follows from the transient appearance of an Sn–H band at 1795 cm.$^{-1}$, characteristic of the group H–Sn–Sn.

With dialkyltin dihydrides as reductants, cleavage occurs usually at room temperature and faster than the preceding addition, ensuring smooth reaction under mild conditions. Both stages are very much slower with organotin monohydrides; e.g. reduction of benzaldehyde with triphenyltin hydride requires several hours' heating at 100°C (140°C with tri-*n*-butyltin hydride); reduction with dialkyltin dihydrides proceeds at room temperature with heat evolution.

Dialkyltin dihydrides—generally the dibutyltin or di-*i*-butyltin compounds—have the further advantage that they give rise to non-volatile polymeric dialkyltins, permitting direct distillation of the carbinol in most cases. When the carbinol is poorly volatile or non-volatile, the $(R_2Sn)_n$ can be oxidized atmospherically to insoluble dialkyltin oxide (373, 560). Yields mostly range from satisfactory to excellent (see Table 10-1). $PtCl_4$, $PdCl_2$, Raney nickel, or nickel acetylacetonate may be added as catalysts (373, 560). These substances are added in small amounts to the mixture, where the tin hydride first reduces them to the colloidal metal. Catalysis probably affects only the first stage of the reduction, i.e. hydrostannation of the carbonyl group. In the case of benzophenone, where reduction at 40°C is very slow in the absence of catalysts, the relative effectiveness of the several catalysts rises in the order shown in Figure 10-1.

To avoid elimination of H_2, the tin hydride is best added dropwise as required to the mixture of carbonyl compound and catalyst.

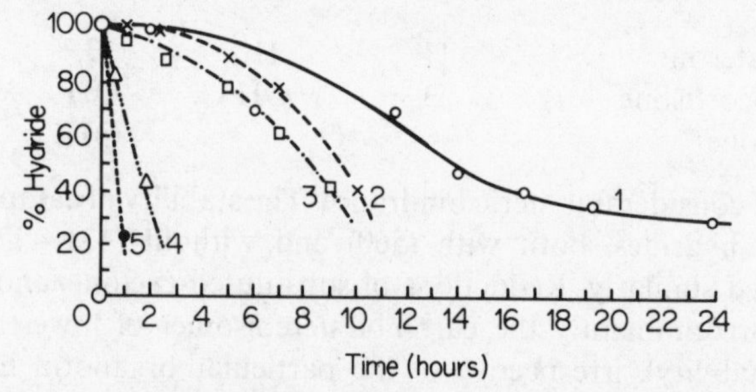

FIG. 10-1 Dependence of reduction rate on catalyst at 40.0°C (373). Benzophenone and Bu_2SnH_2 (10 millimoles each) in 10 c.c. cyclohexane. Curve 1: without catalyst. Curve 2: with Pt from $PtCl_4$. Curve 3: with Raney nickel. Curve 4: with Ni from Ni-acetylacetonate. Curve 5: with Pd from $PdCl_2$. Amount of catalyst in each case about 4–5 mole %

These catalysts also permit considerable control over the course of the reaction between organotin hydrides and unsaturated carbonyl compounds; e.g. 5-hexen-2-one (allylacetone) suffers hydrostannation of the C=C group in the presence of radical initiators [azobisisobutyronitrile (AIBN)], but reduction of the carbonyl group in the presence of $PtCl_4$.

$$MeCOCH_2CH_2CH{=}CH_2 \xrightarrow[AIBN]{Et_3SnH} MeCOCH_2CH_2CH_2CH_2SnEt_3$$

$$MeCOCH_2CH_2CH{=}CH_2 \xrightarrow[PtCl_4]{Bu_2SnH_2} MeCH(OH)CH_2CH_2CH{=}CH_2 + 1/n\,(Bu_2Sn)_n$$

While it is not yet clear to what extent these hydrogenations are stereospecific, there are at least some examples of pronounced selectivity; e.g. it is interesting that progesterone* (**1**) is reduced by diphenyltin dihydride at C-3 about 6.5 times faster than at C-20 (72), thus providing an excellent route from the 3-ketone to the 3-hydroxy-compound. (In contrast, $NaBH_4$ reduces preferentially at C-20.) Hydrocortisone (**2**) gives 63% reduction at C-20 as well as 14% reduction at both C-3 and C-20, while cortisone (**3**) gives 50% reduction at C-3 in addition to 25% at both C-3 and C-20. The C-11 carbonyl in cortisone is not attacked, probably

R2 R3 R1 11 C20—CH2—R4 O 3 O

	R_1	R_2	R_3	R_4
(**1**) Progesterone*	H	H	H	H
(**2**) Hydrocortisone	H	OH	OH	OCOMe
(**3**) Cortisone	=O		OH	OCOMe

because of considerable steric hindrance. The stability of camphor against organotin hydrides, both with (560) and without (431, 432) catalysts, is explained similarly. Reductions of substituted cyclohexanones appear to yield predominantly the carbinol stereoisomer of lowest energy (i.e. greatest stability), irrespective of the particular organotin hydride employed (mono-, di-, or tri-hydrides; butyl- or phenyl-tin hydrides). This means that the 2- and 4-methyl derivatives, and even more so the 4-*t*-butyl derivative, give mainly the *trans*-isomer, while the 3-methyl derivative produces very little of this isomer (431, 432, 858). In this reaction, the

* Inadvertently cited in the original literature as pregnenolone.

organotin hydrides appear to be about as selective as $NaBH_4$, but slightly less selective than $LiAlH_4$ (432).

10-2 ALKYL AND ARYL HALIDES

Alkyl and aryl halides are reduced smoothly by organotin hydrides to the respective hydrocarbons. Both reduction and mechanism have been studied in detail (127d, 429, 437). Iodides are reduced faster than bromides, and bromides faster than chlorides; e.g. reduction of aliphatic bromides by tributyltin hydride is exothermic (no solvent needed), chlorides require heating at about 100°C for several hours, while fluorides are not attacked at all. Effectiveness of the various organotin hydrides increases in the order $Bu_3SnH < Bu_2SnH_2 \approx Ph_3SnH < BuSnH_3 \approx Ph_2SnH_2$ (436). Tributyltin hydride is the reductant most commonly used. Table 10-2 gives a selection from the many examples extant.

Table 10-2
Reduction of Alkyl and Aryl Halides with Tributyltin Hydride

HALIDE	REACTION CONDITIONS	YIELD (%)	REF.
$PhCH_2Br$	in ether, exothermic, 12 hr.	68	437
cyclo-$C_6H_{11}Br$	exothermic, 1 hr.	71	437
cyclo-$C_6H_{11}Cl$	40 min., 120°	36	437
n-$C_8H_{17}Br$	exothermic, 1 hr.	80	437
n-$C_6H_{13}CH(Br)Me$	exothermic, 1 hr.	78	437
$PhCH_2CH_2Br$	4 hr., 100°	85	873
$CHBr_3$	exothermic	62	782
Br_3CF	exothermic	69	782
$PhCHCl_2$	20 min., 140°	70	873
CCl_4	exothermic	85	782
$PhCOCH_2Br$[a]	in benzene, 12 hr.	84	437
3-Bromocamphor[a]	in ether, exothermic, 2 hr.	61	437

[a] The carbonyl group is not attacked.

Geminal polyhalides react in stages; e.g. benzotrichloride can be reduced successively in high yield to benzylidene chloride, benzyl chloride, or toluene (873). The same is true for the series CCl_4, $CHCl_3$, CH_2Cl_2, CH_3Cl (483).*

Cyclic geminal dihalides, e.g. dihalocyclopropanes, are also reduced in stages. Substituted dihalides give a mixture of the isomeric monohalides,

* While some authors report quantitative exothermic reaction between CCl_4 and Ph_3SnH (483), others claim that this hydride has been examined spectroscopically in CCl_4 without decomposition (423).

a result explained by the reaction mechanism (see below) (782). α-Haloketones are reduced selectively to the unhalogenated ketones, before any attack on the carbonyl group (436, 437) (see Table 10-2). Dehalogenation of bromohydrins is equally selective; no elimination has been detected (873):

$$BrCH_2CH(OH)CH_2Br + 2\,Bu_3SnH \rightarrow MeCH(OH)Me + 2\,Bu_3SnBr \quad 85\%$$

1,2-Dihalides have however been found to suffer elimination (436); e.g.

$$\underset{(meso)}{PhCH(Br)CH(Br)Ph} + 2\,Bu_3SnH \xrightarrow{100^\circ} \underset{(trans)}{PhCH{=}CHPh} + H_2 + 2\,Bu_3SnBr$$

1,2-Dibromopropane behaves similarly.

Aromatic halides react far more sluggishly, but otherwise like aliphatic halides. Temperatures of 100—150°C are usually required. Alcohols promote the reaction, while the effect of substituents varies: electron-withdrawing groups accelerate (investigated: Cl, Ph); electron-donor groups (Me, OMe) retard (485). An interesting variant of the reaction dispenses with separate isolation of the organotin hydride (429, 873): whereas $LiAlH_4$ reacts only slowly with alkyl halides, it reduces organotin halides very rapidly to the hydrides (see Chapter 8-1). If the alkyl halide is now mixed with $LiAlH_4$, e.g. in boiling ether, and a little organotin halide added (2.5 mole % suffices), the latter acts as a "hydride carrier". It is first reduced by $LiAlH_4$ to the organotin hydride which in turn reduces the alkyl halide, and is then re-hydrogenated by more $LiAlH_4$. This elegant technique holds great preparative promise.

An S_N2 mechanism can be ruled out on several grounds.

(a) Bridgehead halides are reduced successfully (448).

(b) The same mixture of butene isomers results from the reduction of crotyl chloride or α-methallyl chloride. A radical or ion capable of allylic mesomerism must therefore be an intermediate (526). Formation of both allene and propyne from propargyl bromide is explained similarly:

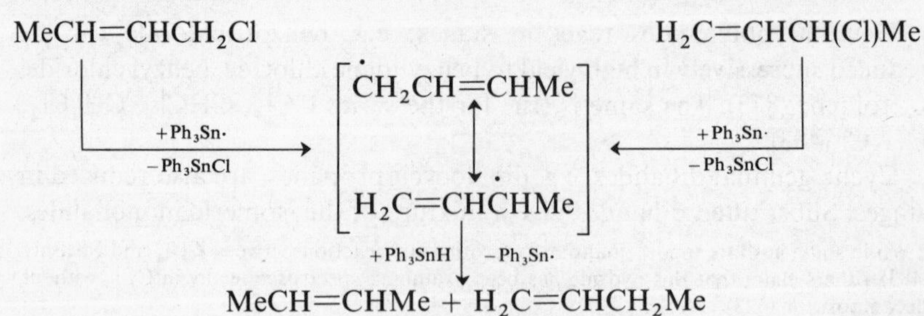

(c) Reduction of 1-phenyl-1-chloroethane with Ph_3SnD affords the racemate (526):

$$(+)\text{-PhCH(Cl)Me} \xrightarrow[-Ph_3SnCl]{+Ph_3SnD} (\pm)\text{-PhCH(D)Me}$$

These and a variety of other arguments prove beyond doubt that dehalogenation proceeds by a radical mechanism (431, 432, 526, 782). Radical sources promote reaction (especially important for the rather

Table 10-3
Relative Rates of Reduction of Alkyl and Aryl Halides by Tributyltin Hydride at 45°C (429)

HALIDE	REL. RATE	HALIDE	REL. RATE
$PhCH_2Cl$	0.05	$H_2C{=}CHCH_2Br$	31
n-BuBr	1.00	$PhCH_2Br$	34
n-$C_8H_{17}Br$	1.10	$BrCH_2COOEt$	50
cyclo-$C_6H_{11}Br$	1.46	*n*-$C_7H_{15}I$	61
cyclo-C_5H_9Br	2.37	CCl_4	76
2-BrC_8H_{17}	2.63	$BrCHCl_2$	112
2-BrC_4H_9	2.99	$HC{\equiv}CCH_2Br$	139
t-BuBr	7.00	$BrCCl_3$	226
$BrCH_2Cl$	26.00		

unreactive aryl halides), while hydroquinone inhibits (526). All available facts lead to the conclusion that a radical-chain reaction is operating:*

$$\text{snH} + \text{R}'\cdot \rightarrow \text{sn}\cdot + \text{R}'\text{H}$$

$$\text{sn}\cdot + \text{RHal} \rightarrow \text{snHal} + \text{R}\cdot$$

$$\text{R}\cdot + \text{snH} \rightarrow \text{sn}\cdot + \text{RH}$$

(R′· is a free radical—stemming from an initiator—or a molecule capable of abstracting a hydrogen atom from the tin hydride)

As expected, the reaction rate depends on the ease of formation of the radical R· from the halide (see Table 10-3) (431, 432). The sequence of reactivities with organotin hydrides is seen to be much the same as that found for other radical reactions (197).

* There seemed to be arguments for a heterolytic mechanism (485), but these were later seen to be invalid (642).

10-3 ACYL HALIDES

Acid halides are easily reduced by trialkyltin hydrides (429), but the yield of aldehyde is largely dependent on reaction partners and conditions. With bulky, e.g. highly branched, groups R on the carbonyl compound, and in dilute solution, about 90% of the theoretical yield of aldehyde is obtained with tributyltin hydride. The remainder, rather surprisingly, consists of the corresponding ester. The amount of ester increases with concentration and is highest when R is small; e.g. 95% of the theoretical yield of ester is obtained in the absence of a solvent when R = Me. Treatment of benzoyl chloride or *p*-toluyl chloride with triphenyltin hydride under various conditions affords the ester, but no aldehyde at all (449).* Acid bromides give a higher yield of aldehyde than the corresponding chlorides (429, 441).

The reaction proceeds by a radical mechanism. Systematic investigation (441) indicates the following competition for the primary acyl radical:

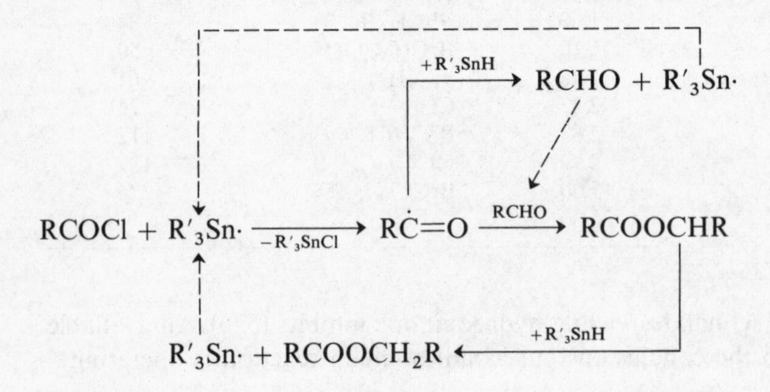

This shows that under suitable conditions some of the aldehyde formed can function as an even better radical trap than the proven scavenger Sn–H (see Chapter 8-4-1). The relationship between radical and trap must therefore be highly specific, and any attempt to derive a general scale of trapping efficiency is futile.

The mechanism can be tested by adding an unrelated aldehyde to the mixture of acyl halide and organotin hydride. In this experiment benzaldehyde and isobutyryl chloride are found to yield benzyl isobutyrate, but no isobutyl benzoate. Conversely, reduction of benzoyl chloride in the presence of isobutyraldehyde yields isobutyl benzoate only, and no benzyl isobutyrate. There is sound evidence for the intermediacy of the acyl radical, e.g. from the fact that triphenylacetyl chloride on reduction

* Earlier authors reported (366) finding benzaldehyde, but did not give amounts, or conditions, or yields.

with tributyltin hydride at 100°C affords (amongst other products) CO and triphenylmethane, as well as the expected aldehyde (441).

$$Ph_3C\dot{C}{=}O \rightarrow Ph_3C\cdot + CO$$

$$Ph_3C\cdot + R_3SnH \rightarrow Ph_3CH + R_3Sn\cdot$$

The behavior of dicarboxy-dihalides is interesting. The radical resulting from halogen-abstraction appears to become trapped by its own carbonyl group in an intramolecular process; e.g. treatment of succinyl (di)chloride with tributyltin hydride yields γ-chloro-γ-butyrolactone; phthalyl (di)-chloride affords the analogous 3-chloro phthalide (428).

$$\begin{array}{l} H_2CCOCl \\ | \\ H_2CCOCl \end{array} + R_3SnH \rightarrow \begin{array}{l} H_2C{-}CO \\ |\qquad\ \ \backslash \\ |\qquad\quad O \\ |\qquad\ \ / \\ H_2C{-}CHCl \end{array} + R_3SnCl$$

10-4 OTHER COMPOUNDS

At present there are only isolated reports concerning reactions of other reducible organic compounds with organotin hydrides; e.g. some aromatic nitro compounds are reduced smoothly to the amino derivatives, leaving other reactive groups such as Br, CH_2Br, CHO more or less intact (Table 10-4).

Aromatic isocyanates yield arylformamides ($LiAlH_4$ reduces right down to the arylmethylamine), and anils the corresponding monosubstituted anilines. Diphenylchlorophosphine gives a good yield of diphenylphosphine (775), and *N*-bromosuccinimide an almost quantitative yield of the parent imide (450). A few reactions of organic sulfur compounds with triphenyltin hydride have been reported (642). Thiophenols and mercaptans yield stannyl aryl(or alkyl) sulfides with elimination of H_2, but aryl disulfides suffer S–S bond scission to give again R_3Sn–S–aryl. On the other hand, (di)benzyl sulfide and disulfide, in which the C–S bond is the equivalent of an allyl C–C, are reduced to toluene, and thiobenzophenone similarly to diphenylmethane. Diphenyl sulfide, sulfoxide, and sulfone do not react.

Carboxylic esters (e.g. benzoates) give hydrocarbons with radical initiators as catalysts and under rather vigorous conditions (370a):

$$RCOOR' + R''_3SnH \xrightarrow{80\text{—}130^{\circ}C} RCOOSnR''_3 + R'H$$

Table 10-4
Hydrogenation of Various Compounds

COMPOUND	HYDRIDE	CATALYST	REACTION CONDITIONS	REDUCTION PRODUCT	YIELD (%)	REFERENCES
m-Nitrobenzaldehyde	Ph_2SnH_2	—	in moist Et_2O, 20°	*m*-Aminobenzaldeyde	65	431, 432
Nitrobenzene	Ph_3SnH	—	?	Aniline	38	617
p-Bromonitrobenzene	Ph_2SnH_2	—	in moist Et_2O, 20°	*p*-Bromoaniline	55	433
Phenyl isocyanate	Ph_3SnH	—	4 hr., 100°	*N*-Phenylformamide	55	484
Naphthyl isocyanate	Ph_3SnH	—	1 hr., 80°	*N*-Naphthylformamide	41	484
Benzylideneaniline	Ph_3SnH	—	22 hr., 124°	*N*-Benzylaniline	35	484
Benzylideneaniline	Bu_2SnH_2	Pt	in Et_2O, 10 hr., 0°	*N*-Benzylaniline	91	373
Benzophenone anil	Bu_2SnH_2	Ni	in Et_2O, 10 hr., 0°	*N*-(Diphenylmethyl)aniline	93	373

Numerous metal halides and oxides, e.g. of Ag, Zn, Cd, Hg, Fe, Pd, Pt, Ge, Sn, Pb, As, Sb, and Bi, are reduced smoothly by triethyltin hydride to the elemental metal (27). In this connection, one should note again the catalytic effect of such metal compounds on various reactions of organotin hydrides.

11 Alkali and alkaline earth stannylmetallics

11-1 STANNYL-ALKALI COMPOUNDS

A number of compounds R_3SnM and R_2SnM_2 are known, all of them deeply colored in solution (315, 875). R may be aliphatic or aromatic; M may be Li, Na, or K. These derivatives are poorly characterized, usually amorphous solids. They are highly reactive, even unstable, sensitive to air and moisture, and often decompose even at room temperature in the absence of air. For example, Me_3SnNa when warmed is transformed rapidly into a mixture of hydrocarbons and tin–sodium alloy; triphenylstannyllithium when boiled in tetrahydrofuran yields tetraphenyltin (245). These alkali-metal compounds are therefore used immediately after they have been prepared. Their great reactivity makes them suitable starting materials for many other organotin compounds.

The commonest method of preparation starts from organotin halides (reactions 11-1 and 11-2).

$$R_3SnBr + 2\,Na \rightarrow R_3SnNa + NaBr \qquad (11\text{-}1)$$

$$R_2SnCl_2 + 4\,Na \rightarrow R_2SnNa_2 + 2\,NaCl \qquad (11\text{-}2)$$

Lithium (757, 839, 840) and potassium compounds can be made by the same method. Apart from organotin halides, suitable starting materials include hexaalkylditins (99, 245, 838, 839) (reaction 11-6) and polymeric dialkyltins (reaction 11-8).

At one time liquid ammonia was commonly employed as the solvent (875). This has now largely gone out of use because of inconvenience of handling and undesirable participation in several reactions. Tetrahydrofuran is quite suitable, but may suffer scission of the ether linkage (240), especially when warm. Ethylene glycol dimethyl ether is rather better, particularly in view of subsequent preparative procedures. An elegant and rapid technique involves the use of monosodium-naphthalene in place of metallic sodium (73). The deep green solution of this adduct in one of the above ethers can be run directly into the aliphatic or aromatic organotin halide or ditin, where it is decolorized until the equivalence point is reached.

Separation of the naphthalene from the reaction products can cause difficulty. In such cases the appropriate amount of sodium metal is first put in one of the ethers together with 0.2—0.5% of the naphthalene theoretically required, and the aliphatic or aromatic ditin added dropwise as required. Conversion into the stannylsodium compound is generally satisfactory even at 20°C (426a).

Compounds R_3SnLi may also be made from $SnCl_2$ (73) (reaction 11-3).

$$\begin{aligned} 3\,PhLi + SnCl_2 &\rightarrow Ph_3SnLi + 2\,LiCl \\ 3\,BuLi + SnCl_2 &\rightarrow Bu_3SnLi + 2\,LiCl \end{aligned} \tag{11-3}$$

Ethyl and *n*-butyl compounds dissociate forming equilibrium systems (73, 250):

$$R_3SnLi \rightleftharpoons R_2Sn + RLi \tag{11-4}$$

Reaction products with Me_3SiCl and with CO_2 are explicable only on the assumption that RLi acts as an intermediate:

$$Bu_3SnLi + ClSiMe_3 \rightarrow 1/n\,(Bu_2Sn)_n + BuSiMe_3 + LiCl$$

$$Bu_3SnLi \xrightarrow[\text{Hydrol.}]{CO_2} \underset{(42\%)}{BuCOOH} + \underset{(32\%)}{BuCOBu} + 1/n\,(Bu_2Sn)_n$$

The equilibrium analogous to 11-4 does not seem to exist for the phenyl compound (73), which can be made from polymeric diphenyltin and phenyllithium (245), tetraphenyltin being formed as a by-product.

The constitution of stannyllithium compounds presents more of a problem than might at first be thought. One aspect of this concerns the bonding between alkali metal and stannyl moiety. Many of these substances show noticeable or considerable conductance in liquid ammonia. They have therefore been regarded as salts consisting of M^+ cations and R_3Sn^- or R_2Sn^{2-} anions. However, this view seems to do justice merely to one facet of their behavior. The character of the alkali metal must correspond more to that in a metal alkyl, say Li in butyllithium, or even Na in sodium-naphthalene. (The latter comparison seems particularly apt because of possible delocalization of electrons contributing to the bonding of the stannyl moiety.) This view also accords with the observations that (a) stannyl-alkali compounds can be used for metalation of hydrocarbons, and (b) their reaction with sulfur and its homologs is analogous to that of the alkali-metal alkyls (755).

The chemistry of these substances just cannot be explained in terms of "salt-like constitution" and simple ionic mechanisms. The simple equations used to describe their formation represent only the overall reaction. In reality, reaction 11-1 proceeds in at least two stages, the first of which (11-5) corresponds to a Wurtz reaction and yields the hexaalkylditin (99). [The same is true for lithium (839, 840).] The mechanism of this reaction and of the subsequent cleavage of the Sn–Sn link by more alkali-metal (11-6) is not known for certain, but a homolytic mechanism must be borne in mind. Reactions 11-7 and 11-8 represent similar stages for dialkyltin compounds.

$$2\,R_3SnBr + 2\,Na \rightarrow R_3SnSnR_3 + 2\,NaBr \qquad (11\text{-}5)$$

$$R_3SnSnR_3 + 2\,Na \rightarrow 2\,R_3SnNa \qquad (11\text{-}6)$$

$$n\,R_2SnCl_2 + 2n\,Na \rightarrow (R_2Sn)_n + 2n\,NaCl \qquad (11\text{-}7)$$

$$(R_2Sn)_n + 2n\,Na \rightarrow R_2SnNa_2 \qquad (11\text{-}8)$$

The versatility of the R_3SnM and R_2SnM_2 compounds, and their resemblance to alkali-metal alkyls, may be illustrated by some examples; others will be found in the literature (315, 875). Triphenylstannyllithium will metalate fluorene in ether, not as well as the analogous silyl and germyl compounds but better than the plumbyl derivative (252). It attacks epoxides (reaction 11-9) as does the Na-compound (reaction 11-10) (252).

$$Ph_3SnLi + H_2C\underset{\diagdown O \diagup}{\text{——}}CHCH_2Cl \xrightarrow{\text{Hydrol.}} Ph_3SnCH_2CH(OH)CH_2Cl$$

$$Ph_3SnNa + H_2C\underset{\diagdown O \diagup}{\text{——}}CH_2 \xrightarrow{\text{Hydrol.}} Ph_3SnCH_2CH_2OH$$

Ethers may suffer cleavage, e.g. tetrahydrofuran (240):

$$R_3SnLi + \overline{CH_2CH_2CH_2CH_2O} \xrightarrow{\text{Hydrol.}} R_3SnCH_2CH_2CH_2CH_2OH$$

This happens more quickly with aliphatic stannylmetallics than with aromatic ones, and has to be watched when tetrahydrofuran is used as solvent. In other reactions, too, stannyl-alkali compounds often behave like rather milder alkyl-alkali compounds: e.g. interaction with organic halides to give tetraorganotins occurs readily and is usually quantitative (73, 245, 341, 839) (reaction 11-11); alkyl phosphates are cleaved by reaction 11-12 (839).

$$Ph_3SnNa + ClCH_2Ph \rightarrow Ph_3SnCH_2Ph + NaCl \qquad (11\text{-}11)$$

$$Ph_3SnLi \xrightarrow{(BuO)_3PO} Ph_3SnBu \qquad (11\text{-}12)$$

Ph_3SnLi in tetrahydrofuran attacks the elements sulfur, selenium, and tellurium; solutions of Ph_3SnXLi (X = S, Se, Te) are obtained fairly simply by this method (755) (see also Chapter 17).

In a number of metal exchange reactions the alkali-metal atom and a halogen atom exchange places in a homolytic process; e.g. formation of a ditin in reaction 11-13 (341) might be explained by intermediate formation of Me_3SnCl, which would at once react with more Me_3SnNa to give ditin (see reaction 11-11). Attempts to prepare octaphenyltritin by reactions 11-14 run into difficulties, probably again because of rapid metal–halogen exchange; only small amounts of the tritin are obtained, most of the product consisting of hexaphenylditin and polymeric diphenyltin (567). In any case it would seem impossible for the Na to have come off as Na^+ and the Cl as Cl^-.

$$2\,Me_3SnNa + HCCl_3 \xrightarrow[NH_3]{liq.} Me_3SnSnMe_3 + NaHCCl_2 + NaCl \qquad (11\text{-}13)$$

$$\left.\begin{array}{l} Ph_3SnNa + ClSn(Ph_2)Cl + NaSnPh_3 \xrightarrow{-2\,NaCl} \\ Ph_3SnCl + NaSn(Ph_2)Na + ClSnPh_3 \xrightarrow{-2\,NaCl} \end{array}\right\} Ph_3Sn{-}SnPh_2{-}SnPh_3 \qquad (11\text{-}14)$$

Metal exchange reactions have also been observed in side-reactions of some other processes (315, 599).

As expected, hydrolysis of Ph_3SnLi with $1N$ HCl affords mainly Ph_3SnH (reaction 11-15) (839):

$$Ph_3SnLi \xrightarrow{H_2O/HCl} Ph_3SnH + LiCl \qquad (11\text{-}15)$$

The corresponding ethyl and butyl compounds behave similarly (840). These reactions are now carried out in complexing organic solvents like tetrahydrofuran or diglyme and are important e.g. for convenient preparation of organotin deuterides from D_2O (427a). Me_3SnNa and $Me_3B{\cdot}NH_3$ interact to give $>90\%$ of hexamethylditin (110).

Formation of a metal ketyl with benzophenone (73) is rather interesting; this again cannot be explained in terms of metal ions.

$$Ph_3SnNa + PhCOPh \rightarrow Ph\dot{C}(ONa)Ph + \tfrac{1}{2} Ph_3SnSnPh_3$$

O_2 and triphenylstannylsodium give an 80% yield of hexaphenylditin and Na_2O_2. Reaction with CO_2 is also unusual: the products are oxalic acid and hexaphenylditin (73). (Na-amalgam too reduces CO_2 to oxalic acid.)

$$2\,Ph_3SnNa + 2\,CO_2 \rightarrow (COONa)_2 + Ph_3SnSnPh_3$$

Here it is quite clear that the sodium reacts as the atom and not as the ion.

11-2 STANNYL-ALKALINE EARTH COMPOUNDS

This group of substances has had little attention. What appears to be the only known example contains Mg and may be prepared like the alkali compounds (Chapter 11-1) via hexaphenylditin or starting from hexaphenylditin. The reaction works in tetrahydrofuran but not in diethyl ether (842); this may account for earlier failures.

$$2\,Ph_3SnCl + 2\,Mg \xrightarrow[\text{(EtBr)}]{\text{THF, room temp.}} (Ph_3Sn)_2Mg + MgCl_2$$

The compound decomposes on warming, but on hydrolysis with NH_4Cl–H_2O the authors obtained a good yield of triphenyltin hydride.

12 Stannylboron compounds

It is curious that compounds containing a Group III element bonded to tin were completely unknown until recently. Colorless compounds with covalent Sn–B bonds have now been made from triethylstannyllithium and the appropriate dimethylaminoboron chloride in ether (624):

$$Et_3Sn\text{–}B(NMe_2)_2 \qquad (Et_3Sn)_2B\text{–}NMe_2$$

They may be distilled in a high vacuum, but decompose when heated to 100°C since they are rather less stable than the analogous silicon derivatives. They are easily hydrolyzed to the trialkyltin hydride, boric acid, and the secondary amine. Halogens cleave the Sn–B bond, while HCl preferentially attacks the B–N bond. The B–N bonds clearly stabilize the bonding system of these molecules (as in the analogous silylboron compounds), since derivatives of the type Et_3SnBR_2 are unstable.

Stannyl compounds containing boron as substituent in the alkyl moiety have been described (762).

13 Stannyl–silicon, –germanium, and –lead compounds

Little is known of such compounds, although their preparation should generally present few difficulties. The usual methods appear to be suitable; e.g. an Sn–Si link is formed in the reaction:

$$Ph_3SnLi + ClSiPh_3 \rightarrow Ph_3SnSiPh_3$$

Admittedly, the ditin and the disilane are formed as by-products, and this indicates rapid metal–halogen exchange before the actual condensation. The tributyl-trimethyl derivative is obtainable similarly (839, 840). In this series the compound $Ph_3SnSi(GePh_3)_3$ might also be mentioned (893).

Sn–Ge bonds are formed by completely analogous reactions. Examples are $Ph_3SnGePh_3$ and $Me_3SnGePh_3$ (242, 325). These compounds can also be made from the corresponding germylpotassium (109a). Condensation of stannylamines with triphenylgermanium hydride provides a mild approach to the Sn–Ge linkage, but the triethyl hydride and organosilicon hydrides do not react with stannylamines (591, 815):

$$Bu_3SnNEt_2 + HGePh_3 \rightarrow Bu_3SnGePh_3 + HNEt_2$$

The aforementioned trimethylstannyl derivative is formed smoothly by this method. A number of phenyl derivatives have been prepared in the same way or by the analogous condensation of triphenyltin hydride with triphenylgermanium diethylamide. Difunctional derivatives give longer chains of metal atoms. Some interesting intermediates and their reaction products with water, phenylacetylene, and phenol have also been described (159b).

Trialkyltin methoxides can be used instead of stannylamines, but only at much higher temperatures, e.g. 180°C (864a).

$Ph_3SnPbPh_3$ has been obtained from triphenyltin chloride and triphenylplumbyllithium (888), after several earlier failures to prepare it. The compound decomposes at about 110°C, can be recovered unchanged from chloroform, but is hydrolyzed in aqueous tetrahydrofuran.

An interesting group of materials comprises substances analogous to, and usually isomorphous with, $(Ph_3Sn)_4Sn$ [see Chapter 14-3, formula (7)]. These have been explored mainly by van der Kerk and his group (889), and all contain a central metal atom tetra-coordinated by similar atoms. Compounds containing tin include $(Ph_3Sn)_4Ge$, $(Ph_3Ge)_4Sn$, $(Ph_3Pb)_4Sn$, and $(Ph_3Sn)_4Pb$. They are colorless or yellow (the penta-lead compound is red) crystalline solids and have been well characterized.

14 Compounds with Sn–Sn bonds

14-1 HEXAORGANODITINS

Ditins $R_3Sn–SnR_3$ were observed as by-products in the earliest attempts to prepare aliphatic and aromatic tetraorganotins, yet their precise modes of formation can be difficult to interpret even today. Their multiplicity is almost an embarrassment to the preparative chemist (315, 416),* and shows that there is a considerable tendency towards formation of ditins. These may result e.g. from ligand exchange or metal exchange (Chapter 11-1), or from interaction of tin tetrahalides with Grignard compounds carrying bulky alkyl groups. They may be the main product or the only product. (The Grignard compound here reduces the triorganotin halide intermediate, especially at higher temperatures.)

Highly reactive metals or organometallic compounds have been in use for some time for specific syntheses (315, 416).

$$2\,R_3SnCl + 2\,Na \rightarrow R_3SnSnR_3 + NaCl \qquad (14\text{-}1)$$

$$6\,PhMgBr + 3\,SnCl_2 \xrightarrow[\text{heat}]{(PhMgBr)} Ph_3SnSnPh_3 + Sn + 6\,MgBrCl \qquad (14\text{-}2)$$

Reaction 14-1 proceeds in stages (for details see Chapter 11-1). The multi-stage reaction 14-2 may be carried out in the same reaction vessel without isolation of intermediates, starting from the organic halide, and can be recommended particularly for ditins with bulky groups.

A further range of syntheses, employing very mild conditions, has been opened up recently by the discovery that stannyl compounds with

* A review, "Catenated Organic Compounds of Silicon, Germanium, Tin, and Lead," containing many references up to early 1965, has been published (238a).

nucleophilic groups, such as stannylamines, stannyl alkoxides, or stannoxanes will condense easily with organotin hydrides, e.g. as in reactions 14-3—14-5. (Reaction mechanisms are discussed in Chapter 8.)

$$R_3SnNR'_2 + HSnR_3 \longrightarrow R_3SnSnR_3 + HNR'_2 \quad (158, 591) \qquad (14\text{-}3)$$

$$R_3SnOR' + HSnR_3 \longrightarrow R_3SnSnR_3 + HOR' \quad (724) \qquad (14\text{-}4)$$

$$\tfrac{1}{2}\,R_3SnOSnR_3 + HSnR_3 \longrightarrow R_3SnSnR_3 + \tfrac{1}{2}\,H_2O \quad (591) \qquad (14\text{-}5)$$

In reaction 14-3, the fastest rates are obtained with the now easily accessible stannyldialkylamines (Chapter 15) with small groups R′ (Me or Et), reaction being usually exothermic. Stannylhydrazines and formylamines react more slowly, but are often useful (158, 489).

Hexaorganoditins are also formed easily from triorganotin hydrides by elimination of H_2, e.g. catalyzed by $PdCl_2$, or by treatment with H_2-acceptors (see Chapter 10):

$$2\,R_3SnH \longrightarrow R_3SnSnR_3 + H_2$$

Reactions 14-3 and 14-5 are suitable also for preparation of unsymmetrical ditins $R_3SnSnR'_3$, by using a substituent R on the hydride different from that on the reaction partner. All the other syntheses yield mixtures of the unsymmetrical ditin and the two symmetrical ones, owing to exchange of functional groups between the reaction partners (see Chapters 8 and 11).

A large number of symmetrical and unsymmetrical ditins is now known, variously carrying aliphatic, cyclic aliphatic, or aromatic groups. A selection is given in Table 14-1.

Ditins with small aliphatic groups are colorless liquids which may be distilled *in vacuo* without decomposition. They are insoluble in water, sparingly soluble in lower alcohols, but easily soluble in all other common organic solvents. They can suffer autoxidation. The peroxide $R_3SnOOSnR_3$ (R = Et) (see Chapter 17-5) is formed first, and reacts with unoxidized ditin to yield the stannoxane $R_3SnOSnR_3$. Some peroxide also decomposes by way of $R_3SnO\cdot$ radicals to give R_3SnOR and R_2SnO. Autoxidation of ditins thus affords a mixture of these products. Autoxidation can also be induced by light, and proceeds then to the same products even at −30°C (11, 12).

Partially or fully arylated ditins are usually crystalline at room temperature and are little or not at all subject to autoxidation, except in the presence of sodium methoxide, ethyl bromide, and a few other substances. Free radicals $R_3Sn\cdot$, e.g. $Ph_3Sn\cdot$, are again assumed to be intermediates (856a). Of course, Sn–Sn groups are oxidized by common strong oxidants.

Table 14-1
Ditins $R_3Sn–SnR_3$ and $XR_2Sn–SnR_2X$

FORMULA	B.P. (°C/mm.)	M.P. (°C)	n_D^{20}	METHOD OF PREPARATION
$Me_3SnSnMe_3$	182	23.3	—	413, 591
	62—63/12			
$Et_3SnSnEt_3$	161—162/23	—	1.5380	270, 453, 591
	149—150/12			
$Bu_3SnSnBu_3$	147—150/0.2	—	1.513	491, 591
i-$Bu_3SnSn$$i$-$Bu_3$	179/3.5	43.8	—	270, 591
	129—131/0.15	55—56		
(cyclo-$C_6H_{11})_3SnSn$(cyclo-$C_6H_{11})_3$	—	decomp. >300	—	417, 591
$Ph_3SnSnPh_3$	—	237 (corr.)	—	248, 415[a], 591
		232.5		
$Me_3SnSnEt_3$	235/748	—	1.5388	406, 413, 591
	106—108/12			
$Et_3SnSn$$i$-$Bu_3$	116—118/0.3	—	1.525	591
$Me_3SnSnPh_3$	—	106	—	160, 406, 591, 840
		108		
$Et_3SnSnPh_3$	187/0.0001	16	1.631	160, 591, 611
i-Bu_3SnSn(cyclo-$C_6H_{11})_3$	—	172	—	591
$Bu_2Sn(Cl)Sn(Cl)Bu_2$	—	25—27	—	236, 725
i-$Bu_2Sn(Cl)Sn(Cl)$$i$-$Bu_2$	—	—	1.550	580, 816
$Bu_2Sn(OCOMe)Sn(OCOMe)Bu_2$	—	−7 to −4	1.5060 (26°)	726, 728
$Bu_2Sn(OCOPh)Sn(OCOPh)Bu_2$	—	31.5—32.5	—	726, 728
$Ph_2Sn(OCOPh)Sn(OCOPh)Ph_2$	—	172—179	—	726, 728
$Ph_2Sn(OCOCH_2Cl)Sn(OCOCH_2Cl)Ph_2$	—	150	—	726, 728
i-$Bu_2Sn(CH_2CH_2CN)Sn(CH_2CH_2CN)$$i$-$Bu_2$	173—175/0.005	—	1.539	593
i-$Bu_2Sn(CH_2CH_2COOMe)Sn(CH_2CH_2COOMe)$$i$-$Bu_2$	148—150/0.0001	—	1.521	593

[a] Reference 415 describes substituted aromatic ditins.

The structure of hexaorganoditins was the subject of debate for quite some time. Although Ladenburg (452) had shown as early as 1870 that the ethyl compound in the vapor state had the formula $Et_3SnSnEt_3$, it was believed later that cryoscopic measurements indicated dissociation of the Sn–Sn bond similar to that of the C–C bond in hexaphenylethane (for reviews see refs. 315 and 416). However, no sign of unpaired electrons could be found by either magnetic balance (537) or ESR (583) measurements. We now know that the explanation lies in a specific failure of cryoscopic measurement.* Osmometry shows no sign of dissociation even in $10^{-3}M$ solution (583). The final proof of the thermal stability of the covalent Sn–Sn bond lies in the resistance of the unsymmetrical ditins to thermal decomposition. Trimethyltriethylditin can be distilled at 235°C without any disproportionation; it remains gas-chromatographically pure even after hours of heating at 170°C. Conversely, the two corresponding symmetrical ditins are stable as a mixture without any sign of comproportionation (583). The dissociation energy of the Sn–Sn bond in the hexaethyl compound has been given as 50 kcal./mole (681a).

The Sn–Sn bond is highly reactive. It is cleaved not only by oxygen (see above for details) but also by sulfur, halogens, alkyl halides, peroxides, and many other compounds:

$$R_3SnSnR_3 + XY \rightarrow R_3SnX + R_3SnY$$

(X can be the same as, or different from, Y)

Both polar and radical mechanisms have been noted, but in many cases the mechanism is not known with certainty. On the other hand, the Sn–Sn bond is stable against water, sodium hydroxide in tetrahydrofuran (but not in wet acetone), and sulfuric acid in tetrahydrofuran (856a).

Additions of Sn–Sn bonds to olefins have been reported also:

$$R_3SnSnR_3 + \;>C{=}C< \;\rightarrow\; -\underset{R_3Sn}{\overset{|}{\underset{|}{C}}}-\underset{SnR_3}{\overset{|}{\underset{|}{C}}}-$$

Table 14-2 contains examples from both categories.

Hexaorganoditins may decompose in more complicated ways, depending on conditions. Fierce heating leads to fission of C–Sn as well as

* The apparent particle weight can sink to less than half the molecular weight of the ditin. However, such "ultradissociation" is impossible, since such particles would not reunite to the ditin on evaporation of the solution. Apart from ditins, the phenomenon is also shown by triphenyltin chloride and many other undissociated substances. No explanation is currently available.

Table 14-2
Some Reactions of Hexaorganoditins $R_3Sn–SnR_3$

Reaction	REF.
$R_3SnSnR_3 + Hal_2 \rightarrow 2\,R_3SnHal$ (R = aliph. or arom.)	87, 315, 835
$Me_3SnSnMe_3 + 2\,HCl \xrightarrow[MeOH]{25°} 2\,Me_3SnCl + H_2$	835
$Et_3SnSnEt_3 + CCl_4 \xrightarrow{20°} Et_3SnCl$ (22%) + Et_2SnCl_2 (15%) + other products	685a
$Me_3SnSnMe_3 + F_3CI \xrightarrow{UV} Me_3SnCF_3 + Me_3SnI$	142
$Et_3SnSnEt_3 + Hg(CH_2COOMe)_2 \xrightarrow[-Hg]{170°} 2\,Et_3SnCH_2COOMe$	496
$Bu_3SnSnBu_3$ + *t*-BuOO*t*-Bu → 2 Bu_3SnO*t*-Bu[a]	584
$Et_3SnSnEt_3 + PhCOOOCOPh \rightarrow 2\,PhCOOSnEt_3$[b]	693
$Et_3SnSnEt_3$ + RCOOO*t*-Bu → $RCOOSnEt_3$ + Et_3SnO*t*-Bu	865
$Et_3SnSnEt_3 + 2\,Ph_3CCl \rightarrow 2\,Ph_3SnCl + Ph_3CCPh_3$	685
$Et_3SnSnEt_3 + BrCH_2CH_2Br \xrightarrow{180°} 2\,Et_3SnBr + H_2C{=}CH_2$	688
$Me_3SnSnMe_3 + F_2C{=}CF_2 \rightarrow Me_3SnCF_2CF_2SnMe_3$	70
$Me_3SnSnMe_3 + F_3CC{\equiv}CCF_3 \xrightarrow[UV]{150°}$ *trans*-$F_3CC(SnMe_3){=}C(SnMe_3)CF_3$	162

[a] About 50% side-reaction.
[b] Radical cleavage of C–Sn also produces some $(PhCOO)_2SnEt_2$.

Sn–Sn bonds. The end-products are tetraorganotins and tin metal; polymeric materials are formed as intermediates (see Chapter 14-3). Irradiation produces the same result (695, 870) as does catalysis by e.g. BF_3, $AlCl_3$, or $AlBr_3$ (110, 695, 870) in a heterolytic process. Stoichiometric amounts of metal halides, e.g. $AlCl_3$ or $AuCl_3$, cleave to organotin halides (695, 834). Hydrogenation of the Sn–Sn bond by means of $LiAlH_4$ is only partially successful (yields of 4—19%); tarry by-products are formed at elevated temperatures (63). [Older reports of successful hydrogenation (616) require correction (63), since the starting material is now known to have been a stannoxane.]

14-2 SUBSTITUTED ORGANIC DITINS

As shown in Chapter 14-1, the ditin grouping Sn–Sn is quite reactive. But mild substitutions need not lead to cleavage of the Sn–Sn link. Ditins also remain as preferred fragments in radical degradation of polymeric diphenyltin by means of benzoyl peroxide (726):

$$\left(-\underset{Ph}{\overset{Ph}{|}}{Sn}-\underset{Ph}{\overset{Ph}{|}}{Sn}- \right)_n + n \begin{array}{c} OCOPh \\ | \\ OCOPh \end{array} \longrightarrow n\, PhCOOSn(Ph_2)Sn(Ph_2)OCOPh$$

Thermal decomposition of diisobutyltin dihydride (and presumably of other dialkyltin dihydrides) also involves intermediate formation of ditin moieties (750).

Preparation mostly starts from organotin hydrides, particularly from those carrying a negative substituent on the tin (see Chapter 8). (For the experiments of C. A. Kraus 1925—1929, see refs. 315 and 416.) Many substituted ditins XR_2SnSnR_2X are known. A selection is listed in Table 14-1.

The acetic esters (**2**) (728, 729) (R′ = Me) have been studied thoroughly. They can also be obtained from equimolar amounts of R′COOH and R_2SnH_2, which is not surprising when one remembers that the acetoxyhydride (**1**) is in equilibrium with dihydride and diacetate (Chapter 8), and that the latter is formed from free acetic acid and the dihydride. Numerous aliphatic and aromatic carboxylic acids give this reaction (728, 729).

Tetraorgano-1,2-dihaloditins (**3**) are formed quantitatively and exothermically by treatment of diorganohalotin hydrides with pyridine or other amines (580). (**3**) may also be obtained from stannyl esters (**2**). Reduction affords the dihydride (**4**). These and some other transformations (580, 730) are shown in the following reaction scheme.

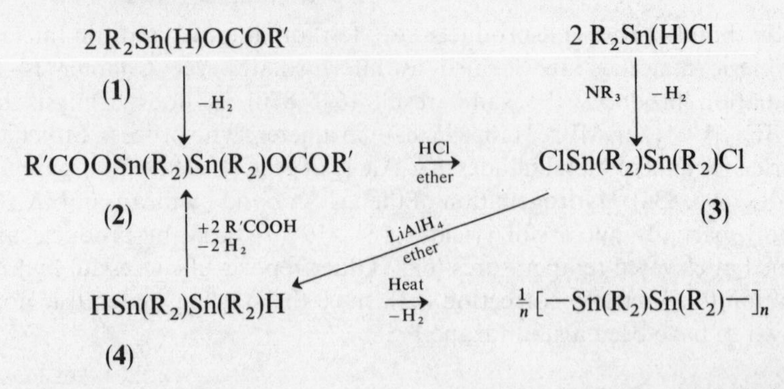

In these reactions the Sn–Sn group is thus relatively stable. Nevertheless, in the stannyl esters (**2**) it can be cleaved by carboxylic acids when R is aliphatic. When R = Ph, the ester (**2**) is stable in this respect, but disproportionates more rapidly when heated than the aliphatic derivatives.

The Sn–Sn linkage also remains intact when all the phenyl groups in hexaphenylditin are replaced in boiling glacial acetic acid by acetoxy groups. The latter may then be replaced stepwise by chlorine (882).

The fully hydrogenated ditin Sn_2H_6 decomposes rapidly even at 20°C (331). Compounds obtained from dibutyltin dihalides by treatment with sodium ethoxide in ethanol were regarded at one time as ditins of structure XR_2SnSnR_2X (327, 328). In reality, they are stannoxanes containing the grouping Sn–O–Sn (see Chapter 17-1).

14-3 LINEAR AND BRANCHED POLYTINS

In 1917 G. Grüttner concluded from his results (270): "This shows that tin is able, just like silicon, to form catenated compounds." Such an attractive hypothesis should really have provided the impetus for polytin chemistry. Instead, this sector of organotin chemistry remained until recently unexplored territory and a hiding place for time-honored but unproven concepts.

The most important of these, and an obstacle to progress for several decades, was the concept of "organic compounds of divalent tin" or supposedly stable "radicals of type R_2Sn." This dates back to the middle of the last century, and until recently arguments were still being sought to explain the electron arrangement of the tin and the evident stability. (For reviews see refs. 315, 416, and 460.) At the same time the available experimental material was inadequate and full of contradictions.

For example, red, yellow, or colorless oils, and colorless or yellow solids were for decades described as "diethyltin" (see refs. 551 and 579). The materials were usually described as polymers, but their molecular weight generally remained undetermined. There was no suggestion as to how the end-group valencies of the supposed polymers might be saturated. There was disagreement between the findings of different investigators and mere speculation regarding molecular structure (314, 460). G. E. Coates (143) therefore concluded in 1960 that pure diethyltin had not yet been prepared. The various products were indeed shown later to be anything but uniform (579, 686); analysis by selective cleavage of Sn–Sn bonds with halogen (579) showed up to 44 mole % triethyl and up to 31 mole % monoethyl moieties. The products were therefore in reality complicated mixtures of polytins, mainly with structures (**5**) and (**6**). Other samples contained halogen or large amounts of diethyltin oxide. (For detailed discussion see refs. 551 and 579.)

$$R_3Sn\text{–}(SnR_2)_n\text{–}SnR_3 \quad \textbf{(5)} \qquad (R_3Sn)_2Sn(R)\text{–}(SnR_2)_m\text{–}SnR_3 \quad \textbf{(6)}$$

The situation concerning "diphenyltin" remained equally confused until 1961 (detailed discussion in refs. 551 and 564). However, the older results and ideas of supposed gradual polymerization of "monomeric diphenyltin" (131, 415) were not able to withstand the test of modern analytical methods. The presence of monomer was disproved in every case. Both the chemical results of the Giessen team (551, 564) and the X-ray structure determination by R. E. Rundle and D. H. Olson (714) show that earlier preparations of "diphenyltin" really yield samples of polytins with true Sn–Sn bonds, consisting mostly of complicated mixtures of types (**5**) and (**6**) (with R = Ph); i.e. they contain tetra-covalent tin. There is no evidence for reversible dissociation into R_2Sn moieties or for "carbene-like behavior." All the chemical reactions of polytins (for a review see ref. 540a) must be understood as reactions of covalent Sn–Sn bonds (see Chapter 14-1) or of covalent C–Sn bonds.*

At the moment, not a single compound is known that contains organic groups covalently or ionically (714) attached to divalent tin, i.e. which might be assigned the structure R–Sn–R.† Such species may, however, occur as short-lived and highly reactive intermediates of some reaction mechanisms, at high temperatures, or in the mass spectrometer (427b) (see Chapter 23-4). [Dicyclopentadienyltin appears to be a special case. It is considered to have an angular sandwich structure involving delocalized covalent bonds (171).] The first attempts at deliberate synthesis

* In spite of this, polytins have sometimes been called "organometallic analogs of carbenes" (540a), presumably for the sake of simplicity.

† In spite of all this, some recent reviews still refer uncritically to old and discredited work, and take no account of the present state of knowledge.

of methyl- and ethyl-substituted tri-, tetra-, and penta-tins were made by C. A. Kraus and his co-workers who studied the interaction of organotin alkali compounds and organotin halides (411, 412). Because of ligand exchange (see Chapter 11-1), which was not then considered, they appear to have obtained mixtures of homologous polymers (**5**) ($n = 0, 1, 2, 3, \ldots$). This follows from discontinuities in the distillation curves, and from the NMR spectra of the products (101). (Aliphatic tri- and tetra-tins in any case distil unchanged *in vacuo*; see below.) Well defined linear polytins have become known only in the last few years; a selection is given in Table 14-3. Octaphenyltritin, which crystallizes uniformly as colorless octahedra (decomp. 280°C), is prepared by reaction 14-6. The sodium compound is first made from the monochloride by treatment with sodium-naphthalene in dimethoxyethane, and the dichloride added subsequently (567).

$$2\,Ph_3SnNa + Ph_2SnCl_2 \xrightarrow{(CH_2OMe)_2} Ph_3Sn\text{–}SnPh_2\text{–}SnPh_3 + 2\,NaCl \qquad (14\text{-}6)$$

Here again, competing reactions arise through metal exchange, yielding hexaphenylditin and polymeric diphenyltin as by-products. These become the sole products under different experimental conditions.

An elegant synthesis of tritins has been developed by the teams in Giessen and Utrecht. It involves condensation of organotin hydrides with organotins carrying nucleophilic groups. [Principle and mechanisms are discussed in Chapter 8 (158, 160, 816)]. The best results are obtained with stannylamines, particularly those with strongly nucleophilic nitrogen, e.g. dimethylamino- and diethylamino-compounds (816) (see reaction 14-7).

$$2\,R'_3SnNEt_2 + HSnR_2H \rightarrow R'_3Sn\text{–}SnR_2\text{–}SnR'_3 + 2\,HNEt_2 \qquad (14\text{-}7)$$

R and R′ may be varied within wide limits, since the low-boiling diethylamine can be removed easily. In other cases, use may be made of organotins carrying other amino groups (158), or –OMe or $-OSnR'_3$ groups (816). A large number of tritins can be prepared in this way; Table 14-3 lists a selection. All have sharp melting or boiling points.

Ligand exchange can be a nuisance in these reactions, and has to be watched particularly with rather more sluggish reactants. For example interaction of dialkyltin dihydride and hexaalkylstannoxane yields exclusively dialkyltin oxide and monohydride, in place of the expected tritin.

$$R'_3SnOSnR'_3 + R_2SnH_2 \begin{cases} \not\longrightarrow R'_3Sn\text{–}SnR_2\text{–}SnR'_3 + H_2O \\ \longrightarrow R_2SnO + 2\,R'_3SnH \end{cases}$$

Again, reaction of stannylenediamines with monohydrides may be rapid and exothermic, but instead of the expected pure tritin R'_3Sn–SnR_2–SnR'_3 one finds an almost inseparable mixture of homologous polymeric polytins R'_3Sn–$(SnR_2)_n$–SnR'_3 ($n = 0, 1, 2, 3, \ldots$). The result can be explained only by a very rapid ligand exchange preceding the condensation.

$$R'_3SnH + R_2Sn(NEt_2)_2 \rightleftharpoons R'_3SnNEt_2 + R_2Sn(H)NEt_2$$

The newly formed reagents can then condense either with each other or with remaining original reagent (detailed discussion in refs. 591 and 816).

Ligand exchange on organotins has often caused difficulty; in the circumstances it is accorded far too little attention even in some of the newer literature.

Linear tetratins too are accessible by the same synthetic route (816) (see Table 14-3).

$$2\,R'_3SnNEt_2 + HSnR_2SnR_2H \longrightarrow R'_3SnSnR_2SnR_2SnR'_3 + 2\,HNEt_2$$

These materials are colorless, and stable up to 150°C, many of them to 200°C. Those with small alkyl groups may therefore be distilled *in vacuo* without decomposition. Some are crystalline and have a sharp melting point.

Linear tetra-, penta-, and hexa-tins (Table 14-3) may be obtained from polytin monohydrides, which in their turn are prepared by condensation of organotin hydrides with *N*-phenylformamido-tin compounds (158). All the above remarks concerning possible ligand exchange with tritins apply equally here.

Apart from these specific syntheses there are also some non-specific ones yielding mixtures of homologous polymeric, and in part branched, polytins.

Treatment of hexaethylditin with di-*t*-butyl peroxide affords much decaethyltetratin as well as higher polytins. It is worth noting that the peroxide cleaves not only Sn–Sn linkages (triethylstannyl *t*-butoxide is one of the products) but also some C–Sn bonds (see Chapters 4-4 and 14-1).

Polytins are also formed as intermediates in the complicated UV- (686), $AlCl_3$- (686), or BF_3-catalyzed (110) disproportionation of hexaalkylditins. The overall reaction (for R = Et) is something like:

$$(n + 1)R_3SnSnR_3 \longrightarrow n\,R_4Sn + R_3Sn\text{–}(SnR_2)_n\text{–}SnR_3 \qquad (n = 1, 2, 3, \ldots)$$

The mixture of polytins at once decomposes further to R_4Sn and tin metal. Higher and branched polytins likewise disproportionate in this way (870).

Table 14-3
Linear Tri-, Tetra-, Penta-, and Hexa-tins

FORMULA	B.P. (°C) AT 10^{-4} mm.	M.P. (°C)	n_D^{20}	METHOD OF PREPARATION
$Me_3Sn–SnMe_2–SnMe_3$	66—68	—	1.5898	816
$Et_3Sn–SnEt_2–SnEt_3$	124—127	—	1.5802[a]	158, 815, 816
$Bu_3Sn–SnBu_2–SnBu_3$	175—177	—	1.5380[b]	158, 815, 816
i-$Bu_3Sn–Sn$$i$-$Bu_2–Sn$$i$-$Bu_3$	169—172	—	1.5431	815, 816
$Ph_3Sn–SnPh_2–SnPh_3$	—	280 (decomp.)[d]	—	567[d]
$Et_3Sn–SnBu_2–SnEt_3$	139—141	—	1.5641	815, 816
$Et_3Sn–SnPh_2–SnEt_3$	—	—	1.6202	160
i-$Bu_3Sn–SnPh_2–Sn$$i$-$Bu_3$	—	28	1.5840 (supercooled melt)	816
$Ph_3Sn–SnBu_2–SnPh_3$	—	83[e]	[e]	160, 816
(cyclo-$C_6H_{11})_3Sn–Sn$$i$-$Bu_2–Sn$(cyclo-$C_6H_{11})_3$	—	168	—	816
$Et_3Sn–(SnEt_2)_2–SnEt_3$	165—172[f]	—	1.6061[c,f]	158, 593, 870
$Bu_3Sn–(SnBu_2)_2–SnBu_3$	—	—	1.5543	158
$Me_3Sn–(Sn$$i$-$Bu_2)_2SnMe_3$	150—155	~ 90	—	816
$Ph_3Sn–(Sn$$i$-$Bu_2)_2–SnPh_3$	—	144	—	816
(cyclo-$C_6H_{11})_3Sn–(Sn$$i$-$Bu_2)_2–Sn$(cyclo-$C_6H_{11})_3$	—	207	—	816
$Et_3Sn–(SnPh_2)_2–SnEt_3$	—	—	1.6877	158
$Et_3Sn–(SnEt_2)_3–SnEt_3$	—	—	1.6433	158
$Et_3Sn–(SnEt_2)_4–SnEt_3$	—	—	1.678	158

[a] Without redistillation (158): 1.5894.
[b] Without redistillation (158): 1.5342.
[c] Without redistillation (158): 1.6283.
[d] The identity of another product (160), M.P. 150—153°C, with octaphenyltritin has not been established.
[e] Reference 160 describes it as a colorless liquid, n_D^{20} 1.6452.
[f] Reference 870 gives B.P. 166—189°C/1 mm. and n_D^{20} 1.5777—1.5844.

A mixture of α,ω-dichloropolydialkyltins is formed when the dichloride R_2SnCl_2 (R = Me, Et, *n*-Bu) is treated with insufficient sodium powder. One particular compound may be the principal product, e.g. 1,4-dichlorooctaethyltetratin (M.P. 179—182°C from pentane) (919). α,ω-Dihalo-polytins can also arise from ring-opening of cyclic polytins (see Chapter 14-4).

Several polytins of general formula $(-R_2Sn-)_n$ but uncertain end-group constitution have been prepared. They include examples with R = Me (n = 12—20) (101), with R = Et or *n*-Bu (n = 7.5—8.3) (915), and with R = Ph [n variable (439) or 13—14 (564)]. Similar compounds are formed in the hydrogenation of carbonyl compounds with R_2SnH_2 (Chapter 10), and in the reaction of organomercury compounds with R_2SnH_2 (Chapter 18). Polymers $(i\text{-BuSnCl})_n$ and $(EtSnBr)_n$ (223) have been described also. All these materials may be partly composed of cyclic polytins (see Chapter 14-4).

Straight-chain polytins are colorless, at least up to the tetratin. It is not yet clear why mixtures of polytins are often yellow, red, or even deep red. Analysis of such colored products by selective cleavage of Sn–Sn bonds invariably shows tin atoms connected to three or even four other stannyl moieties (564, 579, 686). The materials thus contain branched tin chains. In some cases the depth and intensity of color appears to be proportional to the extent of branching. One could therefore imagine mobile electrons participating in various mesomeric forms of the branched-chain system. However, the phenomenon awaits final clarification. Discrete branched-chain phenylpolytins, e.g. (7), are colorless. Mesomerism might be inhibited in these cases by the electron-withdrawing phenyl groups. Deep red mixtures of branched phenylpolytins are formed in the co-condensation of triphenyltin hydride, diphenyltin dihydride, and phenyltin trihydride (564). (For details of this reaction see Chapter 14-4.)

$$\begin{array}{c} SnPh_3 \\ | \\ Ph_3Sn-Sn-SnPh_3 \\ | \\ SnPh_3 \end{array} \quad (7)$$

Whilst the existence of branched catenated systems of covalently bonded tin atoms has been inferred right through the preceding paragraph, there are as yet few examples of discrete branched polytins. Böeseken and Rutgers (85) obtained a crystalline colorless compound to which they assigned structure (7) as early as 1923. Their work has been confirmed recently by the groups of Gilman (239) and van der Kerk (889). Several syntheses were discovered, e.g. that by reaction 14-8. The others yielded mostly hexaphenylditin, as a consequence of ligand exchange.

$$3\ Ph_3SnLi + SnCl_2 \xrightarrow[-2\ LiCl]{THF} (Ph_3Sn)_3SnLi \xrightarrow[-LiCl]{Ph_3SnCl} \textbf{(7)}\quad (70\%) \qquad (14\text{-}8)$$

Both nucleophilic and electrophilic reagents effect Sn–Sn bond scission, but fail to give uniform products (239).

Analogous compounds containing other Group IV elements besides tin (see Chapter 13) are mostly isomorphous with (**7**).

The condensation products of phenyltin trihydride and trialkylstannylamines have been assigned structure (**8**) (R = Et or C_8H_{17}) (160). They are red or yellow oils respectively. No evidence for the structure has been published.

$$(R_3Sn)_3SnPh \qquad \textbf{(8)}$$

Some reactions of these polytins have been mentioned already. Their thermal stability, often to over 200°C, is surprising and in no way inferior to that of the ditins. It shows that Sn–Sn bonds retain their strength even in larger molecules. However, catalysts and preparative impurities may appreciably reduce the thermal stability. Aliphatic derivatives are cleaved easily by air and other mild oxidants, whereas cyclohexyl derivatives are attacked only slightly and aromatic ones not at all. Halogens effect instant Sn–Sn scission; titration with I_2 at 20°C is used for quantitative estimation (see Chapter 22-3). Numerous other polar or radical agents display ready reaction. Amongst these might be mentioned benzoyl peroxide, *t*-butyl peroxide, quinone, sulfur, dibenzyl disulfide, H_2O_2, and diphenyltin dichloride (434, 694). *n*-Butyl, phenyl, allyl, and benzyl bromide likewise cleave the Sn–Sn linkage yielding compounds of type R_3SnBr or R_2SnBr_2 depending on the nature of the alkyl group and hence the mechanism (118). Reactions of the Sn–Sn links are thus very similar to those encountered with ditins and cyclic polytins (see Chapters 14-1 and 14-4 respectively). For a review, see ref. 540a.

In some of the reported reactions, the identity of the polytin starting materials is unfortunately in doubt. They may well have been mixtures of types (**5**) and (**6**), in some cases containing cyclic polytins also (see Chapter 14-4). The exact study of these linear polytins remains a major task for the future.

14-4 CYCLIC POLYTINS

It has become apparent in the last few years that both aliphatic and aromatic diorganopolytins $(-R_2Sn-)_n$ have a surprising tendency towards cyclization. Under suitable conditions the products are of uniform ring size. Covalent tin is thus seen to catenate not only into straight and branched chains but also into ring systems, e.g. (**9**) or (**10**) (see Table 14-4). Formally these compounds are actual "dialkyltin" polymers of exact formula $(R_2Sn)_n$. Some light is thus thrown on the old problem of the "tin dialkyls," which was expressed by G. E. Coates in these words (143): "The constitution and the chemistry of these peculiar compounds, R_2Sn, are not at all well understood."

Cyclic polytins $(R_2Sn)_n$, both aromatic and aliphatic, are now recognized as thermodynamically stable. They are built up from R_2Sn units bound together by covalent Sn–Sn bonds, as described in detail in the following two sections. There is no reason for assuming any reversible dissociation into "carbene-like" monomer moieties R_2Sn. Their chemistry, as far as investigated, is the chemistry of ordinary covalent Sn–Sn and C–Sn bonds. (For a review, see ref. 540a.) Hence, the reactions of cyclic polytins are almost the same as those of linear polytins; see Chapters 14-1—14-3.

Kuivila and his collaborators found that diphenyltin dihydride eliminates hydrogen on treatment with amines, with production of polymers (439):

$$n\,R_2SnH_2 \xrightarrow{\text{Amine}} (R_2Sn)_n + n\,H_2$$

Shortly afterwards it was discovered in the laboratory in Giessen (548, 551, 563) that this reaction is, under specified conditions, eminently suitable for preparation of individual aromatic and aliphatic cyclic polytins. It is important to use only very pure dialkyltin dihydrides and to exclude air carefully. Reaction rate, as well as size and uniformity of the resulting ring system depend on the catalyst used. Catalyst action is unexplained, and mere traces of randomly introduced impurities can have a profound effect.

A second method of cyclization was evolved later, again by the team in Giessen. This has proved especially valuable for aliphatic cyclic polytins (582), and is discussed under that heading.

14-4-1 Aromatic cyclic polytins

Diphenyltin dihydride prepared from the dichloride by means of diethylaluminum hydride and distilled at 10^{-4} mm. Hg, eliminates H_2 by reaction 14-9 when dissolved in pyridine. Reaction is noticeable even at room temperature, and becomes quantitative on brief heating to 70°C. Dodecaphenylcyclohexatin (**9**) is precipitated in 98% yield and can be crystallized from toluene as beautiful colorless platelets. Like all aromatic cyclic (and linear, see Chapter 14-1) polytins it is stable to air, but suffers scission of Sn–Sn bonds instantaneously by I_2. Traces of diphenyltin dichloride accelerate the condensation (reaction 14-9). $Ph_2Sn(Cl)H$, an organohalotin hydride (Chapter 8), probably appears as an intermediate and represents the real reactive species.

$$6\,Ph_2SnH_2 \xrightarrow[-6\,H_2]{\text{Pyridine}} \underset{\mathbf{(9)}}{(\text{–}Ph_2Sn\text{–})_6} \xrightarrow{+6\,I_2} 6\,Ph_2SnI_2 \qquad (14\text{-}9)$$

The structure (563, 564) follows from molecular weight determination and selective degradation; formation of diphenyltin diiodide (reaction 14-9) is quantitative. Attempts to obtain α,ω-dihalopolytins by using an excess of halogen (Ruggli–Ziegler dilution principle) are unsuccessful. The catenated system is degraded at once to diphenyltin diiodide, while the bulk of (**9**) remains unchanged indicating that the ring system is relatively more stable than the corresponding open-chain dihalopolytins. The presence of the six-membered ring has been proved by X-ray structure determination. In the solid [2 moles of *m*-xylene of crystallization per

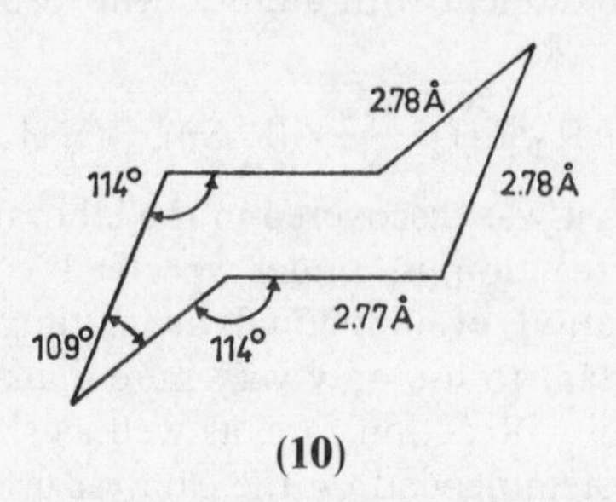

(**10**)

mole of (**9**)] the ring has the chair conformation (**10**). The Sn–Sn distances are exactly twice the covalent radius of tetra-covalent tin, the latter being known from C–Sn bonds (see Chapter 2). The bond angles too are tetrahedral, with small deviations caused probably by the bulky phenyl groups (639). The Sn–Sn bonds are thus purely covalent, the valency distribution is tetrahedral, and the stereochemistry of the ring is that of a scaled-up cyclohexane system. Examination of Stuart molecular models and Dreiding framework models shows that no major steric hindrance is to be expected.

In place of pyridine, other amines too can act as alternative and often more vigorous catalysts for the condensation reaction 14-9. However, condensation then tends to become non-uniform, and a higher polymer [approx. $(Ph_2Sn)_{13-14}$] normally formed in traces becomes more abundant. With methanol as solvent in place of the amine, an intermediate can be isolated and is found to be the α,ω-dihydropolytin (**11**) (564).*

$$6\,Ph_2SnH_2 \xrightarrow{-5\,H_2} H(\text{–}SnPh_2\text{–})_6H \xrightarrow{-H_2} (\mathbf{9})$$
$$(\mathbf{11})$$

Pyridine causes elimination of the last mole of H_2, and gives the six-membered ring (**9**). Reaction is promoted appreciably by traces of alkali. Studies with trialkyltin hydrides (435) show the methoxide anion to be the most likely active agent. This would displace a hydride ion on the tin, giving an intermediate dialkylmethoxytin hydride which would then catenate with elimination of MeOH. The hydrogen evolved would originate from interaction of hydride ion and MeOH.

Elimination of halogen from diphenyltin dichloride, which was used at one time to prepare the alleged "diphenyltin" (see Chapter 14-3) can be made to yield the six-membered ring system (**9**) (amongst other products) with sodium-naphthalene in tetrahydrofuran. Analogous experiments with lithium dust, sodium–potassium alloy, zinc powder, or magnesium (each dispersed or suspended in tetrahydrofuran) indicate some cleavage of C–Sn bonds by these metals. The products are complicated mixtures from which the ring system (**9**) can be recovered in only 5—25% yield by lengthy fractional crystallization. Grignard synthesis from phenylmagnesium bromide and $SnCl_2$ in tetrahydrofuran (soluble!) likewise yields (**9**) in rather variable amounts, besides species of higher molecular weight (564).

Other colorless and usually well crystalline dodecaarylcyclohexatins $Ar_{12}Sn_6$ (with Ar = *p*-tolyl, *p*-ethoxyphenyl, *p*-biphenylyl, and α- or β-naphthyl) can be obtained just like (**9**) from the respective diaryltin dihydride (565).

Diphenyltin dihydride, not distilled before use, with dimethylformamide gave well crystallized decaphenylcyclopentatin (**12**), which proved more soluble than (**9**). The course of the synthesis, and thus also the ring size,

$$(\text{–}Ph_2Sn\text{–})_5 \qquad (\mathbf{12})$$

is obviously influenced by traces of co-catalysts which have so far not been identified (564).

The four-membered ring compound octaphenylcyclotetratin has not been prepared. The difficulties seem to be experimental ones; there is no real reason why it should be incapable of existence. Aliphatic cyclotetratins are known (see below) and so are octaarylcyclotetra-germanes and -silanes (Table 14-4).

* This substance had been isolated previously, but was referred to as "modification B of diphenyltin" (439). Re-examination revealed its identity with (**11**) (personal communication).

14-4-2 Aliphatic cyclic polytins

Cyclization of dialkyltin dihydrides is again the method of choice, with ring size and uniformity of the product again strongly dependent on the catalyst. The mode of catalysis is still a matter of conjecture. For example, pure diethyltin dihydride in benzene–diethylamine catalyzed by ample $ZnCl_2$ affords an excellent and uniform yield of yellow, highly crystalline dodecaethylcyclohexatin (**13**) (582). With a trace of diethyltin dichloride as catalyst and pyridine as solvent (pyridine alone is ineffective in contrast to its behavior in 14-4-1) the uniform and quantitative product is the yellow, crystalline nonamer, octadecaethylcyclononatin (**14**) (579) (reactions 14-10). Diethylchlorotin hydride is most probably one of the reactive intermediates. It is apparent that the catalyst not only accelerates condensation, but also determines the stereochemistry of catenation and especially the process of ring-closure. (**13**) and (**14**) are highly sensitive to air, as are, in principle, all aliphatic compounds with Sn–Sn bonds (see Chapter 14-1).

$$\underset{(\mathbf{14})}{(-Et_2Sn-)_9} \xleftarrow[-n\,H_2]{Et_2SnCl_2/\text{Pyridine}} n\,Et_2SnH_2 \qquad n\,Et_2SnH_2 \xrightarrow[-n\,H_2]{ZnCl_2,\ HNEt_2} (-Et_2Sn-)_6 \tag{14-10}$$

$$3\,Et_2Sn(NEt_2)_2 + 3\,Et_2SnH_2 \xrightarrow[-6\,HNEt_2]{} \underset{(\mathbf{13})}{(-Et_2Sn-)_6}$$

Crystals of the nonamer burst into flames in air. Both are easily soluble in the common organic solvents, but not in the lower alcohols or water. Molecular weights are not dependent on concentration or age of sample. Analysis by specific degradation shows only Et_2Sn moieties; the NMR spectrum shows that all the ethyl groups are equivalent, lending further support to the ring structure. The preparation requires much patience, since even small amounts of impurities can prevent crystallization or lead to compounds of varying ring size. Such mixtures were indeed obtained in the early stages of the investigation (548).

The six-membered ring system (**13**) is also formed exothermically and quantitatively by condensation of Et_2SnH_2 with the stannylenediamine $Et_2Sn(NEt_2)_2$ (582) (see scheme 14-10). This condensation of organotin hydrides with stannyl compounds carrying nucleophilic groups was highly fruitful for linear polytins (see Chapter 14-3). It is now seen to be equally useful for preparing uniform cyclic polytins. The principle of the condensation is discussed in Chapter 8.

Diisobutyltin dihydride yields the nonamer (**15**) by elimination of H_2 under the same conditions which convert the ethyl compound into (**14**). (**15**) can also be obtained by condensation of $i\text{-}Bu_2Sn(NEt_2)_2$ with $i\text{-}Bu_2SnH_2$, or by a third independent route (reactions 14-11) from tetraisobutylditin 1,2-dihydride (582).

$$9\ i\text{-}Bu_2SnH_2 \xrightarrow[-9\,H_2]{R_2SnCl_2/\text{Pyridine}} i\text{-}Bu_{18}Sn_9 \quad (-i\text{-}Bu_2Sn-)_9 \quad (\mathbf{15})$$

$$n\ i\text{-}Bu_2Sn(NEt_2)_2 + n\ i\text{-}Bu_2SnH_2 \xrightarrow{-2n\,HNEt_2} i\text{-}Bu_{18}Sn_9$$

$$3\ i\text{-}Bu_2Sn(NEt_2)_2 + 3\ i\text{-}Bu_2Sn(H)Sn(H)i\text{-}Bu_2 \xrightarrow{-6\,HNEt_2} i\text{-}Bu_{18}Sn_9 \qquad (14\text{-}11)$$

The corresponding six-membered ring compound is formed by treatment of diisobutyltin dichloride with magnesium in tetrahydrofuran. In the case of the *n*-butyl compounds, the six-membered ring system is obtained both by H_2-elimination from the dihydride in pyridine and by cyclization of pure di-*n*-butyltin dihydride with the corresponding diethylamino- or dimethoxy-derivative (582).* All these cyclic polytins are well defined crystalline compounds. Polymeric dimethyltin exists in a rubbery form (101, 582) which awaits further investigation, and also in a six-membered ring modification (101).

The occurrence of nine-membered ring systems in aliphatic cyclic polytins may seem surprising. However, examination of both Stuart and Dreiding models, e.g. for the ethyl compound (**14**), shows that no steric hindrance and only insignificant Pitzer strain is to be expected.† Eight and seven-membered cyclic polytins have not been isolated pure. They may be contained in the reaction products $(R_2Sn)_n$ of variable molecular size which are formed with certain catalysts (548, 551).

The only non-aromatic cyclic pentatin prepared is decacyclohexylcyclopentatin (582).

On the other hand there are two examples of four-membered rings. Octa-*t*-butylcyclotetratin (**16**) is formed in 56% yield by dehalogenation of the dialkyltin dichloride with *t*-butylmagnesium chloride in boiling tetrahydrofuran (203). Alternatively it can be made by condensation of the dihydride with the stannylenediamine (582). The structure has been

* Sawyer (724) too obtained a polymeric dibutyltin from the dimethoxy-derivative, but does not give either molecular weight or structure.

† Nine-membered carbocyclic and nitrogen-heterocyclic ring systems too are no longer unusual and occur naturally, e.g. in caryophyllene and nocardamine (42).

confirmed by X-ray analysis.

$$\begin{array}{ccccc} & Me_3C & & CMe_3 & \\ & | & & | & \\ Me_3C- & Sn & - & Sn & -CMe_3 \\ & | & & | & \\ Me_3C- & Sn & - & Sn & -CMe_3 \\ & | & & | & \\ & Me_3C & & CMe_3 & \end{array} \quad \textbf{(16)}$$

Table 14-4

Cyclic Compounds $(-R_2M-)_n$ *of Group IV Elements*[a,b]

M	NUMBER OF MEMBERS IN RING (n) 4		5		6		9
C[c]	Me	(44a)	—		—		—
Si[d,e]			Me	(127b)	Me	(827)	
	Ph	(254, 372)	Ph	(254)	Ph	(254)	
	p-MeC_6H_4	(372, 703a)	p-MeC_6H_4	(372, 703a)	p-MeC_6H_4	(703a)	
Ge					Me	(541)	
	Ph	(570)	Ph	(570)	Ph	(570)	
	p-MeC_6H_4	(703a)	p-MeC_6H_4	(703a)	p-MeC_6H_4	(703a)	
Sn[f]					Me		
					Et		Et
					Bu		
					i-Bu		i-Bu
	t-Bu						
			cyclo-C_6H_{11}				
			Ph		Ph		
	p-MeC_6H_4				p-MeC_6H_4		
					p-$EtOC_6H_4$		
					p-PhC_6H_4		
					α-Naphth		
					β-Naphth		

[a] For mass-spectrometric data on the phenyl compounds see ref. 427b; for those on the p-tolyl compounds see ref. 703a.
[b] Analogous cyclic lead compounds are not known at present.
[c] Hexamethylcyclopropane has been prepared also (349a).
[d] For a recent review see ref. 254a.
[e] The seven-membered ring $Me_{14}Si_7$ has been prepared also (127b). Five-membered rings of four Si atoms and one Ge or Sn atom are known (290a).
[f] See Chapter 14-4. Detailed discussion and references in the text.

Octabenzylcyclotetratin is formed in good yield when dibenzyltin dihydride (which is extremely susceptible to autoxidation) is treated in dimethylformamide with a trace of dibenzyltin dichloride (566).

Aliphatic cyclic polytins are degraded by halogens, even by iodine, right down to the dialkyltin dihalide. Under Ruggli–Ziegler dilution conditions it is frequently possible to isolate α,ω-dihalopolytins. This is particularly easy with the t-butyl derivative (**16**), which consumes only

25 % of the calculated amount of iodine rapidly and the rest very slowly. This means that the strained four-membered ring is opened easily [similar behavior is observed with octaphenylcyclotetragermane (570)], but that the sterically protected linear chain $I(t\text{-}Bu_2Sn)_4I$ is relatively stable. This situation is totally different from that in the aromatic cyclic polytins, in which the ring is more resistant to cleavage by halogen than the resulting phenylated straight chain (see Chapter 14-4-1). No valid conclusions can be drawn until cyclic polytin chemistry has been explored more fully; at present it is still in its infancy* (for a review, see ref. 540a).

Interpretation of the color of cyclic polytins must be tentative for the same reasons. All known aromatic cyclic polytins are colorless. Their UV spectra usually show, apart from aromatic bands, two specific absorptions at 220 and 280 nm. (564). It could be concluded that electron mobility in the tin ring is diminished by the presence of the electron-withdrawing aryl groups.

This effect is absent in aliphatic cyclic polytins, which are almost all yellow. The corresponding tetraorganotins are colorless. A study of the ethyl compounds shows that increasing length of the tin chain barely affects the position of the UV absorption maximum, but greatly influences the molar extinction coefficient ε; the value of ε increases from that for tetraethyltin (1530 at λ_{max} 219 nm.) through hexaethylditin (1970 at λ_{max} 222 nm.) to many times that size for the nonatin (**14**) (approx. 100,000 at λ_{max} 217 nm.). All three substances show only a single absorption maximum above 200 nm., but in the ring system (**14**) the tail of the relevant band extends into the visible (579). There is obviously some interaction between the tin atoms, like that established for polysilanes (291) (see also Chapter 23-1).

* Some reactions have unfortunately been carried out with ill-defined or impure materials (see Chapter 14-3).

15 Stannylamines

It is surprising that compounds with Sn–N bonds attracted so little attention for quite a long time. Until recently, a few stannyl azides, isocyanates, isothiocyanates (see Chapter 6), *N*-stannyl-carboxylic acid amides (175c, 491) and -sulfonamides (145) were the only known compounds of this class. Some of the numerous complexes of organotin halides with amines (see Chapter 2-3) may also contain Sn–N links, but there is no definite proof at present.

From 1960, however, syntheses and reactions have followed each other in rapid succession. The stannylamines* now constitute one of the most intensively studied areas in organotin chemistry (review in ref. 336). They can be divided into several sub-groups, depending on whether the nitrogen carries alkyl or aryl groups or hydrogen in addition to the one or several stannyl groups. Stannylamines are mostly colorless liquids, which are sensitive to moisture but can be distilled *in vacuo* without decomposition. Some are crystalline solids. Some examples are listed in Table 15-1. Hydrolytic sensitivity is decreased by a second nitrogen in the molecule acting as electron donor. The resulting association, involving pentacoordinate tin, leads to greater stability. This is so e.g. for *N*-trialkylstannyl-imidazoles (see Chapter 2-2).

Tertiary amines of the type $R_3SnNR'R''$ are best prepared by treatment of trialkyltin halides with lithium-, sodium-, or magnesium-amides in ether or benzene (2, 332, 334, 797, 851, 883), e.g. by reaction 15-1:

$$R_3SnCl + LiNEt_2 \rightarrow R_3SnNEt_2 + LiCl \qquad (15\text{-}1)$$

Compounds with two, three, or four dialkylamino groups attached to tin are prepared in a similar manner from R_2SnCl_2, $RSnCl_3$, or $SnCl_4$

* For simplicity, all compounds with Sn–N bonds are here called stannylamines.

Table 15-1
Stannylamines

FORMULA	B.P. (°C/mm.)	M.P. (°C)	n_D^{20}	METHOD OF PREPARATION
$MeSn(NMe_2)_3$	50/1	—	—	481, 883
$Me_2Sn(NMe_2)_2$	138 45/1	—	1.4463	332, 334, 481
Me_3SnNEt_2	140/720 43/8	—	1.4618	334, 444, 883
Et_3SnNEt_2	114—117/23 72/2	—	1.4724	334, 797, 883
$Et_3SnNHPh$	100/0.2	—	—	334
Bu_3SnNH*t*-Bu	124/1	—	1.4773	334
$(Me_3Sn)_2NEt$	93/15	—	1.4968	332, 334
$(Me_3Sn)_3N$	70/0.2 130/14	22—24 26—28	1.5331	332, 334, 461, 732, 798
$(–SnMe_2–NEt–)_3$[a]	104/0.05	—	—	332, 334
Bu_3SnN (pyrrole ring)	139—141/0.62	—	1.4952	360, 490, 493
Pr_3SnN (imidazole ring, =N)	—	152—154	—	360, 490, 493
$Et_3SnN(cyclo\text{-}C_6H_{11})\text{-}CH{=}Ncyclo\text{-}C_6H_{11}$	126/0.0001	—	1.5233	559
$Ph_3SnN(Ph)NHPh$	—	~130	—	611
$Et_3SnN(Ph)CHO$	171—172/13	50—53 63	—	592, 614, 615
$Me_3SnN(SiMe_3)_2$	58—59/1 55/1	20—22	—	481, 734
$(Me_3Si)_2NSn(Me)(NMe_2)_2$	78/0.1	>20	—	481
Me_2Sn (ring: N(Me)–$CH_2CH_2CH_2$–N(Me))	88—90/18	1—3	—	733
Me_2Sn (ring: $N(SiMe_3)$–CH_2CH_2–$N(SiMe_3)$)	121—123	14—16	—	733

[a] Six-membered ring

respectively.* [In some cases, tin alkoxides or stannoxanes have been used instead of organotin halides (732).]

Reaction of organotin oxides with amines or amides provides yet another route to organotin-substituted tertiary amines (175c, 311, 493). The equilibrium normally lies well over towards the starting materials; the water formed must therefore be removed continuously.

$$(R_3Sn)_2O + 2\,HNR'_2 \rightleftharpoons 2\,R_3SnNR'_2 + H_2O$$

$$R_2SnO + 2\,HNR'_2 \rightleftharpoons R_2Sn(NR'_2)_2 + H_2O$$

The reaction is mainly of importance for the preparation of organotin-substituted imidazoles and similar compounds containing penta-coordinate tin (493) (see Chapter 2-2).

Synthesis of organotin-substituted tertiary amines by reaction 15-1 can be supplemented by transamination (reaction 15-2) in which a lower-boiling secondary amine is displaced by a higher-boiling one (334, 733):

$$R_3SnNR'_2 + HNR''_2 \rightarrow R_3SnNR''_2 + HNR'_2 \qquad (15\text{-}2)$$

When the transamination is effected by a primary instead of a secondary amine, the product should be a secondary stannylamine, R_3SnNHR'. This is actually the case when R′ is a bulky group, e.g. *t*-butyl or phenyl (334). However, when R′ is one of the lower alkyl groups—say Me or Et—the reaction leads to bis(trialkylstannyl)alkylamines under even the mildest conditions. Disproportionation, as in reaction 15-3, is probably responsible (332, 334). Attempts to prepare the secondary amine R_3SnNHR' by a reaction analogous to 15-1 from trialkyltin halide and lithium methylamide or ethylamide fail similarly, and afford the bis-(trialkylstannyl)alkylamine instead. The disproportionation reaction 15-3 can also be viewed as a transamination, one molecule of secondary stannylamine displacing H_2NR' from another molecule.

$$2\,R_3SnNHR' \rightarrow (R_3Sn)_2NR' + H_2NR' \qquad (15\text{-}3)$$

Synthesis of a secondary stannylamine has been accomplished with the aid of a silylamine (2):†

$$Me_3SnBr + Me_3SiNHEt \rightarrow Me_3SnNHEt + Me_3SiBr$$

A crystalline complex is formed as an intermediate.

*Explosions have been reported when warming $Me_2Sn(NMe_2)_2$ with chloroform, and during distillation of $H_2Sn(NR_2)_2$ (684a).

† Other authors (334), however, observed spontaneous disproportionation even with this stannylamine.

The analogous ammonia derivatives R_3SnNH_2 transform by various intermediate stages into tris(trialkylstannyl)amines $(R_3Sn)_3N$ (332, 334). The same compounds are accessible directly from trialkyltin halides and lithium amide or sodamide (334, 798) or quite simply from lithium nitride (461):

$$3\,R_3SnCl + Li_3N \xrightarrow{THF} (R_3Sn)_3N + 3\,LiCl$$

Quite complicated compounds, e.g. (733)

Si(CH3)3 | N — benzo-fused ring — Sn(CH3)2 — N | Si(CH3)3

or cyclic stannazanes (332, 334), can be prepared by transamination, starting from bis(dialkylamino)dialkyltins and primary amines, e.g.

$$3\,Me_2Sn(NMe_2)_2 + 3\,H_2NEt \xrightarrow[-6\,HNMe_2]{} (\text{–}Me_2Sn\text{–}EtN\text{–})_3$$

Other possible syntheses of Sn–N links are based on hydrostannation of azomethines, azo-compounds, and isocyanates (see Chapter 9).

NMR studies lead to the conclusion that the nitrogen–tin bond is purely covalent (444). Certain observations seem to indicate some $d\pi$–$p\pi$ double bonding (445, 481), but other authors regard this as out of the question (755).

Stannylamines are generally very reactive. The Sn–N bond is cleaved rapidly by water, with production of organotin hydroxides or oxides and amine. Preparation and purification of stannylamines therefore calls for strict exclusion of moisture. Alcoholysis to the amine and stannyl alkoxide is likewise extremely rapid. The displacement reaction 15-2 also succeeds with equimolar amounts of stannylamines and other organo-substituted Group V hydrides. In this way stannyl-arsines, -phosphines, and -bismuthines can be obtained in good yield (Chapter 16).

The known reactions of the stannylamines may be divided into two main types.

(a) Condensation with proton-donor compounds (Chapter 19). Organotin hydrides also come under this heading, the products being compounds with Sn–Sn bonds and secondary amines (see Chapter 8-4-4).

(b) Addition of the Sn–N group to unsaturated systems (see Chapter 19).

16 Stannyl-phosphines, -arsines, -stibines, and -bismuthines

The rapid development of stannylamine chemistry and the surprising properties of stannylamines (see Chapters 15 and 19) created further interest in P-, As-, Sb-, and Bi-analogs. These were unknown only a short time ago; Et_3SnPPh_2, reported in 1959 (425), was the first compound of this class, and $(Me_3Sn)_3P$ the second (105). The number of known compounds is now quite sizeable. H. Schumann and M. Schmidt (755) give a critical review and tabular summaries of many of these.

Preparative methods are in principle the same as for stannylamines. Treatment of the appropriate stannyl lithium compound with a suitable phosphorus halide in tetrahydrofuran affords representatives of the series R_3SnPR_2, $(R_3Sn)_2PR$, and $(R_3Sn)_3P$ (125, 737, 752, 755), and also some complicated tin–phosphorus ring compounds. Analogous arsines and stibines are accessible similarly (126, 754, 755). Condensation of organotin halides with alkylphosphorus alkali compounds, e.g. MePHLi in ether at $-20°$, is feasible in some isolated instances (56a).

Some syntheses fail, probably because of rapid metal–halogen exchange:

$$R_3SnNa + ClPR'_2 \rightleftharpoons R_3SnCl + NaPR'_2 \qquad \text{(ref. 127)}$$

The "displacement reaction" (333) mentioned earlier (Chapter 15) proceeds smoothly and under very mild conditions; yields are almost quantitative:

$$Me_3SnNMe_2 + HPPh_2 \rightarrow Me_3SnPPh_2 + HNMe_2$$

$$Me_3SnNMe_2 + HAsPh_2 \rightarrow Me_3SnAsPh_2 + HNMe_2$$

Condensation of the appropriate organotin chloride with the free phosphine in the presence of amines (752, 755) provides a general synthetic pathway to compounds of types R_3SnPR_2, $R_2Sn(PR_2)_2$, and $RSn(PR_2)_3$, e.g.

$$RSnCl_3 + 3\,HPR'_2 \xrightarrow[-3\,Et_3N\cdot HCl]{+3\,Et_3N} RSn(PR'_2)_3$$

$Sn(PR'_2)_4$ too has been prepared in this way. The versatility of the method is shown by the production of cyclic compounds of type (**1**) from R_2SnCl_2 (R = Me, Bu, Ph), $PhPH_2$, and stoichiometric amounts of triethylamine using the Ziegler–Ruggli procedure. Polymers are formed as by-products. Similar reaction of $PhSnCl_3$ with PH_3 gives, besides polymeric material, a deep yellowish compound in 10% yield. IR spectra indicate T_d symmetry and suggest a formula $(PhSnP)_4$ and cubane-like structure (850a).

The tris(triphenylstannyl) derivatives of As, Sb, and Bi are colorless solids that may be recrystallized from pentane (754, 755, 864b).

$$3\,Ph_3SnLi + MCl_3 \xrightarrow{THF} (Ph_3Sn)_3M + 3\,LiCl \qquad (M = As, Sb, Bi)$$

Phenyl- and diphenyl-arsine chlorides react in the same way as MCl_3 (753a). Tris(triethylstannyl)-antimony and -bismuth derivatives have been prepared from the triethyls and triethyltin hydride (685b).

Elemental phosphorus is arylated by tetraphenyltin in a similar way to elemental sulfur. The final products are Ph_3P and tin phosphides. However, if the reaction is carried out in a sealed tube at 235—250°C, a colorless ring compound (**1**) and a yellow diphosphine (**2**) may be isolated in poor yield (752, 753, 755):

$(–Ph_2Sn–PhP–)_3$ $\qquad$ $(Ph_3Sn)_2P–P(SnPh_3)_2$

(**1**) M.P. 66° $\qquad$ (**2**) M.P. 110°

The compound $(R_2P)_2SnR_2$ (R = Ph) is orange-red (127), whereas stannylphosphines with only one Sn–P group are colorless. Air-free water is without effect (755)* while ethyl iodide or sodium in liquid ammonia react rapidly. Atmospheric oxidation first gives phosphine oxides (737), and subsequently stannyl phosphinates or phosphonates. Oxidation is quantitative with H_2O_2 in ethanol (755).

The nature of the Sn–P bond is now a matter of considerable interest. The bond is covalent, very easily oxidized,† but stable to at least 200°C in the absence of air.

* Ready hydrolysis has been reported several times, but seemingly always in the presence of air (425).

† Autoxidation is retarded when there is appreciable shielding by phenyl groups, as in $(Ph_3Sn)_3P$ or $(Ph_2P)_4Sn$.

This and the small dipole moment of Et_3SnPEt_2 (~1.0 D) has been attributed to overlap of the lone pair of the phosphorus with a vacant *d*-orbital of the tin. The intensive absorption of this last compound at 214 nm. is explained similarly (127). [The possibility of $p\pi$–$d\pi$ participation in the Sn–P bond is criticized in ref. 755.]

The chemistry of these materials is only in its early stages, and rapid development must surely be expected. The Sn–P link will certainly add to polar multiple bonds in much the same way as stannylamines (see Chapter 19): triphenylstannyldiphenylphosphine adds to allyl chloride, styrene, and phenylacetylene with the aid of radical initiators (751), i.e. by a radical mechanism. With appropriate reactants, polar reactions are possible: adducts are formed without catalyst at 80—100°C with thiourea, phenyl isocyanate and isothiocyanate, CS_2, $CSCl_2$ and COS, but there is no reaction with CO_2 (750b).

Stannyl-arsines, -stibines, and -bismuthines have been mentioned. Not much is known about these, although a number of methyl, ethyl, propyl, butyl, and phenyl derivatives have been well characterized (126, 753a, 755). R_3Sn–As and R_3Sn–Sb links are cleaved by methyl iodide with formation of R_3SnI, while oxidation of stannylarsines yields dimers or tetramers of esters $R_3SnOAs(O)R_2$ (126). Et_3Sn–Sb and Et_3Sn–Bi bonds react with e.g. oxygen, benzoyl peroxide, and alkyl bromides (685b).

17 Organotin oxygen, sulfur, selenium, and tellurium compounds (except stannyl esters)

Some of these substances had been regarded for a time as well known and structurally understood. In the last few years, however, they have become the subject of renewed and vigorous discussion. The structural picture now seems more complex, and some of the older ideas have had to be revised or extended considerably. In addition, several representatives of this class have attained industrial importance (see Chapter 25). The number of new patents is therefore considerable and increasing rapidly.

Results to-date at least show that tin is capable of forming stable covalent bonds with the elements of Group VI just like carbon, silicon, germanium, and lead.

17-1 TRIORGANOTIN HYDROXIDES AND STANNOXANES $R_3SnOSnR_3$

These two classes must be discussed together, since they are related to each other by the equilibrium 17-1, which can adopt all possible positions depending on the species. Triorganotin hydroxides are obtained easily

$$2\,R_3SnOH \rightleftharpoons R_3SnOSnR_3 + H_2O \qquad (17\text{-}1)$$

by hydrolysis of the halides. Equilibrium 17-2 is established even in aqueous dioxane, in most cases quite rapidly (677).

$$\begin{aligned} R_3SnCl + H_2O &\rightleftharpoons R_3SnOH + HCl \\ R_3SnCl + KOH &\rightarrow R_3SnOH + KCl \end{aligned} \qquad (17\text{-}2)$$

In the presence of alkali, reaction is quantitative (315). The hydroxides (or by equation 17-1 the stannoxanes) are formed readily also by hydrolysis of stannyl esters (Chapter 7), alkoxides (Chapter 17-4), and amines (Chapter 15).

Preparation involves shaking a solution of the halide in an organic solvent (or a liquid halide without solvent) with alkali (315, 416).

Many of the hydroxides are solids possessing a sharp melting point (Table 17-1). However, those with larger aliphatic groups (from C_4 upwards) are liquid and perhaps not to be regarded as definite compounds, but rather as solutions of H_2O in the respective stannoxane. Trimethyltin hydroxide (487) differs appreciably from its homologs. In contrast to these, it is dehydrated (see below) only with difficulty (e.g. by reaction with metallic sodium); it sublimes readily and without decomposition, and is easily soluble in water and alcohol, less so in benzene. Solubility in organic media rises rapidly with increasing size of the alkyl group, and solubility in water decreases. Whereas Me_3SnOH is stable in air, Et_3SnOH absorbs carbon dioxide with formation of the carbonate $(Et_3SnO)_2CO$. Hexapropylstannoxane and tripropyltin hydroxide behave similarly; on heating, CO_2 is evolved and the stannoxane re-formed (798). In contrast hexabutylstannoxane and its hydroxide are unaffected by CO_2 (317).

All triorganotin hydroxides render water slightly alkaline, depending on the nature of the organic groups and on a number of other factors which are not yet fully understood. Neither data nor interpretations (317, 606, 853) are in complete agreement. It must be assumed that the dissociation is not simply into R_3Sn^+ and OH^- but that it involves complex or solvated stannyl ions depending largely on the experimental conditions. Because of these difficulties, no dissociation constants are quoted here.

Older measurements already suggested that Me_3SnOH is associated in benzene. Existence of a dimer over a wide range of concentrations in several boiling organic solvents is now quite certain. Okawara and his co-workers (638) have assigned the dimer structure (**1**) with penta-

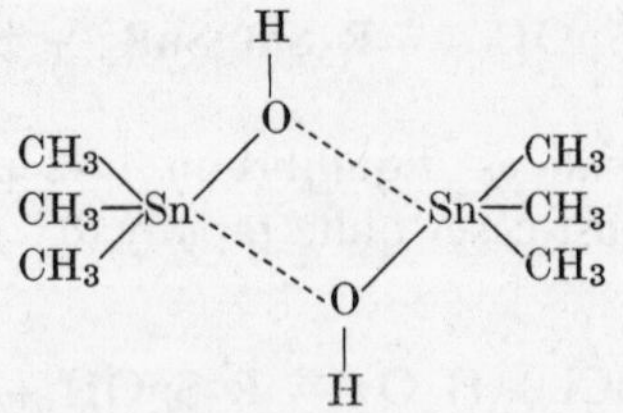

H
O
CH_3 CH_3
CH_3—Sn Sn—CH_3
CH_3 CH_3
O
H

(**1**)

coordinate tin on the basis of the IR spectrum. For the solid the results of the same group of workers suggest a polymeric structure (**2**) with planar Me_3Sn groups and again penta-coordinate tin (345) (see also Chapter 2).

$$\left[\cdots Sn(CH_3)_3 \cdots O(H) \cdots Sn(CH_3)_3 \cdots O(H) \cdots \right]_n$$

(**2**)

Raman and IR spectra certainly indicate association via Sn–O–Sn bridges (422). There are no definitive results for compounds with other alkyl groups.

All compounds R_3SnOH are in equilibrium with the corresponding stannoxane $R_3SnOSnR_3$ (equation 17-1). The methyl derivative loses water only with great difficulty (sodium metal is required); the corresponding stannoxane is consequently hygroscopic in air. On the other hand, hexabutylstannoxane and other stannoxanes with larger alkyl groups are formed already by warming the hydroxide or even directly from the hydrolysis of the chloride. In some cases it is doubtful altogether whether one is dealing with a properly defined triorganotin hydroxide R_3SnOH or with a stannoxane hydrate $(R_3Sn)_2O{\cdot}H_2O$ (277, 295). Certainly the substance generally described as triethyltin hydroxide (M.P. 56°C) loses water spontaneously when dissolved (in the absence of CO_2) in benzene, cyclohexane, carbon tetrachloride (295), or nonane (12). The water condenses out as droplets, and the molecular weight (by cryoscopy) and IR spectrum are then those of hexaethylstannoxane (12). Triphenyltin hydroxide behaves similarly. Its spectrum in solution in CS_2 corresponds to that of the stannoxane $Ph_3SnOSnPh_3$. Definite proof of a real hydroxide is however supplied for the solid phase by absorption bands (Nujol mull) of the Sn–O–H group at 896 and 3606 cm.$^{-1}$, and confirmed by similar bands for the Sn–O–D group of the deuteride at 662 and 2664 cm.$^{-1}$. These Sn–O–H bands can even be used for quantitative determination of the hydroxide in the presence of the stannoxane (222). The question of distinction between hydroxide and stannoxane hydrate, and the assignment of Raman and IR bands (see Chapter 23) has been the subject of several enquiries (421, 451, 913). Heating can change triorganotin hydroxides and stannoxanes into diorganotin oxide plus tetraorganotin (18). The reaction may take place in air at 80°C (295);

with trimethyltin hydroxide it starts at 100°C, and with a certain vinyl compound already at 20°C (152) by reaction 17-3:

$$Bu_2Sn(Cl)CH{=}CH_2 \xrightarrow[-Cl^-]{+OH^-} Bu_2Sn(OH)CH{=}CH_2 \xrightarrow[\text{several days}]{20^\circ}$$

$$Bu_2SnO + H_2O + (H_2C{=}CH)_2SnBu_2 \qquad (17\text{-}3)$$

Hexaphenylstannoxane decomposes similarly on melting (748).

Stannoxanes from the ethyl compound upwards, e.g. the technically important hexabutylstannoxane ("tributyltin oxide") can be distilled *in vacuo* without decomposition (Table 17-1). They are not associated, at least in dilute solution (653).

Apart from the stannoxanes $R_3SnOSnR_3$ there is also a series of disubstituted stannoxanes $XR_2SnOSnR_2X$ (**2a**) formed by partial hydrolysis of diorganotin compounds or by their interaction with diorganotin oxides (Chapter 17-2) as in reactions 17-4.

$$\left.\begin{array}{l} 2\,R_2SnX_2 \xrightarrow{\text{partial hydrolysis}} \\ R_2SnX_2 + R_2SnO \end{array}\right\} \longrightarrow \underset{(\mathbf{2a})}{XR_2SnOSnR_2X} \qquad (17\text{-}4)$$

(X = e.g. Oalkyl, Oaryl, OCOMe, OCOPh, $OSiR_3$, Cl, I)

(18, 19, 151, 486, 633, 637, 658, 687, 830, 844, 916, 917)

Further hydrolysis transforms them into diorganotin oxides $(R_2SnO)_n$ (Chapter 17-2).

Under appropriate conditions, intermediates can be isolated; e.g., controlled basic hydrolysis of (**2a**) (X = Cl) gives $Cl{-}(Bu_2SnO)_n{-}SnBu_2{-}Cl$ (532a).

Some of the earliest organotin compounds (486) belong to the type (**2a**). Composition and preparations were described as early as 1862 by A. Strecker (830), and later by P. Pfeiffer (658) in 1914. Later still, there was temporary confusion when some representatives of the series, obtained by another route, were thought to be ditins (328). In the end, the original stannoxane structure (658) was confirmed once more (14, 326).

In contrast to the hexaorganostannoxanes (above) these compounds are associated in solution (627, 658), as shown also by NMR measurements (147). [For a review see Okawara and Wada (635a).] When R = Bu

Table 17-1
Triorganotin Hydroxides and Stannoxanes

FORMULA	B.P. (°C/mm.)	M.P. (°C)	n_D^{20}	METHOD OF PREPARATION
Me_3SnOH	subl. >80°	118—119	—	384, 407, 422, 487
$Me_3SnOSnMe_3$	84/22	—	—	278, 422
	77—79/12			
Et_3SnOH	—	56[a]	—	295, 384, 610
$Et_3SnOSnEt_3$	272	—	1.4982	32, 384, 610
	146—147/12			
$Bu_3SnOSnBu_3$	210—214/10	—	1.4871	610
	164—174/4			
i-$Bu_3SnOSn$$i$-$Bu_3$	197—198/12	—	1.4859	591, 902
	142—146/0.001			
(cyclo-$C_6H_{11})_3SnOH$	—	220—222	—	417
$(PhCH_2)_3SnOSn(CH_2Ph)_3$	—	120	—	56
Ph_3SnOH	—	119—120	—	222, 417, 451, 748
$Ph_3SnOSnPh_3$	—	123—124	—	222, 417, 451, 748
$Et_2Sn(Cl)OSn(Cl)Et_2$	—	175	—	19, 658
		178		
$Bu_2Sn(Cl)OH$	—	105—107	—	236
$Bu_2Sn(Cl)OSn(Cl)Bu_2$	—	112—114	—	14, 236, 326
		114.5—115.5		
$Bu_2Sn(SCN)OSn(SCN)Bu_2$	—	84	—	17, 19
$Bu_2Sn(OCOMe)OSn(OCOMe)Bu_2$	—	58—60	—	19, 382, 917
$Et_2Sn(OCOPh)OSn(OCOPh)Et_2$	—	215—217	—	687, 867
$Me_2Sn(OPh)OSn(OPh)Me_2$	—	190—197	—	151, 275
$Bu_2Sn(OC_6H_4OMe$-$p)OSn(OC_6H_4OMe$-$p)Bu_2$	—	86—89	—	151

[a] Earlier statements of melting points between 40 and 50°C can be traced to contamination with triethyltin carbonate, formed by the action of air on the stannoxane (295, 798).

and X = Cl or Br, the dimer is present even in boiling CCl_4 (17).

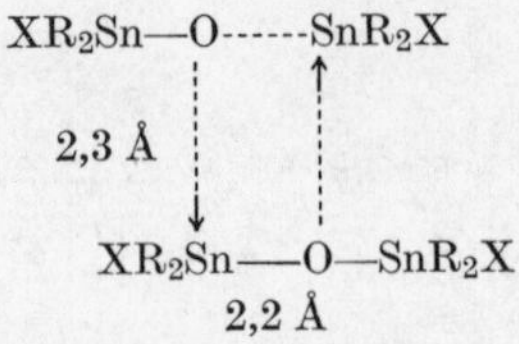

$$\begin{array}{ccc} XR_2Sn\text{—}O & \cdots\cdots & SnR_2X \\ \vdots\ 2{,}3\ \text{Å} & & \uparrow \\ \downarrow & & \vdots \\ XR_2Sn\text{——}O & \text{—} & SnR_2X \\ & 2{,}2\ \text{Å} & \end{array}$$

(3)

X-Ray structure determinations and experiments with the ^{119}Sn isotope have shown that the association takes the form (**3**) with an almost square four-membered ring (627). The ring has been assigned a skeletal vibration at 400 cm.$^{-1}$ (786). X = Cl can be partially or completely replaced by other groups (635). Thus, treatment with formic or glacial acetic acid yields dialkylchlorostannyl carboxylates $R_2(Cl)SnOCOR'$ (872).

Stannoxanes may be employed for a number of other reactions. Their condensation with proton donors will be discussed separately in Chapter 19. $LiAlH_4$ or diorganoaluminum hydrides effect hydrogenation to the triorganotin hydride, often with good yields (149, 589). It should be

$$Me_3SnOSnMe_3 \xrightarrow[-Me_4Sn]{+MeLi} Me_3SnOLi \xrightarrow[-LiCl]{+Et_3SiCl} Me_3SnOSiEt_3 \qquad (17\text{-}5)$$

remembered, though, that at higher temperatures stannoxanes can react with organotin hydrides (see Chapter 14). Methyllithium likewise produces cleavage (reaction 17-5). The resulting compound lends itself to synthesis, e.g. of stannosiloxanes (738). These materials are also accessible simply from stannoxanes and silanols (383), or from simultaneous hydrolysis of triorganotin and triorganosilicon halides. They are monomeric in solution (634) (as far as they have been examined) and have, like the analogous germanium and lead compounds, proved of quite varied interest (740, 849).

17-2 DIORGANOTIN OXIDES

Preparation and reactions of these compounds are very similar to those in the previous section. Diorganotin oxides are obtainable by hydrolysis of e.g. diorganotin dihalides, diamides, dicarboxylates, or dialkoxides. The hydrolysis is often reversible; e.g. hydrogen halides reconvert the oxides into the dihalides as in reaction 17-6. This is the commonest method not only of preparing diorganotin oxides (R may be aliphatic or aromatic) but also of purifying or isolating diorganotin dihalides.

$$R_2SnX_2 \underset{+2\,HX,\ -H_2O}{\overset{+2\,NaOH,\ -2\,NaX,\ -H_2O}{\rightleftharpoons}} R_2SnO \qquad (X = \text{e.g. Cl, OCOMe}) \tag{17-6}$$

The reactions really take place in several stages, with stannoxanes $XR_2SnOSnR_2X$ among the intermediates.

Hydrolysis 17-6 should yield initially diorganotin dihydroxides, but these generally eliminate water spontaneously and change into the oxides. The dihydroxide has been identified in exceptional cases only, e.g. with the di-*t*-butyl, di-*t*-amyl, and recently the di-(*o*-phenoxyphenyl) derivatives (668). As with the triorganotin compounds, hydrolysis involves some solvation of the ions; association via oxygen bridges as in (**4**) is considered most likely (606).

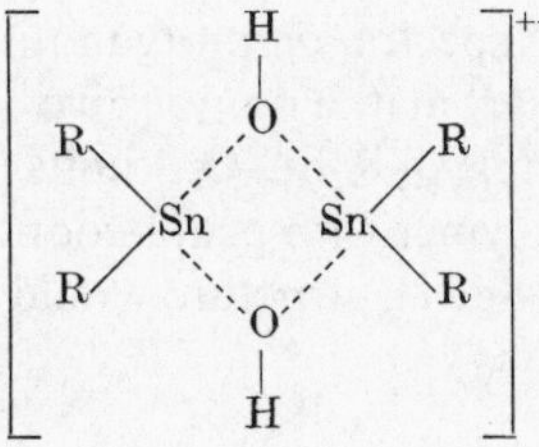

(**4**)

Diorganotin oxides are colorless, amorphous or microcrystalline powders. They mostly decompose on heating (62), without melting, sometimes forming hexaorganostannoxanes. (The same is true for diphenyltin oxide.)* They are practically insoluble in water, but react smoothly with both organic and inorganic acids, as mentioned above (reaction 17-6 and Chapter 7). When freshly prepared they are sometimes soluble in alkalis also. Old samples are not, so there must be some change of modification on ageing. However, precise details are lacking. Compounds described as diorganotin oxides could well be entirely or partially $HO(R_2SnO)_nH$, $Cl(R_2SnO)_nH$, or $R'COO(R_2SnO)_nCOOR'$, depending on the method of preparation.

Diorganotin oxides are unquestionably polymeric, so their formula ought really to be given as $(R_2SnO)_n$. Association occurs via Sn–O bridges (669) which show up clearly in the IR spectrum (786). The value

* The statement that it melts at 38°C (315) is traceable to an error in *Chem. Abstr.*, **53**, 17030 (1959). In the original paper (678) this melting point refers to diphenyltin dibromide.

of n is still in doubt.* Distinct solubility in ethers and other polar organic solvents and generally good reactivity, e.g. of dibutyltin oxide with acetylacetone, leads one to suspect that n cannot be large and that it may often be no larger than 3 (as in the analogous thio-compounds; Chapter 17-6). This is thought to be so e.g. for dibutyltin oxide (14), and the compounds are then supposed to have the ring structure (**5**).

$$\begin{array}{ccc} & O & \\ R_2Sn & & SnR_2 \\ | & & | \\ O & & O \\ & \underset{R_2}{Sn} & \end{array}$$

(**5**)

However, Mössbauer spectra of diorganotin oxides (R = aliphatic, aromatic, or substituted aromatic) do not give any indication of a structure built up from simple $(\text{–}R_2Sn\text{–}O)_n$ units. They suggest rather a reticular coordination polymer with penta-coordinate tin (260). Structure (**6**) is suggested (18, 146). Such a structure would also account for the low

$$\begin{array}{l} \text{—}R_2Sn\text{——}O\text{—}R_2Sn\text{—}O\text{—}R_2Sn\text{——}O\text{—} \\ \quad\;\uparrow \qquad\quad \downarrow \qquad\qquad\qquad \uparrow \qquad \downarrow \\ \text{—}O\text{—}R_2Sn\text{—}O\text{—}R_2Sn\text{——}O\text{—}R_2Sn\text{—} \end{array}$$

(**6**)

frequency of the Sn–O absorption in the IR (666) and for the high intermolecular orientation deduced from X-ray structure analysis (697). It seems clear that one and the same diorganotin oxide can exist in several modifications, depending on the conditions of preparation, and that these modifications may be interconvertible. In view of the considerable interest aroused by these substances, it is remarkable that there is so little accurate information on this point.

Diorganotin oxides are formally analogous to silicones. There have therefore been many attempts to prepare corresponding high molecular

*The isolated observation of the monomeric nature of di-*n*-dodecyltin oxide requires confirmation; the sample may have contained water (810).

weight diorganotin compounds. As far as is known at present, only partial success has been achieved (see Chapter 21). The obstacle appears to be the tendency of Sn–O compounds to cyclize.

17-3 STANNONIC ACIDS

Organotin trihalides are hydrolyzed even more easily than mono- or di-halides (they usually fume in moist air). However, the expected tri-hydroxides immediately eliminate water to form so-called "stannonic acids." These are the products isolated from reaction 17-7 (R = alkyl or aryl).

$$RSnX_3 \underset{+3\,HX,\ -2\,H_2O}{\overset{+3\,NaOH,\ -3\,NaX,\ -H_2O}{\rightleftharpoons}} RSn(O)OH \qquad (17\text{-}7)$$

The name was meant originally to emphasize the analogy with German "carbonsäuren," i.e. carboxylic acids. It is a poor guide to the real character of these compounds. Reaction 17-7 can be reversed with hydrogen halides, particularly smoothly with concentrated hydriodic acid. At one time this constituted an important synthesis of organotin trihalides, since these were inaccessible or awkward to prepare by other routes. The required stannonic acids can be obtained from potassium or sodium stannite by treatment with alkyl halide in cold alkali (454, 530, 659, 660) (reaction 17-8).

$$MeI + Na_2SnO_2 \rightarrow MeSn(O)O^-Na^+ + NaI \qquad (17\text{-}8)$$

Hydrolysis 17-7 is of course a multi-stage reaction. Aliphatic trichlorides $RSnCl_3$ yield $RSn(OH)_2Cl$, $RSn(OH)Cl_2{\cdot}H_2O$, $RSn(OH)Cl_2$, or $[RSn(O)Cl]_n$ depending on the experimental conditions and the nature of R. Hydrolysis of aryltin trichlorides yields no properly defined intermediate products (489a).

Stannonic acids are soluble in caustic alkalis, but not in alkali carbonate solutions. They are precipitated from alkali as colorless infusible precipitates by passing in CO_2. Stannonic acids with large alkyl groups are easily soluble in common organic solvents. They are now prepared usually from the easily accessible organotin trihalides (Chapter 5). Surprisingly little work has been done on their structure. The scarcity of reliable data is even greater than in the case of the diorganotin oxides (Chapter 17-2). It is probable that the stannonic acids are cyclic hydroxy-stannoxanes. There are some good arguments for the trimeric structure (7) (454), and this would also accord with the IR spectrum (920). In addition, as with the diorganotin oxides (Chapter 17-2) there is a strong possibility of additional coordinate bonding O-→Sn giving penta-coordinate tin. Methylstannonic acid forms well defined crystalline pentaacylstannoxanes (**8**) with some car-

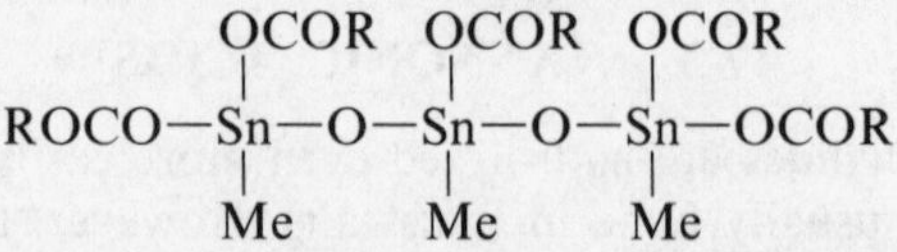

$$\mathrm{ROCO{-}\underset{Me}{\overset{OCOR}{Sn}}{-}O{-}\underset{Me}{\overset{OCOR}{Sn}}{-}O{-}\underset{Me}{\overset{OCOR}{Sn}}{-}OCOR}$$

(**8**)

boxylic acids (454), and this reaction supports the six-membered ring structure. The free acids eliminate water rapidly above 120°C (920), when moist already above 50°C (464). This polycondensation reaction produces linear or network polymers (**9**); e.g. with R = Bu, heating for 3 hours at 150°C results in a largely linear polymer, a butyl-polystannoxane of average molecular weight 8000.

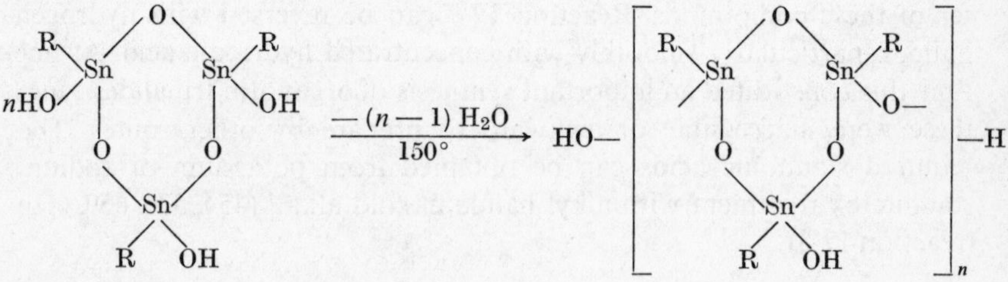

$$n\,\text{(7)} \xrightarrow[150^\circ]{-(n-1)\,H_2O} \text{(9)}$$

(**7**): six-membered $(RSn(OH)O)_3$ ring — R, Sn, O, Sn, R; nHO, OH; O, O; Sn; R, OH

(**9**): HO—[ring: R, Sn, O, Sn, R; —O—; O, O; Sn; R, OH]$_n$—H

The polymer is colorless, infusible, soluble in chloroform and carbon tetrachloride, capable of swelling in benzene, and reasonably stable to hydrolysis and to heating up to 250°C (920). (Almost all C–Sn bonds are ruptured at around 250°C; see Chapter 4.)

Like all monoorganotin compounds, stannonic acids decompose relatively easily with complete elimination of the organic moiety; e.g. heating of all aliphatic and some aromatic stannonic acids with hydrogen halides affords the corresponding hydrocarbon and tin tetrahalide (464, 663). Boiling with caustic alkali again gives the hydrocarbon, but also some R_2SnO and/or R_3SnOH (315, 464). Stannonic acids with large alkyl groups, e.g. *n*-octyl, are rather more stable to disproportionation than those with small alkyl groups (464, 489a).

Finally, it should be pointed out that the chemistry of organotin hydroxides and oxides is seen more and more to be that of stannoxanes, i.e. of compounds with SnOSn groups. All these compounds should therefore now be viewed from a common standpoint, even though our present knowledge reveals considerable gaps in quite fundamental areas.

17-4 ORGANOTIN ALKOXIDES

Compounds R_3SnOR' and $R_2Sn(OR')_2$ will be discussed together, since their properties are very similar and the methods of preparation the same. [Properly defined compounds $RSn(OR')_3$ appear to be unknown at present.] Preparation involves elimination of water from the appropriate hydroxide, oxide, or stannoxane and alcohol or phenol, the water being distilled off as the benzene or toluene azeotrope (reaction 17-9). For diphenoxides, boiling tetralin has been recommended (696a).

$$\begin{aligned} \tfrac{1}{2}\,R_3SnOSnR_3 + HOR' &\rightarrow R_3SnOR' + H_2O \\ R_2SnO + 2\,HOR' &\rightarrow R_2Sn(OR')_2 + H_2O \end{aligned} \qquad (17\text{-}9)$$

Heating of dialkyl carbonates with stannoxanes (evolution of CO_2) is another useful method (175d).

Alternatively, one can start from the alkali-metal alkoxide (or phenoxide) and the organotin halide (416), and this can be of advantage in some cases, e.g.,

$$Ph_3SnCl + NaOC_6H_4Cl\text{-}o \xrightarrow[\text{boil}]{\text{THF}} Ph_3SnOC_6H_4Cl\text{-}o + NaCl \quad (43)$$

$$n\text{-}Bu_2SnCl_2 + 2\,NaOPh \xrightarrow[\text{boil}]{n\text{-Heptane}} n\text{-}Bu_2Sn(OPh)_2 + 2\,NaCl \quad (150)$$

$$Et_3SnCl + RMgOCH(Me)C{\equiv}CH \xrightarrow[\text{boil; hydrol.}]{\text{Ether}} Et_3SnOCH(Me)C{\equiv}CH \quad (791)$$

This last method was widely used at one time, but has now been largely abandoned in favor of reaction 17-9.

Organotin alkoxides and free alcohols establish an equilibrium as in equation 17-10. Lower-boiling alcohols can therefore be displaced by higher-boiling ones (315).

$$R_3SnOR' + HOR'' \rightleftharpoons R_3SnOR'' + HOR' \qquad (17\text{-}10)$$

Another possible route to organotin alkoxides involves alcoholysis of stannylamines (75, 851), e.g. (22) as in reaction 17-11.

$$Ph_3SnNEt_2 + MeOH \rightarrow Ph_3SnOMe + HNEt_2 \qquad (17\text{-}11)$$

Hydrostannation of carbonyl compounds (see Chapter 9) and cleavage of Sn–C bonds by phenols (see Chapter 4) should be mentioned as other feasible approaches.

Most organotin alkoxides are liquids which can be distilled without decomposition. A few are crystalline solids (see Table 17-2). Their IR

Table 17-2
Organotin Alkoxides[a]

FORMULA	B.P. (°C/mm.)	M.P. (°C)	n_D^{20}	METHOD OF PREPARATION
Me_3SnO*t*-Bu	subl. at 0.1	decomp. >45	—	23
Bu_3SnOMe	97—97.5/0.06	—	1.4745	15, 546
i-$Bu_3SnOCH(Me)Et$	88/0.4	—	1.474	558
Ph_3SnOMe	—	65—66	—	15, 22
Et_3SnOPh	260	—	1.5422	355, 719, 721
	115—116/1		1.5415	
$Et_3SnOC_6H_4(CH_2OH)$-*o*	—	64	—	558
$Et_3SnO(CH_2)_4OSnEt_3$	184—185/4.5	—	1.4975	790
$Et_3SnOCH_2CH_2C{\equiv}CH$	157—158/3	—	—	791
$Ph_3SnOC_6H_4Cl$-*o*	—	59—60	—	43
$Me_2Sn(OEt)_2$	81/0.1	—	1.4775	234
$Bu_2Sn(OMe)_2$	126—128/0.05	—	1.4880	15, 234
$(C_8H_{17})_2Sn(OBu)_2$	182—188/0.05	—	1.4735	234
$Bu_2Sn(OPh)_2$	—	45—48	—	150
$Bu_2\overline{SnOCH_2CH_2O}$	—	195—200	—	115
$Ph_2Sn(OMe)_2$	—	270 (decomp.)	—	505
$Bu_2Sn(O\text{-}8\text{-quinolyl})_2$	—	155—156	—	74, 234

[a] Many more compounds are named in the patent literature, but often without adequate characterization.

spectra have been discussed more than once (116, 420, 526a) (see also Chapter 23). Molecular weights in solution (as far as they have been determined) are always those of the monomer. For undiluted liquid compounds $R_2Sn(OR)_2$, however, a dimeric structure with SnO(R)···Sn bridging has been concluded from IR measurements (526b).

Dimerization through oxygen donors seems likely also when electron-withdrawing groups are located on the tin atoms. Compounds of this type arise by redistribution reactions (175a):

$$R_2Sn(OMe)_2 + R_2SnX_2 \rightarrow 2\,R_2Sn(X)OMe$$

(R = Me, Et, Pr, Bu; X = F, Cl, Br, I, SCN, OCOMe, OSO_2-camphor)

The Sn–OR group is highly reactive: water effects immediate hydrolysis (reversal of reaction 17-9), alcohols lead to equilibrium 17-10, and carboxylic acids or acid chlorides yield stannyl esters in an exothermic reaction (see Chapter 7). Carboxylic esters suffer transesterification (857):

$$RCOOR' + R_3SnOR'' \rightleftharpoons RCOOR'' + R_3SnOR' \qquad (17\text{-}12)$$

while alkyl halides give the ether and an organotin halide (673). Organosilicon halides likewise cleave the Sn–O bond, affording Si–OR compounds and organotin halides (672). Amongst other reaction partners one might just mention BCl_3, which yields $RO\text{–}BCl_2$ (235). Addition of organotin alkoxides to unsaturated systems and condensation with acidic compounds are discussed in Chapter 19. Reduction to organotin hydrides was dealt with in Chapter 8.

17-5 ORGANOTIN PEROXIDES

Organosilicon and organogermanium peroxides have been studied for quite some time (reviews in refs. 16 and 705). Work on analogous tin compounds first appeared in 1958 (704), but has expanded rapidly ever since. The compounds are all much less stable than alkyl peroxides. The reason for this remains largely unknown, as does the decomposition mechanism (16). Anyway, the Sn–OO bond is almost entirely covalent (705) and the decomposition of radical type, since organotin peroxides catalyze polymerization of olefins such as styrene and methacrylic esters (9, 705).

Stannyl hydroperoxides R_3SnOOH are the simplest type. These are known with R = Me, Et, and Ph. They are formed from the corresponding stannoxane or hydroxide by treatment with hydrogen peroxide and removal of the water formed (11, 170), e.g.

$$Et_3SnOSnEt_3 + 2\,H_2O_2 \xrightarrow{-60^\circ} 2\,Et_3SnOOH + H_2O$$

They are fairly stable at room temperature. The kinetics of decomposition in solution have been studied (170). The triethyltin compound forms a complex $(R_3SnOOH)_2 \cdot HOOH$ that dissociates in solution (11). (For structural problems, see below.)

Trialkylstannyl alkyl peroxides R_3SnOOR' are easily accessible by several routes which resemble those for the corresponding alkoxides (Chapter 17-4) (16, 704, 705).

$$(\text{cyclo-}C_6H_{11})_3SnCl + NaOOCMe_2Ph \xrightarrow[96\,\%]{20^\circ} (\text{cyclo-}C_6H_{11})_3SnOOCMe_2Ph + NaCl$$

$$Ph_3SnCl + HOOCMe_3 + NH_3 \xrightarrow[83\%]{20°} Ph_3SnOOCMe_3 + NH_4Cl$$

$$Bu_3SnOMe + HOO(9\text{-Decalyl}) \rightarrow Bu_3SnOO(9\text{-Decalyl}) + MeOH$$

$$R_3SnNR''_2 + HOOR' \rightarrow R_3SnOOR' + HNR''_2$$

$$R_3SnH + HOOR' \xrightarrow[-H_2O]{\text{exotherm.}} R_3SnOR' \xrightarrow[-R'OH]{HOOR'} R_3SnOOR'$$

A convenient method involves elimination of water from stannoxanes (or trialkyltin hydroxides) and alkyl hydroperoxides (80a):

$$(R_3Sn)_2O + 2\,HOOR' \rightarrow 2\,R_3SnOOR' + H_2O$$
$$(R = Bu, Ph; R' = t\text{-Bu}, CMe_2Ph)$$

The water can be distilled off directly *in vacuo* (R = Bu) or by azeotropic distillation with benzene (R = Ph), in both cases at room temperature. Trimethyltin hydroxide reacts in a similar way with $MgSO_4$ as dehydrating agent. Trimethyltin *t*-butyl peroxide is partially associated in benzene solution, presumably as a dimer of structure (**9a**) (80a). This emphasizes once again the donor strength of oxygen atoms bonded to tin (cf. the R_3SnOOH complex mentioned above).

```
                 t-Bu
                /
            O—O
           /    \
     Me3Sn        SnMe3
           \    /
            O—O
           /
       t-Bu
```

(**9a**)

The compounds are decomposed rapidly by moisture and by acids and alkalis. Most of them deflagrate on rapid heating, but are otherwise reasonably stable. Those with small alkyl groups have been distilled without decomposition and show little sensitivity to impact or friction.

Trialkylstannyl acyl peroxides are formed by reaction with per-acids (16, 832), e.g.

$$Ph_3SnOH + HOOCOR' \xrightarrow{0°} Ph_3SnOOCOR' + H_2O$$

They are rather unstable, and both aromatic and aliphatic representatives undergo an internal redox reaction even at room temperature.

$$R_2Sn(R)\text{—}O\text{—}O\text{—}COR' \longrightarrow R_2Sn(OR)(OCOR')$$

A similar rearrangement is also observed in bis(trialkylstannyl) peroxides (9, 16, 705). The phenyl derivative of this series decomposes spontaneously and exothermically even at room temperature, yielding phenol and diphenyltin oxide as well as some Ph_3SnOH. It is obtained as follows:

$$2\,Ph_3SnCl + H_2O_2 + 2\,NH_3 \xrightarrow[-5°]{\text{Ether}} Ph_3SnOOSnPh_3 + 2\,NH_4Cl$$

Bis- and tris-(alkylperoxy)-compounds $R_2Sn(OOR')_2$ and $RSn(OOR')_3$ are obtainable by methods analogous to those described. They are mostly rather unstable (16, 705). Preparation of the stannoxane-peroxide (**10**) (174) and of the photo-peroxides $R_3Sn\text{–}R'\text{–}OOR''$ and $R_3Sn\text{–}R'\text{–}OOH$ (see Chapter 3-4) deserves brief mention.

$$R'OOSn(R_2)OSn(R_2)OOR' \qquad \textbf{(10)}$$

17-6 ORGANOTIN SULFUR COMPOUNDS*

Compounds with Sn–S bonds show extensive parallels with the corresponding oxygen derivatives (Chapters 17-1–17-4). A few peculiarities will be commented upon. The synthetic possibilities also are largely the same. Compared with other sectors of organotin chemistry, that of organic tin–sulfur compounds lay altogether neglected for a long time. Most of these substances were first prepared from 1953 onwards. Table 17-3 gives some examples. Some Sn–S compounds have recently acquired technical importance (see Chapter 25); the number of patents is therefore considerable.

Triorganotin hydrosulfides $R_3Sn\text{–}SH$ have not been characterized unequivocally. They appear to transform spontaneously into compounds $R_3SnSSnR_3$ (see below). A crystalline lithium derivative has however been prepared. This is dimeric in benzene, and probably has the four-membered ring structure (**11**). It is sensitive to solvolysis and oxidation; at elevated temperatures it splits into $R_3SnSSnR_3$ and Li_2S. It is obtained under quite mild conditions from Ph_3SnLi and sulfur, when the S_8-molecules appear to be degraded in stages by an S_N2 mechanism (755, 756).

* For a recent review see ref. 1a.

Table 17-3
Organotin Sulfur Compounds

FORMULA	B.P. (°C/mm.)	M.P. (°C)	n_D^{20}	METHOD OF PREPARATION
$Me_3SnSSnMe_3$	233—235 118/18	6	—	279, 413
$Et_3SnSSnEt_3$	187—188/20 133—137/1	—	1.5468	279
$Ph_3SnSSnPh_3$	—	145.5—147	—	446, 698, 699
$(Me_2SnS)_3$	—	149.5—151	—	279, 698, 699
$(Et_2SnS)_3$	219—221	24	—	279
$(Pr_2SnS)_3$	254/16	—	—	280
$(Bu_2SnS)_3$	—	63—69	—	698, 699
$(Ph_2SnS)_3$	—	182—184	—	196, 666, 697, 698, 699
Me_3SnSMe	163	—	1.5306	1
$Me_3SnSt\text{-}Bu$	42/0.1	—	1.5083	1
Et_3SnSBu	88—91/1	—	1.5133	719
Et_3SnSPh	150/1.7	—	1.5828 1.5794	172
Ph_3SnSPh	—	102—103	—	172
$Me_2Sn(SPr)_2$	74/0.1	—	1.5498	1
$Pr_2Sn(SPh)_2$	226—230/1	—	1.6298	721
$Ph_2Sn(SPh)_2$	—	65—65.5	—	52, 172
$MeSn(SEt)_3$	90/0.05	—	1.5972	1
$EtSn(SMe)_3$	66/0.001	—	1.6232	1
$Me_2\overline{SnSCH_2CH_2S}$[a]	113—115/5	82—83	—	1, 884

[a] See also Chapter 20.

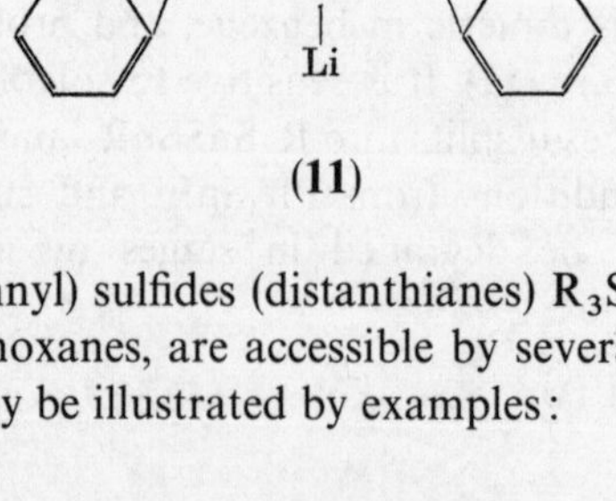

(11)

Bis(triorganostannyl) sulfides (distanthianes) $R_3SnSSnR_3$, the analogs of hexaorganostannoxanes, are accessible by several reactions (315, 446, 699, 719), which may be illustrated by examples:

$$2\,R_3SnCl + H_2S \rightarrow R_3SnSSnR_3 + 2\,HCl$$

$$Et_3SnOSnEt_3 + H_2S \rightarrow Et_3SnSSnEt_3 + H_2O$$

$$Ph_3SnOSnPh_3 + COS \rightarrow Ph_3SnSSnPh_3 + CO_2$$

$$Ph_3SnSLi + ClSnPh_3 \rightarrow Ph_3SnSSnPh_3 + LiCl$$

In the first equation, the halide can be replaced by an alkoxide R_3SnOR' (491) or by a hydroxide; also, Na_2S_2 or NaHS have been recommended in place of H_2S. In the last reaction the organotin chloride can be replaced e.g. by the analogous Ge or Pb chloride, to yield mixed sulfides $Ph_3SnSGePh_3$ or $Ph_3SnSPbPh_3$ respectively (755).

The sulfides are soluble in many organic solvents, and monomeric in solution (279, 719). Most of them are liquids and may be distilled without decomposition. With non-volatile acids, or acids stronger than H_2S, they react analogously to stannoxanes, e.g. (719)

$$Et_3SnSSnEt_3 + 2\,HOPh \rightarrow 2\,Et_3SnOPh + H_2S$$

Copper powder, freshly prepared, forms ditins:

$$R_3SnSSnR_3 + Cu \rightarrow R_3SnSnR_3 + CuS \qquad (R = Et, Bu)$$

if oxygen is excluded. In the presence of oxygen, however, stannoxanes are formed, and halides R_3SnCl in the presence of CCl_4 (685a).

Diorganotin sulfides $(R_2SnS)_n$ are associated like the corresponding oxides. They are however crystalline and soluble in organic solvents, and most have sharp melting points. They are not hydrolyzed even by boiling water. In all cases investigated they have been found to be cyclic trimers (**12**)* (279, 698, 755, 759).

$$(\text{–}R_2Sn\text{–}S\text{–})_3 \quad (\mathbf{12})$$

They are prepared by exchange of a reactive substituent for sulfur, e.g. (279, 315, 434, 698, 699)

$$Ph_2SnO + NaSH \xrightarrow{\text{Acid}} Ph_2SnS$$

$$R_2SnO + CS_2 \rightarrow R_2SnS + COS$$

$$R_2SnCl_2 + H_2S \rightarrow R_2SnS + 2\,HCl$$

$$1/n\,(Ph_2Sn)_n + S \rightarrow Ph_2SnS \qquad \text{(ref. 434)}$$

* The monomeric formulation has been retained for simplicity's sake.

The six-membered ring of diorganotin sulfides is also obtained by cleavage of C–Sn bonds, e.g. in $(p\text{-}MeC_6H_4)_3SnCH_2CH{=}CH_2$, by means of H_2S_5 (759). Organotin polysulfides appear as intermediates.

At 200°C tetraphenyltin reacts smoothly with elemental sulfur (88, 742, 755). The Ph–Sn bond appears to suffer polar bond scission under the polarizing influence of some $S^{\delta+}\text{–}(S)_n\text{–}S^{\delta-}$ units. The sulfur chains are then degraded subsequently. The reaction would be expected to yield the thiophenoxide Ph_3SnSPh, but this reacts further with sulfur to give the equally unstable bis(thiophenoxide) and affords ultimately a mixture of (**12**) and diphenyl sulfide:

$$3\ Ph_2Sn(SPh)_2 \rightarrow (\mathbf{12};\ R = Ph) + 3\ PhSPh$$

The behavior of tetrabutyltin is entirely analogous, but reaction proceeds already at 150°C to yield (**12**; R = Bu) and BuSBu.

The corresponding dithiols $R_2Sn(SH)_2$ are unknown. However, a highly reactive lithium derivative $Ph_2Sn(SLi)_2$ has been obtained from Ph_2SnLi_2 and sulfur (755).

The thio-analogs of the stannonic acids are represented by the compounds $(RSnS)_2S$ which are formally thiostannonic anhydrides (315, 659, 660). The methyl and butyl derivatives are crystalline solids which decompose without melting; they are practically insoluble in water or common organic solvents but react with acids and alkalis. The aromatic representatives are easily soluble in pyridine, aniline, or nitrobenzene.

For more than six decades there was little reliable information about their molecular weight and structure.* Recently, the methyl compound has been shown by mass spectrometry to be dimeric. A three-dimensional X-ray structure determination showed an adamantane-type structure (**12a**). In the crystal, Sn–Sn distances were found to be 3.8 Å, Sn–S 2.35 Å, and Sn–C 2.1 Å (see other data in Chapter 2-1). The true formula of the compound is therefore $(MeSn)_4S_6$ (177a). It is reasonable to assume similar structures for the butyl and other compounds of this group.

The compounds $(RSn)_4S_6$ are formed, in analogy with stannonic acids, by treatment of organotin trihalides with Na_2S in aqueous solution. Precipitation is quantitative:

$$4\ BuSnCl_3 + 6\ Na_2S \rightarrow (BuSn)_4S_6 + 12\ NaCl$$

* A review (315) suggests the formula $R\overset{\overset{S}{\uparrow}}{Sn}S\overset{\overset{S}{\uparrow}}{Sn}R$ for these compounds, without quoting any source or advancing any arguments for this suggestion. Ref. 847 quotes earlier work (659, 660) when advancing the formula $R\overset{\overset{S}{\|}}{Sn}S\overset{\overset{S}{\|}}{Sn}R$, but there is no mention of such a formula in the work quoted. Molecular weight determination is not mentioned in either.

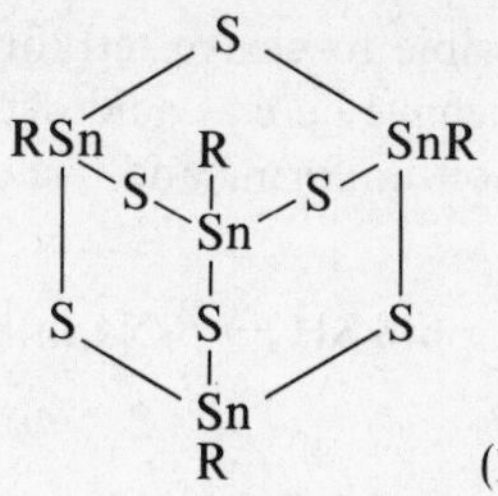

```
            S
          /   \
     RSn    R    SnR
      |  S  |  S  |
      |    Sn     |
      |     |     |
      S     S     S
        \   |   /
           Sn
           R            (12a)
```

Stannyl alkyl and aryl sulfides of the series R_3SnSR' and $R_2Sn(SR')_2$ are liquids or crystalline solids respectively. Most of them can be distilled (315). The third series $RSn(SR')_3$ comprises liquids and solids which cannot be distilled, but which are nevertheless monomeric in solution (Table 17-3).

Formation of thioalkoxides is analogous to that of alkoxides, but usually both smoother and simpler (1, 172, 719, 721, 823), e.g. as in the following:

$$Et_3SnOSnEt_3 + 2\,HSPh \rightarrow 2\,Et_3SnSPh + H_2O$$

$$Et_3SnSEt + HS\textit{i}\text{-}Pr \rightarrow Et_3SnS\textit{i}\text{-}Pr + HSEt$$

$$R_2SnCl_2 + 2\,HSR' \xrightarrow{\text{Bases}} R_2Sn(SR')_2 + 2\,HCl$$

$$R_2Sn(OR'')_2 + 2\,HSR' \rightarrow R_2Sn(SR')_2 + 2\,R''OH$$

$$RSn(O)OH + 3\,R'SH \rightarrow RSn(SR')_3 + 2\,H_2O$$

$$RSnBr_3 + 3\,R'SH \xrightarrow{NaOH/H_2O} RSn(SR')_3 + 3\,HBr$$

It will be seen that other substituents can be displaced by mercaptans ("transesterification"), sometimes spontaneously even at room temperature, and that lower-boiling mercaptans can be displaced by higher-boiling ones. On the other hand, the –SR group is displaced by stronger or poorly volatile acids (e.g. RCOOH). Understandably, it is not attacked by water. (For an explanation, see below.) Phosphorus trichloride effects cleavage:

$$3\,Me_3SnSMe + PCl_3 \rightarrow 3\,Me_3SnCl + P(SMe)_3$$

Elemental halogens cleave into organotin halide and R′SSR′ (434), providing an iodometric determination of the Sn–SR′ link (1). Treatment with methyl iodide affords a sulfonium iodide $(Me_3SnS^+Me_2)I^-$ which is stable to hydrolysis (1).

In principle it is possible to cleave tetraorganotins with thioalcohols or thiophenols, which behave here as acids (see Chapter 4). The products however tend to be non-uniform, and yields are only moderate, e.g. (720, 722):

$$Et_4Sn + \beta\text{-Naphth SH} \rightarrow \beta\text{-Naphth SSnEt}_3 + C_2H_6 \quad 40\%$$

Interesting tin–sulfur heterocycles are formed when the above-mentioned thiolysis of diorganotin dihalides is carried out with 1 molecule of a dithiol instead of 2 molecules of a monothiol (see Chapter 20).

Stannyl esters of thioacids are easily accessible. (For a general discussion of stannyl esters see Chapter 7.) The following examples indicate some of the possible approaches:

$$Bu_2SnO + 2\,HSC(S)R' \rightarrow Bu_2Sn[SC(S)R']_2 + H_2O$$

$$Ph_3SnSLi + ClCOPh \rightarrow Ph_3SnSCOPh + LiCl \quad (755)$$

$$Ph_3SnBr + NaSC(S)NHCH_2Ph \rightarrow Ph_3SnSC(S)NHCH_2Ph \quad 94\% \quad (446)$$

17-7 ORGANOTIN SELENIUM AND TELLURIUM COMPOUNDS

Our knowledge of these compounds is still largely incomplete. From what little information there is, we may conclude that the general relationships and the various categories are in principle the same as with organotin oxygen compounds (Chapter 17-1—17-4). Elemental selenium and tellurium react by similar mechanisms as sulfur (see above). They react with triphenylstannyllithium in tetrahydrofuran at room temperature (755) to give solutions of $Ph_3SnSeLi$ and $Ph_3SnTeLi$ respectively. These compounds cannot be isolated, but the solutions may be used for further reactions; e.g. treatment with the appropriate triphenyl-metal halide yields the highly crystalline compounds: $Ph_3SnSeGePh_3$, M.P. 145°C; $Ph_3SnSeSnPh_3$, M.P. 148°C; $Ph_3SnSePbPh_3$, M.P. 138°C; $Ph_3SnTeGePh_3$, M.P. 145°C; $Ph_3SnTeSnPh_3$, M.P. 150°C; $Ph_3SnTePbPh_3$, decomp. 136°C. The selenium compounds are very pale yellow, and the tellurium compounds yellow. The bonds are covalent (755). The good solubility in benzene, dioxane, tetrahydrofuran, and chloroform is noteworthy, as is the thermal and—in most cases—solvolytic stability. Methyl derivatives are obtained similarly (755). Compounds $(R_2SnX)_n$ are known only for selenium. Like the analogous sulfur compounds (see above) they form cyclic trimers; e.g. (**13**) is obtained from dimethyltin dichloride and sodium selenide (743, 877).

$$3\,Me_2SnCl_2 + 3\,Na_2Se \xrightarrow[\text{Benzene}]{80°,\ 75\%} (\text{–}Me_2Sn\text{–}Se\text{–})_3 + 6\,NaCl$$

(13)

The corresponding butyl compound is one of the products of the pyrolysis of Bu_4Sn and selenium at 200°C (cf. similar reaction with sulfur, above). The other main product is BuSeBu. The phenyl derivative corresponding to (**13**) cannot be made in this way since the required temperature is at least 240°C and some decomposition occurs before this. The following synthesis, however, has proved successful (755) (R = Ph):

$$2\,R_3SnSeLi + R_2SnCl_2 \xrightarrow[-2\,LiCl]{} R_3SnSeSn(R_2)SeSnR_3$$

$$\longrightarrow R_3SnSeSnR_3 + \tfrac{1}{3}(R_2SnSe)_3$$

The only selenium analogs of stannonic acids prepared to date appear to be $(MeSn)_2Se_3$ and $(EtSn)_2Se_3$. Both are obtained by treatment of the stannonic acid with hydrogen selenide (847). The structure Se=Sn(R)SeSn(R)=Se has been proposed without recourse to molecular weight determination. Analogy with stannonic acids and trimeric dimethyltin selenide (above) indicates that some association is likely.

Stannyl alkyl selenides are formed in the pyrolysis of tetraphenyltin with selenium (755):

$$Se + Ph_4Sn \xrightarrow{200°} Ph_3SnSePh$$

or alternatively by displacement of an alkoxy group from some organotin alkoxides (558):

$$Et_3SnO\textit{i}\text{-}Bu + HSeCH_2Ph \rightarrow Et_3SnSeCH_2Ph + HO\textit{i}\text{-}Bu$$

This selenoalkoxide is stable to water; it is cleaved by dilute hydrochloric acid and, on heating, by acetic anhydride.

18 Organotin compounds containing other elements

In principle it now seems possible to link stannyl groups to any element of the Periodic Table. Depending on the character of the particular element, the bond will be more or less stable, and therefore also less or more reactive. In this chapter only those compounds are discussed which contain tin bonded to elements not mentioned so far, especially sub-group and transition elements.

Relatively few such compounds are known. However, this state of affairs should change rapidly under the impact of better preparative and analytical techniques. Thus, $(Ph_3Sn)_2Zn$, a compound highly sensitive to air, has been prepared in liquid ammonia:

$$2\,Ph_3SnK + ZnCl_2\,(dry) \xrightarrow[-2KCl]{} Ph_3SnZnSnPh_3$$

The corresponding Sn–Cd and Sn–Hg compounds, even less stable, were not accessible by this method (190c).

Another route to compounds of this kind involves reaction of triphenyltin hydride with proper metal alkyls in strongly complexing solvents like dimethoxyethane, tetrahydrofuran, or 2,2′-bipyridyl (176b):

$$2\,Ph_3SnH + Et_2Zn \rightarrow Ph_3SnZnSnPh_3 + 2\,C_2H_6$$

$$2\,Ph_3SnH + Me_2Cd \rightarrow Ph_3SnCdSnPh_3 + 2\,CH_4$$

The products are isolated as coordination complexes. Corresponding halides $Ph_3SnZnCl$ and $Ph_3SnCdCl$ are accessible similarly. These are shown also to occur as intermediates in some other reactions (176b).

The corresponding mercury compound, remarkably unstable and highly reactive, has recently been prepared by another procedure (186b):

$$Ph_3SnH + [(Me_3Si)_2N]_2Hg \rightarrow Ph_3SnHgSnPh_3 + 2\,(Me_3Si)_2NH$$

Rather less is known about trialkyltin analogs. Triethyltin hydride reacts smoothly with diethylzinc (866) forming an intermediate with an Sn–Zn bond. Under the experimental conditions this decomposes at once with separation of zinc metal:

$$Et_3SnH + Et_2Zn \rightarrow Et_4Sn + C_2H_6 + Zn$$

$$2\,Et_3SnH + Et_2Zn \rightarrow Et_3SnSnEt_3 + 2\,C_2H_6 + Zn$$

The Sn–Hg compound is even more unstable, and suffers immediate radical fission at 20°C with separation of Hg. The following reactions occur spontaneously at room temperature:

$$2\,R_3SnH + Et_2Hg \rightarrow R_3SnSnR_3 + 2\,C_2H_6 + Hg \quad (426, 569, 570, 868)$$

$$R_2SnH_2 + Et_2Hg \rightarrow 1/n\,(R_2Sn)_n + 2\,C_2H_6 + Hg \quad (426, 570)$$

$$Et_3SnH + EtHgCl \rightarrow Et_3SnCl + C_2H_6 + Hg \quad (868)$$

$$Et_3SnH + EtHgOCOMe \rightarrow Et_3SnOCOMe + C_2H_6 + Hg \quad (61)$$

At 0°C, there is practically no reaction between triethyltin hydride and diethylmercury, but di-*t*-butylmercury reacts smoothly even at −15°C. $Et_3SnHgSnEt_3$ is formed; it is deep yellow and stable at −30°C under argon in the dark. $Me_3SnHgSnMe_3$ forms deep red crystals (74a).

The analogous silicon and germanium compounds are rather more stable. Those with aliphatic groups can be distilled without decomposition,

$$2\,R_3GeH + Et_2Hg \rightarrow R_3GeHgGeR_3 + 2\,C_2H_6$$
$$(R = Et, Bu, Ph)$$

while the phenyl derivatives have sharp melting points. Decomposition into Hg and the digermane or disilane respectively takes place only on heating to quite high temperatures or on UV irradiation. The same is true for the dihydrides (426, 569, 570):

$$n\,R_2GeH_2 + n\,Et_2Hg \xrightarrow[-2nC_2H_6]{} (\text{–}R_2Ge\text{–}Hg\text{–})_n \xrightarrow[\text{heat}]{\text{UV}} (R_2Ge)_n + n\text{Hg}$$

These M–Hg compounds are versatile reagents for several types of syntheses (870a).

Metal carbonyls containing a covalent tin–metal bond are now known for Cr, W, Mo, Mn, Fe, and Co. In some cases the tin–metal bond has been proved; in others it has merely been postulated. The following equations and formulae should give an indication:

$$Ph_3SnCl + Cr(CO)_6 + Na(C_5H_5) \xrightarrow[-NaCl,\ -3\,CO]{THF} Ph_3Sn\text{–}Cr(CO)_3C_5H_5 \qquad (543)$$

The analogous tungsten compound was obtained similarly (543). $Ph_3SnMn(CO)_5$ (structure proved by X-ray) (875b). $Ph_3SnMn(CO)_4\text{–}PPh_3$ (structure proved by X-ray) (106).

$$Me_2Sn(Cl)Mn(CO)_5 + [C_5H_5Mo(CO)_3]^- \xrightarrow{-Cl^-} (CO)_5MnSnMe_2Mo(CO)_3C_5H_5 \qquad (643)$$

$$R_2Sn(C{\equiv}CR')_2 + Fe_3(CO)_{12} \rightarrow (CO)_4Fe\langle(SnR_2)_2\rangle Fe(CO)_4 \qquad (313)$$

$$BuSnCl_3 + Co_2(CO)_8 \rightarrow BuSn[Co(CO)_3]_3 \qquad (313)$$

$$Bu_3SnCl + Na[Co(CO)_4] \xrightarrow[-NaCl]{} Bu_3SnCo(CO)_4 \qquad (297)$$

Such compounds can even add to olefins:

$$Me_3SnMn(CO)_5 + F_2C{=}CF_2 \rightarrow Me_3SnCF_2CF_2Mn(CO)_5 \qquad (141)$$

This class of organotin compounds has been mentioned here briefly for the sake of completeness. Detailed treatment belongs properly to the field of transition-metal chemistry (768).

19 Reactions of stannylamines, stannoxanes, and organotin alkoxides

Stannylamines, stannoxanes, and organotin alkoxides have in the last few years proved to be versatile and, on the whole, smoothly reacting starting materials for synthetic purposes, and are amongst the most commonly used laboratory organotin compounds, being exceeded in importance only by the organotin hydrides (see Chapters 9 and 10). Their synthetic possibilities have been demonstrated largely by the groups working with A. G. Davies (80, 173) and with M. F. Lappert (336, 454a). All three classes of compound resemble each other in their reactions, justifying the treatment of these reactions in one and the same chapter. Admittedly, there are marked differences in reactivity: the stannylamines (Sn–N group) are the most reactive, followed by the alkoxides, and—some way behind—the stannoxanes (Sn–O group). There are two reaction types: condensation and addition. The potential use of organotin oxides and alkoxides in organic synthesis has been reviewed (173b).

19-1 CONDENSATION REACTIONS

19-1-1 Condensation with proton donor compounds

The general equations for these reactions are:

$$R_3SnX + HY \rightarrow R_3SnY + HX \tag{19-1}$$

$$R_2SnX_2 + 2\,HY \rightarrow R_2SnY_2 + 2\,HX \tag{19-2}$$

$$RSnX_3 + 3\,HY \rightarrow RSnY_3 + 3\,HX \tag{19-3}$$

where X represents NR'_2, OR', or $OSnR'_3$. HY can be any compound with an acidic hydrogen (reviews in refs. 335, 336, and 673a), e.g. a carboxylic acid:*

$$R_3SnOR' + MeCOOH \rightarrow R_3SnOCOMe + HOR'$$

Out of the large number of known or possible reactions of this type one might mention condensation with the NH-group of acyl amides and carbamates (612), e.g.

$$PhNHCOR + Et_3SnOMe \xrightarrow[-MeOH]{} PhN(SnEt_3)COR$$
$$(R = H, Me, OMe)$$

If the residue X attached to the amide group is rather more electronegative, the initial *N*-trialkylstannylation is followed by more complex reactions (175c). Thus, primary amides ultimately yield trialkyltin isocyanates.

$$2\,XCONH_2 + (R_3Sn)_2O \rightarrow 2\,R_3SnNCO + 2\,HX + H_2O$$

while secondary amides form distannyl-ureas

$$2\,XCONHR' + 2\,(R_3Sn)_2O \rightarrow$$
$$R_3SnNR'CONR'SnR_3 + 2\,R_3SnX + CO_2 + H_2O$$

Reactions of type 19-1 have been studied most extensively, and the experience gained can mostly be applied to types 19-2 and 19-3 also.

Special preparative interest attaches to condensations with compounds having an acidic CH function, since these can give new Sn–C bonds in excellent yield. Cyclopentadiene, fluorene, and indene have been stannylated in this way (335), and so have alkynes.

$(C_6H_5)_3Sn—N(CH_3)_2$ + [indene, H_2] ⟶ [1-(triphenylstannyl)indene, H, $Sn(C_6H_5)_3$] + $HN(CH_3)_2$

The group X on the tin, which is eliminated in these reactions, is best chosen to give an easily separated product, e.g. $HNMe_2$ (333, 337), H_2O (612, 561), or MeOH (497, 612). With stannylamines, reaction is mostly quantitative even at 20°C, e.g.

$$RC{\equiv}CH + Me_2NSnMe_3 \rightarrow RC{\equiv}CSnMe_3 + HNMe_2$$

With stannoxanes, the water formed must be removed continuously if reaction is to be quantitative. This can be achieved either by azeotropic

* For degradation of alkoxides, malonic acid is often the acid chosen, since its stannyl ester is precipitated quantitatively from petroleum ether (562).

distillation with benzene, or by using a solvent of suitable boiling point and allowing the condensate to interact with e.g. CaH_2.

Some functionally substituted CH-acidic compounds react smoothly under such mild conditions (373a, 561):

$$2\ p\text{-}MeOC_6H_4C{\equiv}CH + R_3SnOSnR_3 \xrightarrow[-H_2O]{} 2\ p\text{-}MeOC_6H_4C{\equiv}CSnR_3$$

$$2\ MeOCH{=}CHC{\equiv}CH + R_3SnOSnR_3 \xrightarrow[-H_2O]{} 2\ MeOCH{=}CHC{\equiv}CSnR_3$$

$$2\ NCC{\equiv}CH + R_3SnOSnR_3 \xrightarrow[-H_2O]{} 2\ NCC{\equiv}CSnR_3$$

Trihalomethyltin compounds, hardly accessible by other methods, are prepared conveniently by a similar procedure (175b):

$$X_3CH + R_3SnNR'_2 \rightarrow X_3CSnR_3 + HNR'_2$$

If the substituent itself has an acidic hydrogen, this will of course enter into the reaction also; e.g. propargyl alcohol first forms an organotin alkoxide, which then reacts with an excess of stannoxane to give the stannylacetylene (561, 792):

$$2\ HC{\equiv}CCH_2OH + R_3SnOSnR_3 \xrightarrow[-H_2O]{} 2\ HC{\equiv}CCH_2OSnR_3$$

$$R_3SnOSnR_3 + 2\ HC{\equiv}CCH_2OSnR_3 \xrightarrow[-H_2O]{} 2\ R_3SnC{\equiv}CCH_2OSnR_3$$

This means that condensation occurs preferentially at the more acidic hydrogen. Enols yield enolates, e.g. from dibutyltin oxide and acetylacetone (561):

$$2\ MeCOCH_2COMe + Bu_2SnO \xrightarrow[-H_2O]{} [MeCOCH{=}C(Me)O]_2SnBu_2$$

(M.P. 30°C)

Stannylamines are reported to be powerful dehydrochlorinating agents, e.g. in the case of alkyl chlorides (127c):

$$MeCH_2CH_2CH_2Cl + Me_3SnNMe_2 \xrightarrow[4\ hr.]{40°C}$$

$$MeCH{=}CHMe\ \ (88.5\%\ \textit{trans} + 5.2\%\ \textit{cis}) + MeCH_2CH{=}CH_2\ (6.3\%)$$

A group of interesting condensations with these highly reactive stannylamines should be mentioned separately (537a, 537b, 812b). α-Tolunitrile, as expected, yields a *C*-stannyl-substituted product:

$$PhCH_2CN + R_3SnNEt_2 \xrightarrow[-HNEt_2]{} PhCH(CN)SnR_3$$

while alkyl-substituted malononitriles form *N*-stannyl ketenimines, and alkyl-substituted malonic esters, or cyanoacetic esters respectively, form

O-stannyl ketenacetals (for properties of these materials see Chapter 9-8):

$$PhCH_2CH(CN)_2 + R_3SnNEt_2 \xrightarrow[-HNEt_2]{} PhCH_2C\begin{matrix} C{=}N \\ \\ C{=}N \end{matrix}SnR_3$$

$$PhCH_2CH(X)COOR' + R_3SnNEt_2 \xrightarrow[-HNEt_2]{} PhCH_2C(X)\!=\!C(OR')OSnR_3$$

$$(X = CN, COOR')$$

Condensation with aliphatic nitro-compounds affords *O*-stannyl nitronic acid esters (sometimes well crystallized, e.g. with R = Et):

$$R'CH_2NO_2 + R_3SnNEt_2 \xrightarrow[-HNEt_2]{} R'CH{=}N(O)OSnR_3$$

$$(R' = H, Me, Et, Ph)$$

Stannoxanes and carboxylic esters react quantitatively in accordance with reaction 19-4,* effecting a kind of transesterification:

$$R'COOR'' + R_3SnOSnR_3 \rightarrow R'COOSnR_3 + R_3SnOR'' \quad (19\text{-}4)$$

$$[R = Et, Bu; R' = \text{e.g. } Me, Ph; R'' = Et\ (574), CH_2C{\equiv}CH\ (561)]$$

Reactions of stannoxanes with carboxylic acids (yielding stannyl esters) have been described in Chapter 7, and those with hydrogen halides under organotin halides. Stannyl derivatives of still weaker acids, e.g. HCN, are best obtained from stannylamines or alkoxides, rather than from stannoxanes. In this way, bis(dialkylamino)dialkyltins and anhydrous hydrocyanic acid afford organotin dicyanides (479).

The gradations in the reactivities of stannoxanes, organotin alkoxides, and stannylamines are brought out also by the instantaneous, quantitative alcoholysis of stannylamines to alkoxides or hydrolysis to stannoxanes (335, 336), whereas alkoxides are only hydrolyzed fairly rapidly to stannoxanes and alcohol. The reverse reactions, such as formation of an alkoxide from stannoxane and alcohol, or a stannylamine from stannoxane and amine, succeed only if the water formed is removed continuously from

* This reaction had been known for some time, but was formulated (26) at first as

$$2\,R'COOR'' + R_3SnOSnR_3 \rightarrow 2\,R'COOSnR_3 + R''OR''$$

In fact, only 1 mole of alkyl ester is used up, and no dialkyl ether has been detected in any case investigated (175c, 574). The correct equation is therefore 19-4.

the equilibrium mixture (see Chapter 15 on stannylamines, and Chapter 17-4 on organotin alkoxides). In reactions 19-1—19-3, HY can also be a higher-boiling alcohol or higher-boiling amine (336). The stannyl moiety is then displaced from the lower-boiling one.

Finally, one might mention condensation of stannylamines with diazomethane (454a), and also condensations with arsines and phosphines (335) (see Chapter 16) and with organotin hydrides, which behave in this context as proton donors (see Chapter 8-4-4).

19-1-2 Condensations with other compounds

Only a few examples are known at present, but these few already promise an extension of preparative possibilities; e.g. organotin alkoxides react smoothly with acyl halides to yield the esters:

$$R'COCl + R_3SnOR'' \rightarrow R'COOR'' + R_3SnCl$$

With alkyl bromides or iodides they afford ethers (673) (chlorides, with the exception of benzyl chloride, do not react):

$$BuI + Bu_3SnOMe \xrightarrow{65\,\%} BuOMe + Bu_3SnI$$

$$H_2C{=}CHCH_2Br + Bu_3SnOMe \xrightarrow{100\,\%} H_2C{=}CHCH_2OMe + Bu_3SnBr$$

$$BrCH_2CH_2Br + Bu_2Sn(OMe)_2 \rightarrow \begin{cases} BrCH_2CH_2OMe + Bu_2Sn(Br)OMe \\ MeOCH_2CH_2OMe + Bu_2SnBr_2 \end{cases}$$

$$2\,BrCH_2CH_2Br + Bu_2Sn\begin{matrix} OCH_2CH_2O \\ OCH_2CH_2O \end{matrix}SnBu_2 \xrightarrow{80\,\%} 2\ \text{(1,4-dioxane)} + 2\,Bu_2SnBr_2$$

Organotin alkoxides behave here much like alkali-metal alkoxides, mechanisms being undoubtedly polar. They are, however, much milder in their action, and that is their advantage. Various solvents, $ZnCl_2$, $AlCl_3$, and even UV irradiation have no, or only negligible, catalytic effect.

Reaction of stannoxanes with alkyl halides is similar; e.g. α,ω-dihalides react in two stages to give cyclic ethers (673a):

$$\mathrm{Hal(CH_2)_nHal + R_3SnOSnR_3 \xrightarrow[-R_3SnHal]{} Hal(CH_2)_nOSnR_3 \xrightarrow[-R_3SnHal]{} (CH_2)_n\ O}$$

Stannylamines are suitable for introduction of $-NR_2$ groups under mild conditions; e.g. carboxylic esters and anhydrides are transformed into dialkyl-amides, usually in excellent yield (232a).

19-2 ADDITION TO POLAR MULTIPLE BONDS

Addition of Sn–N and Sn–O groups to various unsaturated systems in the manner of Sn–H groups is a fairly recent discovery. Several other metal–N and metal–O compounds react similarly (for reviews see refs. 78, 155, and 454b), but the tin compounds are particularly versatile. These reactions represent a new type, related to the Grignard reactions, and are probably more common than is realized. Even at this early stage it is clear that they have wide preparative application. Furthermore, the resulting compounds are mostly quite reactive themselves so that they are suitable for interaction with other reaction partners. A distinction can be made between the primary addition and possible secondary reactions.

19-2-1 Primary addition

Stannylamines $R_3SnNR'_2$ (232, 333, 335, 336) as well as organotin alkoxides R_3SnOR' and stannoxanes $R_3SnOSnR_3$ (76, 78, 79, 80b, 173, 176) add to unsaturated systems A=B, such as isocyanates, isothiocyanates, CO_2, CS_2, SO_2, carbodiimides, and ketenes, and also to some highly polar aldehydes, nitriles, and ketones.

In these addition reactions, stannylamines are the most reactive partners, but even peroxides give additions which are similar in principle, e.g.

$$\mathrm{RN{=}C{=}O + Ph_3SnOO\mathit{t}\text{-}Bu \rightarrow Ph_3SnN(R)COOO\mathit{t}\text{-}Bu}$$

[The products in this case are crystalline and constitute a new class of organic peroxides (74c).]

Difunctional analogs behave like the monofunctional ones, but may react in stages, e.g. (175a)

$$\mathrm{RN{=}C{=}O + Bu_2Sn(OMe)_2 \rightarrow RN(COOMe)Sn(Bu)_2OMe \xrightarrow{+RNCO} RN(COOMe)Sn(Bu)_2N(COOMe)R}$$

Table 19-1
Addition of Sn–N and Sn–O Groups to Polar Multiple Bonds A=B

A=B	$R_3Sn–NR'_2$ (232, 233a, 333, 336)	$R_3Sn–OR'$ (76, 78, 80b, 173, 173b, 176)	$R_3Sn–O–SnR_3$ (78, 80, 80b, 173, 173b)
RN=C=O	$Me_3SnN(Ph)CONMe_2$	$Et_3SnN(Bu)COOEt$	$Pr_3SnN(Ph)COOSnPr_3$
RN=C=S	$Me_3SnN(Ph)CSNMe_2$	$Bu_3SnN(Ph)CSOMe$	Secondary reactions
RCH=O	—	$Bu_3SnOCH(OMe)CCl_3$	$Bu_3SnOCH(OSnBu_3)CCl_3$
RC≡N	$Me_3SnN{=}C(Ph)NMe_2$	$Bu_3SnN{=}C(OMe)CCl_3$	$Bu_3SnN{=}C(OSnBu_3)CCl_3$
O=C=O	$Me_3SnOCONMe_2$	$Bu_3SnOCOOMe$	$Bu_3SnOCOOSnBu_3$
S=C=S	$Me_3SnSCSNMe_2$	$Bu_3SnSCSOMe$	$Bu_3SnSCSOSnBu_3$
O=S=O	$Me_3SnOSONMe_2$	$Bu_3SnOSOOPh$	$Bu_3SnOSOOSnBu_3$
RN=C=NR	$Me_3SnN(Tol)C(NEt_2){=}NTol$[a]	$Bu_3SnN(Tol)C(OMe){=}NTol$[a]	$Bu_3SnN(Tol)C(OSnBu_3){=}NTol$[a]
$R_2C{=}C{=}O$	$Me_3SnCH_2CONMe_2$	Bu_3SnCH_2COOMe	$Bu_3SnCMe_2COOSnBu_3$
CCOOR ⫼ CCOOR	$ROCOC(SnMe_3){=}C(NMe_2)COOR$ (233)	$ROCOC(SnMe_3){=}C(OMe)COOR$ (497)	

[a] Tol = p-MeC_6H_4.

Reactions are generally rapid, exothermic, and quantitative even at room temperature, but those with nitriles are rather slower than the remainder. It is important that the system A=B must be polar and open to nucleophilic attack. Thus, in aldehydes or ketones, only strongly polarized carbonyl groups act as systems A=B in the above sense, and electron-withdrawing groups such as CCl_3 are essential as substituents. The lone electron pair on the N or O of the attacking organotin probably plays an important part just as in condensations with organotin hydrides (see Chapter 8-4-4). [For stannylamines it has even been suggested that the reaction is controlled thermodynamically through a four-center intermediate state (232).] Unsubstituted olefins and some alkynes appear to be unaffected, but ketenes react at the C=C group. Polar olefins RHC=CR′R″ (R and R′ = H and/or Me; R″ = CN, CHO, or COOMe), react too, at least with the very active stannylamines R_3SnNMe_2 (R = Me, Et) (233a). This behavior resembles that of epoxides (R′ = e.g. Et, Ph) (856b):

$$R'\underset{\diagdown O \diagup}{CH-CH_2} + R_3SnNEt_2 \rightarrow R'\underset{NEt_2}{\underset{|}{C}}HCH_2OSnR_3$$

Acetylenedicarboxylic esters given good addition with both alkoxides (497) and amines (233, 233a).

The products of these reactions include many substituted derivatives of urea, carbamic acid, thiourea, and thiocarbamic acid, also distannyl acetals, amidines, imino-ethers, and derivatives of carbonic acid, dithiocarbonic acid, guanidine, carboxylic acids, and ethylenedicarboxylic acid. The versatility and preparative utility of these new syntheses should be obvious, particularly where amination and alkoxylation are concerned. Table 19-1 lists the systems A=B which have so far been used as substrates, and gives a small selection of the resulting compounds.

Many of these compounds can be distilled *in vacuo* without decomposition, but some of them revert to the starting materials. The addition itself may be reversible, therefore, as in the formation of stannyl and distannyl acetals from aldehydes (176) (R = Me, Pr, Bu, CBr_3):

$$RCHO + Bu_3SnOMe \overset{20°}{\rightleftharpoons} RCH(OMe)OSnBu_3$$

$$RCHO + Bu_3SnOSnBu_3 \rightleftharpoons RCH(OSnBu_3)_2$$

Attempts to distil the adduct from organotin phenoxides and isocyanates regenerate the starting materials, whereas the corresponding alkoxide adducts can be distilled unchanged.

The structure of the adducts (Table 19-1) follows from IR and NMR spectra as well as from chemical reactions. Unambiguous assignment is not always possible, since C=O and C=N absorptions in the region 1630–1695 cm.$^{-1}$ cannot be differentiated with certainty (78); e.g. addition to isocyanates may take form (**1**) or form (**2**) or both (78), i.e. addition may occur at the C=N group or at the C=O group:

$$\begin{matrix} R'NCO \\ + \\ R_3SnOR'' \end{matrix} \rightarrow \begin{matrix} R'N{-}CO \\ | \quad\;\; | \\ R_3Sn \quad OR'' \end{matrix} \; (\mathbf{1}) \quad \text{or} \quad \begin{matrix} R'N{=}C{-}OR'' \\ | \\ OSnR_3 \end{matrix} \; (\mathbf{2})$$

(R = Et, Bu; R' = Me, Et, Bu, Ph, α-Naphthyl; R'' = Me, Et, Ph)

(**1**) and (**2**) may even be in equilibrium via penta-coordinate tin. In any case, all reactions of the adducts can be interpreted in terms of structure (**1**), i.e. of an *N*-alkyl-*N*-stannylcarbamic ester.

Addition of the Sn–N group in organotin azides to the C≡C group is somewhat different, but rather interesting, as mentioned in Chapter 6.

$$Bu_3SnN_3 + \begin{matrix} CCOOEt \\ ||| \\ CCOOEt \end{matrix} \rightarrow \begin{matrix} N{-}{-}CCOOEt \\ || \qquad || \\ N \quad\;\; CCOOEt \\ \diagdown \;\; \diagup \\ N \\ SnBu_3 \end{matrix}$$

19-2-2 Secondary reactions

All the above adducts contain new Sn–N or Sn–O groups, and can therefore interact with the acidic hydrogen of carboxylic acids, H_2O, H_2S, amides, or alcohols and phenols as described in Chapter 19-1. With appropriate co-reactants the original organotin compound may be recovered. The organotin is then acting merely as catalyst for the addition of the acidic compound, e.g. ROH, to the unsaturated substrate. It has been debated whether this is the explanation of the well known catalytic action of tin compounds in polyurethane formation (75, 173).

Addition of the newly formed Sn–N or Sn–O links to other polar multiple bonds is feasible for the same reason. Elimination may occur in subsequent reactions, in which ultimately one acceptor, $A^2{=}B^2$, may displace another, $A^1{=}B^1$ (80b) (X = e.g. OR, NR'_2):

$$R_3Sn{-}A^1{-}B^1{-}X + A^2{=}B^2 \rightarrow R_3Sn{-}A^2{-}B^2{-}X + A^1{=}B^1$$

Further insertion of the Sn–O or Sn–N bond is another possibility, depending on the nature of A=B and on the conditions. If $A^1{=}B^1$ is identical with $A^2{=}B^2$, di-, tri-, or poly-merization may result; e.g. a substance R_3SnOR'', formed by addition of R_3SnOR' to an oxygen-containing system, A=B, is itself capable of such addition. Under suitable reaction conditions this can result ultimately in polymerization (77), e.g.

with chloral (176):

$$Bu_3SnOMe + O{=}CHCCl_3 \rightleftharpoons Bu_3SnOCH(CCl_3)OMe \xrightarrow{n\,O=CHCCl_3} Bu_3Sn\text{–}(OCH(CCl_3))_{n+1}\text{–}OMe$$

Even copolymerization is possible, as has been demonstrated by successive reaction with isovaleraldehyde, chloral, and phenyl isocyanate (173):

$$Bu_3SnOMe + O{=}CH\textit{i}\text{-}Bu + O{=}CHCCl_3 + O{=}C{=}NPh \longrightarrow Bu_3SnN(Ph)COOCH(CCl_3)OCH(\textit{i}\text{-}Bu)OMe$$

Isocyanates by themselves can be polymerized with catalysts containing Sn–N and Sn–O groups (75, 173). The stannoxane-catalyzed system has been investigated carefully (79, 80, 80b). The primary products are stannyl esters of *N*-alkyl-*N*-stannylcarbamic acid (**3**) (see Chapter 19-2-1). These react with more isocyanate to yield the isocyanurate (**6**) (R′ = R″ or R′ ≠ R″). The intermediate stages (**4**) and (**5**) were not isolated.

$$R'\text{—}NCO + R_3Sn\text{—}O\text{—}SnR_3 \longrightarrow R_3Sn\text{—}N(R')\text{—}C(=O)\text{—}O\text{—}SnR_3 \ (\mathbf{3}) \xrightarrow{R''\text{—}NCO}$$

$$[R_3Sn\text{—}N(R'')\text{—}C(=O)\text{—}N(R')\text{—}C(=O)\text{—}O\text{—}SnR_3] \ (\mathbf{4})$$

$$\downarrow R''\text{—}NCO$$

$$[R_3Sn\text{—}N(R'')\text{—}C(=O)\text{—}N(R'')\text{—}C(=O)\text{—}N(R')\text{—}C(=O)\text{—}O\text{—}SnR_3] \ (\mathbf{5}) \longrightarrow \text{cyclo-}[N(R')\text{—}C(=O)\text{—}N(R'')\text{—}C(=O)\text{—}N(R'')\text{—}C(=O)] + R_3Sn\text{—}O\text{—}SnR_3 \ (\mathbf{6})$$

Alternatively, depending on reaction conditions and the nature of R′ and R″, the allophanate (**4**) may decarboxylate and transform into the urea (**7**). This happens already at room temperature when R′ = Ph and R″ = α-naphthyl.

$$(\mathbf{4}) \xrightarrow[-CO_2]{} R_3SnN(\alpha\text{-Naphth})CON(Ph)SnR_3 \qquad (7)$$

This also explains why heating of some carbamates (**3**) and other stannyl carbamates $R'NHCOOSnR_3$ to 150°C leads directly to urea derivatives (79). At this temperature (**3**) is in equilibrium with its starting materials and can therefore add to some of the liberated isocyanate to give (**4**), which then decarboxylates. The last step must be assumed to take place via an intermediate state (**8**) (80) which is strongly reminiscent of that proposed for the decarboxylation of dimethylacetic acid (**9**) (880):

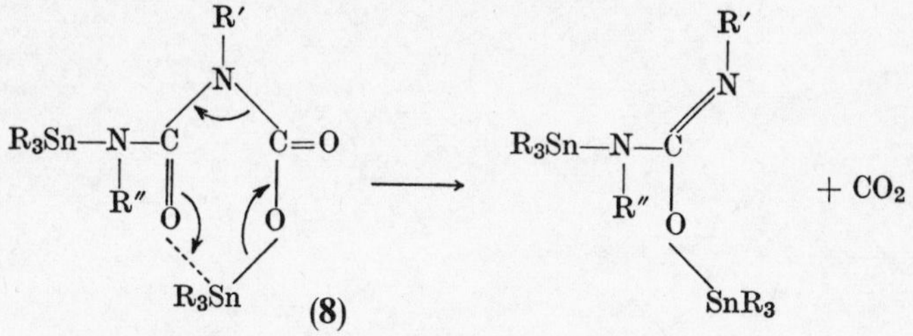

R′ N R₃Sn—N—C C=O R″ O O R₃Sn (**8**) ⟶ R′ N R₃Sn—N—C R″ O SnR₃ + CO_2

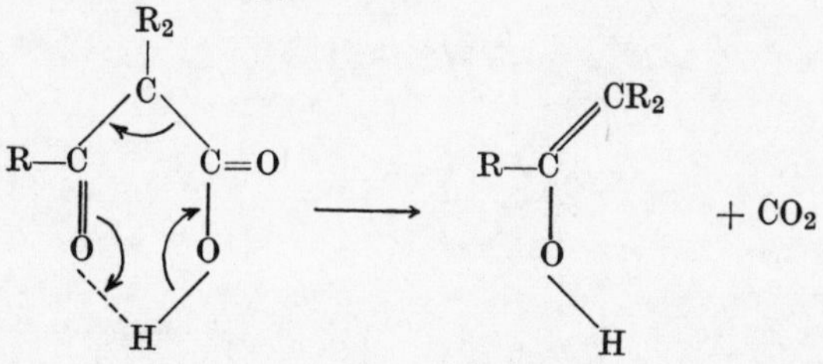

R₂ C R—C C=O O O H (**9**) ⟶ CR₂ R—C O H + CO_2

It should be noted that the imino-product from (**8**) is assumed to rearrange spontaneously to the urea, involving migration of the stannyl moiety from O to N (see also addition to isocyanates, Chapter 19-2-1 and 9-6-4).

These results can be applied to the preparation of unsymmetrically substituted biurets, which were previously unobtainable, e.g. (**10**) (80):

$$MeNCO + R_3SnOSnR_3 \rightarrow R_3SnN(Me)COOSnR_3 \xrightarrow[-CO_2]{PhNCO}$$

$$R_3SnN(Me)CON(Ph)SnR_3 \xrightarrow{\alpha\text{-NaphthNCO}}$$

$$R_3SnN(\alpha\text{-Naphth})CON(Me)CON(Ph)SnR_3 \xrightarrow{H^+}$$

$$\alpha\text{-NaphthNHCON(Me)CONHPh}$$

(**10**)

It is clear that there are appreciable differences in the reactivities of the various Sn–N groups, and that these depend on the other substituents on the nitrogen. Aliphatic groups generally make for faster reaction than aromatic ones. One may conclude that, apart from steric effects, the determining factor is the nucleophilicity of the nitrogen.

It is more than likely that the application of Sn–N and Sn–O compounds as reactants and especially as catalysts in preparative organic chemistry will be extended well beyond the boundaries of present knowledge.

20 Tin heterocycles

Introduction of tin as heteroatom into carbocyclic compounds provides some interesting stereochemical problems, as the large covalent radius of the tin (see Chapter 2) affects its environment. Tin heterocycles are formed surprisingly readily by several methods. The first of these was described by Grüttner and Krause in 1917 (272), but the field then lay largely neglected until recent years. A number of experimental results is now available. However, there remains an almost total lack of systematic stereochemical studies.

20-1 RING SYSTEMS WITH ONE HETEROATOM

The 1,1-diethyl and 1,1-dimethyl derivatives of stannacyclohexane are colorless liquids of supposedly pleasant odor, which may be distilled *in vacuo* without decomposition. They have been characterized quite unambiguously (272, 416). They are obtained even under conditions not altogether favorable for ring-closure, e.g. in 25–30 % yield from the Grignard synthesis 20-1.

$$\begin{matrix} ClMg(CH_2)_5MgCl \\ + \\ Et_2SnBr_2 \end{matrix} \rightarrow \text{(stannacyclohexane ring, } SnEt_2\text{)} \xrightarrow[100^\circ]{Br_2} BrSn(Et)_2(CH_2)_5Br \quad (20\text{-}1)$$

The ring system of these compounds is strained, as shown by their unusually (for tetraorganotins) high refractive indices, viz. $n_D{}^{20} = 1.5067$ for the diethyl and $n_D{}^{23} = 1.5024$ for the dimethyl compound. In contrast to these, tetraethyltin, a normal tetraorganotin of comparable molecular weight, has a refractive index of only 1.4720. The atmospheric oxidation of these ring compounds is also unusual for saturated tetraorganotins. The action of bromine is specific: with the diethyl compound ring-opening is the only effect (see reaction 20-1); no other C–Sn bond is ruptured. Dreiding models, too, confirm that the ring is strained owing to the large covalent radius of the tin. Nevertheless, lead too can be incorporated as heteroatom in cyclohexane (271), in spite of its still greater size.

With the dimethyl compound, bromine again effects mainly ring-opening, but this time also cleaves an Sn–Me bond to give 1-bromo-1-methylstannacyclohexane (272). With the analogous 1,1-diphenyl derivative ($n_D^{20} = 1.6007$) which is obtained from diphenyltin dichloride and 1,5-dilithiumpentane (64), substitution of the phenyl groups becomes the predominant reaction. (Phenyl groups are more easily cleaved from tin by polar mechanisms than are saturated aliphatic groups; see Chapter 4-2.)

$$\text{(I)(Ph)Sn}(CH_2)_5 \xleftarrow[-PhI]{+I_2} Ph_2Sn(CH_2)_5 \xrightarrow[-2PhH]{+2HBr\,(gas)} Br_2Sn(CH_2)_5$$

The halogen derivatives provide a route to a large number of 1-substituted stannacyclohexanes, but this possibility does not appear to have been much explored.

Grignard reactions similar to 20-1 have now been found suitable for general synthesis not only of stannacyclohexanes, but also of corresponding stanna-cyclopentanes and -cycloheptanes (928a):

$$BrMg(CH_2)_nMgBr + R_2SnCl_2 \rightarrow R_2Sn\widehat{\quad(CH_2)_n}$$

$$(n = 4, 5, 6;\ R = \text{Me, Et, Bu, } neo\text{-}C_5H_{11}, \text{Ph})$$

Analogous Si-, Ge-, and Pb-compounds have been prepared and used for spectroscopic comparison (64). The ring modes in the IR are displaced systematically in going from the carbon compound to the lead compound.

An elegant synthesis of the stannacyclohexane system has been achieved (671) by a modification of the preparation of tetraorganotins with alkylaluminums (549) (see Chapter 3-3-2):

$$3\,H_2C{=}CHC(Me_2)CH{=}CH_2 + 2\,i\text{-}Bu_2AlH \xrightarrow[-4i\text{-}C_4H_8]{160^\circ}$$

$$Me_2\langle\ \rangle AlCH_2CH_2C(Me_2)CH_2CH_2Al\langle\ \rangle Me_2$$

$$\xrightarrow[-2\,R'_2O\cdot AlCl_3]{+2\,R'_2O,\ +3\,Bu_2SnCl_2} 3\ \text{(4,4-}Me_2\text{-1,1-}Bu_2\text{-stannacyclohexane)}$$

$$\xrightarrow[-2\,R'_2O\cdot AlCl_3]{+1.5\,SnCl_4,\ +2\,R'_2O} 1.5\ Me_2\langle\ \rangle Sn\langle\ \rangle Me_2$$

The two methyl groups of the 3,3-dimethyl-1,4-pentadiene favor the horseshoe conformation required for ring-closure. The corresponding spiro-compound, a derivative of 1,1′-spirobisstannacyclohexane, is obtained similarly. Its structure has been ascertained by stepwise ring-opening at the C–Sn bonds by means of Br_2.

Derivatives of stannacyclopentane became known only recently (see above), but those of the unsaturated stannacyclopentadiene had been prepared earlier: 1,1-disubstituted stannoles are obtained usually in 40—70% yield from 1,4-dilithiumtetraphenylbutadiene (**1**) (93, 456, 457).

$$\begin{matrix} PhC{=}C(Ph)Li \\ | \\ PhC{=}C(Ph)Li \end{matrix} \xrightarrow[-2\,LiHal]{+R_2SnHal_2} \text{Ph}_4C_4SnR_2$$

(**1**) (**2**)

(R = Ph, Me, $CH{=}CH_2$)

The products are well defined colorless crystalline solids of high melting point. Only the hexaphenylstannole (**2**) (R = Ph) is faintly green and fluoresces when solid (93). One must assume that the success of this synthesis is due to a favorable preferred conformation of the dilithium derivative (**1**) which seems moreover to be present in the *cis-cis*-form. Corresponding Si- and Ge- and many other heterocycles have been obtained by very similar methods (93, 456, 457). The stannole ring (**2**) can be opened at the tin by halogens (Cl_2, Br_2, I_2) under mild conditions.

$$\text{Ph}_4C_4\cdot NiBr_2\ (\mathbf{4}) \leftarrow R_2SnBr\text{–}C(Ph){=}C(Ph)\text{–}C(Ph){=}C(Ph)\text{–}Br\ (\mathbf{3}) \rightarrow \tfrac{1}{2}\ \text{octaphenylcyclooctatetraene}\ (\mathbf{5})$$

(**4**) (**3**) (**5**)

More drastic action leads to 1,4-dihalotetraphenylbutadiene and diorganotin dihalide (217).

The first step involves conversion of (**2**) into the stannyl derivative (**3**), which can be used for some remarkable syntheses (217, 716a). Heating forms tetraphenylcyclobutadiene as a labile intermediate. In the presence of $NiBr_2$ this yields a stable complex (**4**). In the absence of $NiBr_2$ it affords a dimeric product (216) which was for a time regarded as octaphenylcubane (64, 218, 219), but has now been identified as octaphenylcyclooctatetraene (**5**) (852).

Two molecules of the butadiene derivative (**1**) with $SnCl_4$ give octaphenyl-1,1′-spirobistannole:

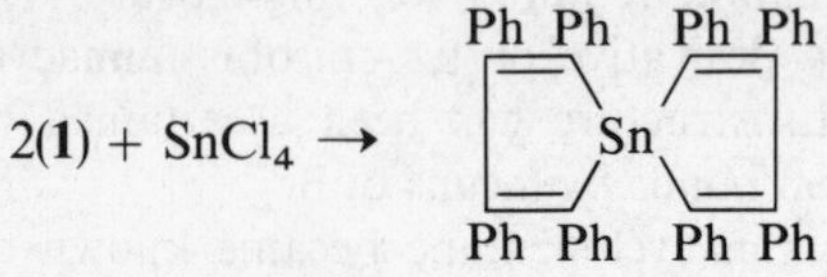

$$2(\mathbf{1}) + SnCl_4 \rightarrow$$

Ph Ph Ph Ph
Sn
Ph Ph Ph Ph

Dibenzostannoles are obtained by the action of the well tried 2,2′-dilithiumbiphenyl (894) on diorganotin dihalides (231). These compounds can also be regarded as derivatives of 9-stannafluorene:

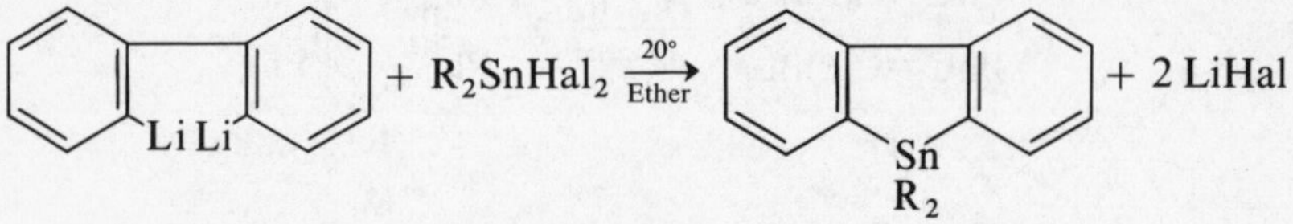

$$\text{(biphenyl-2,2'-diyl)Li Li} + R_2SnHal_2 \xrightarrow[\text{Ether}]{20°} \text{(dibenzostannole)} SnR_2 + 2\,LiHal$$

(R = Et, Bu, cyclo-C_6H_{11}, Ph, p-MeC_6H_4, o-PhC_6H_4)

The structure has been proved by degradative reactions and molecular weight determinations. The analogous reaction with $SnCl_4$ would appear to give the spiro-compound, which is rather more labile than the simple stannoles (231).

The saturated stannacycloheptane system has become known only very recently (like stannacyclopentane, see above), but a benzo-derivative was reported some years ago (473).

1,1-Disubstituted stanna-2,6-cycloheptadienes (stannepines) were obtained as by-products (12—28 %) in the polyaddition of diorganotin dihydrides to 1,5-hexadiynes (619) (see Chapter 21). (For hydrostannation of alkynes, see Chapter 9.)

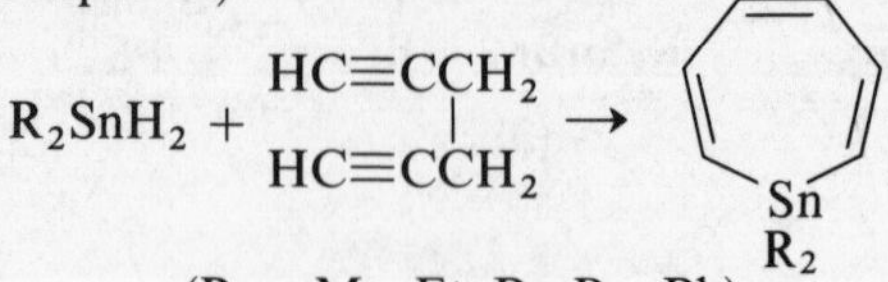

$$R_2SnH_2 + \begin{matrix} HC{\equiv}CCH_2 \\ | \\ HC{\equiv}CCH_2 \end{matrix} \rightarrow \text{(stannepine)}\ SnR_2$$

(R = Me, Et, Pr, Bu, Ph)

The compounds are colorless liquids which can be distilled *in vacuo*; their structure is supported by elemental analysis and IR spectroscopy. A 4,5-benzo-derivative has been obtained similarly (473). Attempts to prepare the corresponding stannacyclodecadienes from 1,8-nonadiyne yielded polymers only (619).

A crystalline dibenzostannepine is formed (besides some polymeric material) by treatment of the lithium dibenzyl derivative with diphenyltin dichloride (430).

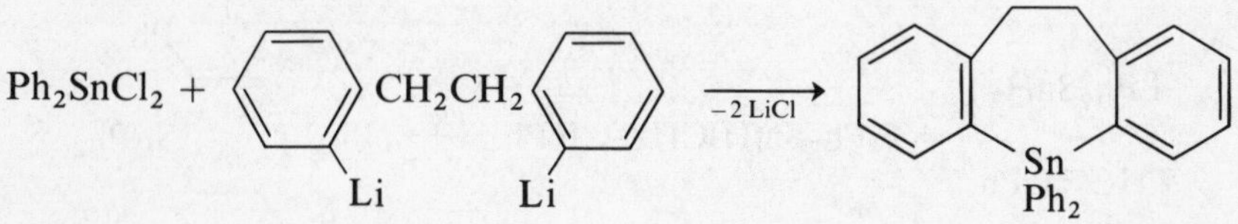

$$Ph_2SnCl_2 + (o\text{-}LiC_6H_4)CH_2CH_2(C_6H_4\text{-}o\text{-}Li) \xrightarrow{-2\,LiCl} \text{dibenzostannepine } (Sn\,Ph_2)$$

This heterocycle exists in two interchangeable modifications of M.P. 136—137° and 146—147°C. The difference may be one of conformation. Derivatives other than the 1,1-diphenyl compound have been prepared also, among them the 1,1-dichloride, the oxide, and the dimethyl compound.

An interesting exchange of the heteroatom has been observed with a dimethyl benzostannepine, in a reaction with phenylboron dichloride (466a) (50% yield of pure product):

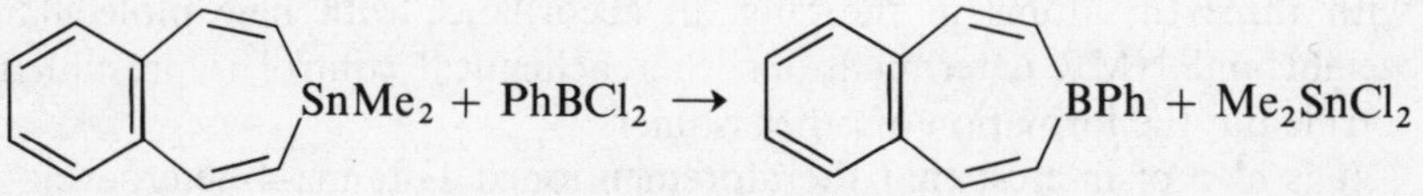

$$\text{benzostannepine-}SnMe_2 + PhBCl_2 \rightarrow \text{benzoborepine-}BPh + Me_2SnCl_2$$

20-2 RING SYSTEMS WITH TWO OR MORE HETEROATOMS

Only a few tin heterocycles of this type are known at present. 1,4-Diheterocyclohexanes are formed by hydrostannation of suitable vinyl compounds (293) (for general discussion of this reaction see Chapter 9):

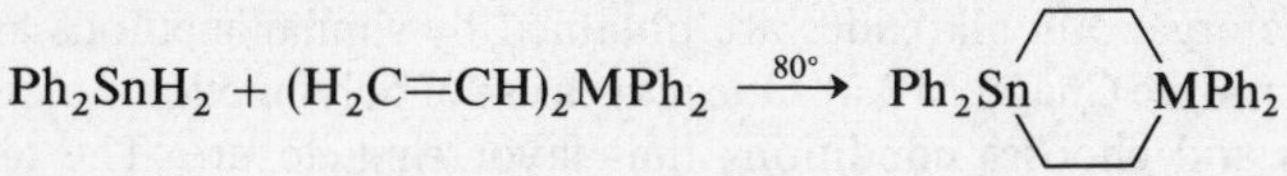

$$Ph_2SnH_2 + (H_2C{=}CH)_2MPh_2 \xrightarrow{80^\circ} Ph_2Sn\langle\rangle MPh_2$$

$$(M = Si, Ge)$$

The Si- and Ge-compounds are both well defined,* crystalline solids of sharp melting point. Yields of pure product are 23% and 17% respectively. These relatively high yields of heterocyclic product may be due to a favorable conformation of the vinyl derivatives caused by the presence of the two phenyl groups. The analogous lead compound is unobtainable

* The structure has been confirmed by NMR spectroscopy (473a).

by this method; lead metal invariably separates. Attempts to prepare the corresponding distannacyclohexane in this way also fail. The intermediate open-chain product, formed by addition of one Sn–H to one vinyl group, either survives as such or reacts further to give insoluble linear polymers. The larger size of the tin atom may possibly encourage free rotation of the vinyl groups and so favor catenation rather than ring-closure. The desired system (**6**) was supposed to be formed almost quantitatively by addition of diphenyltin dihydride to phenylacetylene (293):

$$2\,Ph_2SnH_2 + 2\,HC{\equiv}CPh \rightarrow 2\,Ph_2Sn(H)CH{=}CHPh \xrightarrow{100^\circ} Ph_2Sn\langle CH_2CH(Ph)\rangle_2 SnPh_2 \quad (\mathbf{6})$$

Two isomers were postulated, which could be separated by fractional crystallization from *n*-butanol, and had identical IR spectra, but M.P.s of 70—72 and 144—145°C. However, entirely different results have since been reported (473a). The "lower-melting isomer" was found to be a mixture of fractions softening or melting between 50 and 105°C, with molecular weights of 1200—1800. The higher-melting product does not contain the ring system (**6**) but seems to be $[-Ph_2Sn-C(CH_2Ph)-]_3$ with three tin atoms in the ring, in accordance with new molecular weight and NMR determinations. Stereochemical control is presumed to rule out the formation of other isomers.

It is also of interest that the aforementioned 1-stanna-4-heterocyclohexanes (heteroatom = Si or Ge) form easily isolated 1:1 complexes with cyclic solvents such as benzene, toluene, pyridine, or dioxane (293). Similar complexing has been observed with cyclohexatins (see Chapter 14-4).

Stanna-heterocycles can of course be prepared also by various methods described in other chapters, but none of these appears to have been used extensively. Some examples should suffice.

Cyclic organotin alkoxides are obtained by similar methods as open-chain ones (see Chapter 17-4). One makes use of polyfunctional alcohols or phenols and chooses conditions that favor ring-closure. The tendency

$$Bu_2SnO + \begin{matrix} HO-C\lessdot \\ | \\ HO-C\lessdot \end{matrix} \xrightarrow[-H_2O]{} \begin{matrix} -C-C- \\ O\quad O \\ Sn \\ Bu_2 \end{matrix}$$

$$\text{2,2'-biphenyldiol (OH HO)} + Me_2SnCl_2 \xrightarrow[-2\,NaCl,\ -2\,NH_3]{+2\,NaNH_2/Ether} \text{dibenzodioxastannepine } (O-SnMe_2-O)$$

towards cyclization is again remarkably great. Derivatives of e.g. dioxastannacyclopentane (145a) (including the benzo-compound) and dibenzodioxastannepine (194) have been prepared in this way.

Ethylene glycol similarly gives a dimeric product, a 1,6-distanna-2,5,7, 10-tetraoxacyclodecane.

$$2\,Bu_2SnO + 2\,HOCH_2CH_2OH \xrightarrow[3\text{ hr.},\ -2H_2O]{\text{Boiling xylene}} Bu_2Sn\begin{matrix}OCH_2CH_2O\\ OCH_2CH_2O\end{matrix}SnBu_2$$

86% M.P. 230°

Propylene glycol again forms a monomeric product, apparently because steric conditions here favor the six-membered ring.

$$Et_2SnO + HO(CH_2)_3OH \xrightarrow[3\text{ hr.},\ -H_2O]{\text{Boiling xylene}} Et_2Sn\begin{matrix}O\\ O\end{matrix}(CH_2)_3$$

90% M.P. 169—173°

Other similar cyclic compounds have been synthesized (863a). Dialkyltin alkoxides (splitting off e.g. MeOH) can be used instead of dialkyltin oxides.

These compounds are remarkably stable, both thermally and hydrolytically. Sometimes dimeric heterocyclic products can give the corresponding monomeric species on heating (674a).

Cyclic carboxylates and thio-compounds too are prepared by methods similar to those for open-chain ones, e.g. (885)

$$\begin{matrix}H_2C-SH\\ |\\ H_2C-SH\end{matrix} + Cl_2SnMe_2 \xrightarrow[-2\,Et_3N\cdot HCl]{+2\,Et_3N} \text{(CH}_2\text{)}_2\begin{matrix}S\\ S\end{matrix}SnMe_2$$

Propane-1,3-dithiol reacts analogously to form the six-membered ring. Yields are good. The Si- and Ge-compounds are obtained in the same way (885).

Unsaturated cyclic compounds arise as insoluble solids from reaction of disodium ethylene-1,2-dithiolates with a dialkyltin dichloride in water. They are purified by recrystallization from methanol (R′ = CN) or by sublimation (R′ = H) (2a):

$$(NaS)R'C{=}CR'(SNa) + R_2SnCl_2 \xrightarrow{-2\,NaCl} \text{cyclo-}[R'C{=}CR'\text{–S–}SnR_2\text{–S}] \qquad (R' = H, CN)$$

The monoalkyl thiostannonic anhydrides and their adamantane-like cyclic system should be kept in mind here, too (see Chapter 17-6).

Derivatives of phenostannazine can be obtained by treatment of diorganotin dihalides with dilithiumdiphenylamines (prepared via the dibromides), e.g. (841)

$$RN(C_6H_4Li\text{-}2)_2 + Ph_2SnCl_2 \xrightarrow{-2\,LiCl} RN(C_6H_4)_2SnPh_2$$

Homocyclic systems of tin atoms only were discussed in Chapter 14-4.

21 Macromolecular organotins

Quite a large number of macromolecular organotins are known,* but until now interest has centered on the search for new syntheses rather than on detailed investigation of products. Ascertainable facts are often confined to the bare statement that polymeric materials were obtained. Various questions have now created renewed interest, not least being that of their technological application (see Chapter 25). The literature is therefore extensive and contains a high percentage of patent specifications.

Two main categories can be distinguished: macromolecules containing tin as part of the main chain, and those carrying stannyl groups as substituents on the main chain. Both may be sub-divided further. Preparative methods include those commonly employed in the chemistry of macromolecules, and also some special methods. All of them can be classified under one or other of the two main reaction types leading to polymers: polycondensation and polyaddition.

21-1 POLYMERS WITH TIN ATOMS IN THE MAIN CHAIN†

21-1-1 Polymers with tin inserted at regular intervals in a carbon chain

21-1-1-1 Preparation by Organometallic Methods. The simplest case is that of diorganotin groups connected by methylene bridges. Such compounds were first prepared, with a low degree of polymerization, by C. A. Kraus and W. N. Greer (411).

$$n\,CH_2Cl_2 + n\,R_2SnNa_2 \rightarrow (\text{–}CH_2\text{–}R_2Sn\text{–})_n + 2n\,NaCl\ (n \sim 6)$$

* R. K. Ingham and H. Gilman have reviewed the literature up to early 1962 (314). A comprehensive review up to the beginning of 1967 is given in a Russian monograph by D. A. Kochkin and I. N. Azerbaev (382a).

† Polytins are discussed in Chapter 14-3 and 14-4.

They are liquids which age, and solidify in the process. Nothing is known about the end-groups. Products with R = Bu or Ph, or with substituents on the methylene groups, are resinous, glassy, or rubbery solids (198, 513). Using α,ω-dihalo-hydrocarbons, e.g. $ClCH_2CH_2Cl$, p-ClC_6H_4Cl, p-BrC_6H_4Br, or $Br–(CH_2)_4–Br$ in place of methylene chloride, one obtains resins or elastomers (156, 513). Polymers $(–CH_2–SnCl_2–)_n$ are formed from $SnCl_2$ and diazomethane in benzene (896). An interesting variant of these arises from treatment of styrene with dialkyltin dihalides and lithium (541). Reaction probably proceeds as follows:

$$n\,PhCH{=}CH_2 + n\,ClSn(Me_2)Cl + 2n\,Li \longrightarrow (–CHPh–CH_2–SnMe_2–)_n + 2n\,LiCl$$

It is not clear whether the reaction intermediate is the carbene analog Me_2Sn: or an organolithium compound. Polymers are also obtained from alkylation of tin halides by means of alkylaluminums (see Chapter 3-1-2); e.g. treatment of 1,7-octadiene with di-i-butylaluminum hydride yields the polymeric trialkylaluminum which then affords a linear polymer (749):

$$(–\underset{|}{Al}–C_8H_{16}–\underset{\underset{C_8H_{16}–}{|}}{Al}–C_8H_{16}–)_n + 3n\,Bu_2SnCl_2 + 2n\,R_2O \longrightarrow (–Bu_2Sn–C_8H_{16}–)_{3n} + 2n\,R_2O{\cdot}AlCl_3$$

21-1-1-2 Preparation by Hydrostannation.

(a) Using α,ω-diolefins

Hydrostannation of olefins can be turned into polyaddition by suitable choice of reaction partners (362):

$$n\,H_2C{=}\underset{\underset{R'}{|}}{C}–R''–\underset{\underset{R'}{|}}{C}{=}CH_2 + n\,R_2SnH_2 \longrightarrow (–CH_2–\underset{\underset{R'}{|}}{CH}–R''–\underset{\underset{R'}{|}}{CH}–CH_2–SnR_2–)_n \qquad (21\text{-}1)$$

The reaction can proceed without a catalyst, but is then restricted usually to aromatic organotin hydrides, especially the labile diphenyltin dihydride, or to particularly reactive olefins if aliphatic organotin hydrides are to be used (5, 472, 618, 620, 812). One version starts with the aromatic ditin dihydride (**1**) (472) whose rigid structure inhibits the formation of low molecular weight cyclic by-products.

$$H{-}\underset{CH_3}{\overset{CH_3}{Sn}}{-}C_6H_4{-}\underset{CH_3}{\overset{CH_3}{Sn}}{-}H$$

(1)

Like the simple hydrostannations, these polyadditions can be accelerated by radical initiators such as azobisisobutyronitrile (5, 812). UV light too has been reported as effective (607).

Organometallic catalysis using alkylaluminums (see Chapter 9-1-2) has been found preferable in several cases (550, 576, 750), especially in reactions of the easily accessible aliphatic di- or tri-hydrides with unactivated olefins, e.g. 1,7-octadiene, or compounds of the type

$$H_2C{=}CH{-}(CH_2)_n{-}R_2Sn{-}(CH_2)_n{-}CH{=}CH_2 \quad (n \text{ often} = 1 \text{ or } 2)$$

The alkylaluminum is subjected to a displacement reaction with an excess of diene (928) and diluted with an inert solvent, e.g. cyclohexane. The organotin hydride, also diluted, is then added dropwise at a rate which must not appreciably exceed the reaction rate. The rate can be followed easily by observation of the Sn–H band in the IR spectrum.

Organotin trihydrides (e.g. isobutyltin trihydride) or trienes (e.g. 1,2,4-trivinylcyclohexane) yield network polymers. Thus, reaction of 3 moles of 1,7-octadiene with 2 moles of isobutyltin trihydride (catalyst: i-Bu_2AlH) affords a crystal-clear, insoluble elastomer which swells considerably in benzene (750). Any desired amount of cross-linking can, of course, be attained by admixture of a trifunctional component to a reaction of dihydride and α,ω-diolefin.

Many different products have been obtained by these techniques, as shown in Table 21-1. They include materials with additional Ge or Pb atoms in the main chain.

Some major difficulties remain to be overcome, however. Average molecular weights from linear polyaddition as in reaction 21-1 are often below 10,000 (insofar as they have been measured at all), and catalysts appear to have little effect in this respect. The chain-termination reactions responsible have not been investigated, and there is also no information on the nature of the end-groups. Higher molecular weights are usually found only for products from highly reactive olefins, such as isopropenylbenzene (5) or styrene (618) derivatives. The average molecular weight can be increased by removal of some of the low molecular weight components through some form of after-treatment, e.g. heating *in vacuo* at 140°C, or extraction (472) or fractional precipitation with alcohol from benzene. The possibility of secondary changes in the polymer, especially cross-linking and elimination of certain small fragments, occurring on heating can of course not be excluded. Such changes can sometimes be recognized by the appreciable deviation of elemental analysis results from the theoretical values (472, 749). Introduction of small amounts of a trifunctional component, mentioned above, provides a partial remedy for unduly low molecular weights (750) by compensating through cross-linking for premature chain-termination.

All products reported to-date are transparent and colorless. Depending on the components and on the degree of polymerization, they can be liquid, viscous, rubbery, crumbly, waxy, or glassy (see Table 21-1). Although the tin content may be as high as 50 %, the materials resemble hydrocarbons in many respects, e.g. in their solubility in benzene. Those polymers which have a largely unbranched structure (618) are viscous liquids or amorphous (insofar as they have been investigated) brittle solids of low melting point. Intermolecular forces are thus weak, most likely owing to the large bulk of the organotin groups. Incorporation of amide links in the main chain should intensify these forces, and softening points of 110–120°C are indeed attained in such materials (472). Highly cross-linked products from trifunctional reagents are infusible and practically insoluble, but still capable of considerable swelling.

In the absence of air, many of these polymers are stable up to 250°C. In air, or when impure, they start to decompose slowly above 220°C, phenyltin compounds even below this.

A polymeric material containing phosphorus has been described (607), and also a polymer from divinyl sulfone and diphenyltin dihydride (647). The latter has a higher Sn : S ratio than expected, which may indicate some Sn–Sn bonding. It is not clear whether the hydride in this case reacts with the sulfone group also.

(b) From acetylenes and α,ω-diacetylenes

Generally, addition of organotin hydrides to C≡C groups is far easier than addition to olefinic double bonds (see Chapter 9-4).

With dihydrides, the first reaction results in exothermic formation of a dialkylvinyltin hydride. With continued heating this can be made to undergo polyaddition, as shown by the example of dipropyltin dihydride and phenylacetylene. The following route was the one first suggested (619):

$$n\,\mathrm{PhCH{=}CHSnR_2H} \xrightarrow[24\text{ hr.}]{130^\circ} (\mathrm{-CHPh{-}CH_2SnR_2-})_n$$

$$(\mathrm{R} = \mathrm{Pr};\ n \approx 11)$$

Table 21-1
Linear Polymers with Tin Atoms in the Main Chain, prepared by Hydrostannation of Olefins by Reaction 21-1; Further Examples in the Text

CONSISTENCY OF THE POLYMER	R	R′	R″	CATALYST	$\overline{M}$	REF.
1. Viscous oil	*n*-Bu	H	$-(CH_2)_4-$	*i*-Bu_2AlH	4700—20,000	750
2. Viscous oil	*i*-Bu	H	$-CH_2-(Bu_2Sn)-CH_2-$	*i*-Bu_2AlH	6200	750
3. Solid	Ph	Me	*p*-C_6H_4-*p*	AIBN	2800	5
4. Viscous liquid	Pr	H	*p*-C_6H_4-*p*	—	?	618, 620
5. Hard, crumbly (M = Sn)	Ph	H	*p*-$C_6H_4-(Ph_2M)-C_6H_4$-*p* (M = Ge, Sn, Pb)	—	48,000 (M = Sn)	472, 620
6. Viscous oil	*i*-Bu	H	$-(CH_2)_6-(Bu_2Sn)-(CH_2)_6-$	*i*-Bu_2AlH	2700	749
7. Glass-like	Ph	H	$-CH(O-CH_2)_2C(CH_2-O)_2CH-$	—	19,000	618
8. Viscous oil	Bu	Me	$-COO-(CH_2)_2-OCO-$	AIBN	?	812

It remains to be seen whether the product is contaminated with cyclic compounds of low molecular weight, and whether the stannyl moiety can also be attached to the α-carbon atom (see ref. 472).*

Because of the great difference in reactivity the C≡C groups react quantitatively before there is any measurable attack on the C=C groups. In this way 1,5-hexadiyne yields rubbery or slowly creeping elastomers of which some are soluble and some insoluble in benzene (619):

$$n\,HC{\equiv}CCH_2CH_2C{\equiv}CH + n\,R_2SnH_2 \rightarrow$$

$$(\text{-}CH{=}CHCH_2CH_2CH{=}CHSnR_2\text{-})_n$$

$$(R = Me, Et, Pr, Bu, Ph)$$

As far as it can be measured, n is about 160–180, and the weight average molecular weight about 50,000–75,000.

Some cyclic 1:1 adducts (see Chapter 20) are formed as by-products. These can be distilled off *in vacuo* at 150—250°C or removed by fractional precipitation. The phenyl derivative decomposes in nitrogen above 275°C, but on prolonged heating decomposition sets in as low as 100°C. The other compounds likewise change slowly on heating, as shown by their growing insolubility in benzene. Cross-linking of the highly unsaturated chains appears to be the reason.

Polymers are also formed from 1,8-nonadiyne and from *p*-diethynylbenzene. As far as they are soluble, they show average degrees of polymerization around 130—250. The ditin dihydride (**1**) yields glass-like polymers with phenylacetylene.† With hexadiyne or nonadiyne it gives rubbery or highly viscous products, and with *p*-diethynylbenzene infusible apparently cross-linked products (472).

Reaction of diacetylene with diphenyltin dihydride is incomplete and not particularly uniform; the products are therefore of low molecular weight (458).

21-1-2 Polymeric stannoxanes and stannyl esters

Interest in this group of compounds was awakened quite early by the apparent analogy with the silicones. Hopes of obtaining polymers by hydrolysis of diorganotin and monoorganotin compounds were fulfilled

* Recent NMR studies showed the structural unit $[\text{-}R_2Sn\text{-}CH(CH_2Ph)\text{-}]_n$ to be present (473a).

† Here again two tin atoms are bonded to the same carbon atom: $[\text{-}Me_2Sn\text{-}C_6H_4\text{-}Me_2Sn\text{-}CH(CH_2Ph)\text{-}]_n$ (473a).

only partially:

$$n\,R_2SnCl_2 \xrightarrow{\text{hydrol.}} (-R_2Sn-O-)_n$$

$$n\,RSnCl_3 \xrightarrow{\text{hydrol.}} [-RSn(OH)-O-]_n \xrightarrow[-n/2\,H_2O]{} \left(-RSn \begin{matrix} / \quad\quad O- \\ \backslash \quad / \\ O-RSn \\ \backslash\; O- \end{matrix} \right)_{n/2}$$

(One can also start from alkoxides or carboxylic esters, instead of halides.) n generally remains quite small, e.g. 3, so the intermediate diorganotin oxides and stannonic acids appear to have a great tendency towards ring-formation (see Chapter 17). In the case of the diorganotin oxides the end valencies may be saturated by hydroxy groups [e.g. with R = CH_2Ph and n = 8—9 (807)], or OR groups (503, 715), or OCOR groups (504). Polymeric stannthianes $(-R_2Sn-S-)_n$ are also known (314) (see Chapter 17-6). Compounds of type (**2**) are related to polyesters. They are formed from long-chain dicarboxylic acids, e.g. terephthalic or sebacic acid, and diorganotin oxides (34, 401).

$$(-R_2Sn-OCO-R'-COO-)_n \qquad \textbf{(2)}$$

Maleic acid is here of some importance (681), whereas succinic acid and adipic acid give apparently cyclic condensates of low molecular weight. All these compounds can be classified as stannyl esters (see Chapter 7).

Some further variants are known (314), in which oxygen may be partially or completely replaced by sulfur:

$$n\,R'O-SnR_2-OR' + n\,MeCOO-SnR_2-OCOMe \xrightarrow{-(n-1)MeCOOR'} R'O(-R_2Sn-O-)_nCOMe$$

Polymers are also obtained similarly by heating dialkyltin diacetates with tin tetraethoxide to 140°C (393). Stannosiloxanes, e.g. $(-R_2Sn-O-R'_2Si-O-)_n$, sometimes have remarkably good mechanical and thermal properties. Many types and preparative processes have been described, for which reference should be made to the original papers, e.g. refs 4 and 632, or to a review article (314).

Quite another group of polymers is obtained by addition of organotin halides or esters to ethylene or propylene oxides and subsequent insertion of further epoxide molecules at 160—200°C (178). These are polyethers containing a stannoxane group as part of the main chain; n may vary from 1 to 50, e.g.

$$R_3Sn(OCH_2CH_2)_nCl \quad \text{and} \quad R_2Sn[(OCH_2CH_2)_nCl]_2$$

Polyethers carrying stannyl groups as substituents on the main chain are discussed in Chapter 21-2-2. Partial hydrolysis of dialkyltin diacetates yields at most tetramers (with acetoxy end-groups) (918), but heating of dibutyltin diacetate with dibutyltin dibutoxide has given polymers with molecular weights as high as 4200 (921).

21-2 POLYMERS WITH STANNYL SUBSTITUENTS

Stannylation of polymers (subsequent to polymerization) appears hardly to have been attempted. Introduction of silyl groups into polystyrene, by reaction of halosilanes with poly-(*p*-styrenyllithium), has been accomplished, but the reaction fails with stannyl groups apparently for steric reasons (92). The alternative method, stannylation of monomers before polymerization, has been adopted frequently.

21-2-1 Polymers from olefinic organotin compounds

Large numbers of unsaturated organotins seem to have been prepared in the last few years for the sole purpose of testing their polymerizability (92, 314, 613).

The simplest compounds of this type, vinyltins and allyltins (see Chapter 3-3), appear to be incapable of radical polymerization (391, 395, 397). Slight oligomerization of trimethylvinyltin ($n \approx 3$) has been achieved at 6000 atm. with *t*-butyl peroxide as catalyst (399). Trimethylallyltin resists polymerization, but can be copolymerized with methyl methacrylate. In other cases, radical polymerization of methyl methacrylate, vinyl acetate, and styrene is inhibited strongly by organotins, the effect increasing with the number of allyl or vinyl groups on the tin. By far the most powerful inhibitor is tetraallyltin (395, 397). The effect is understood readily in terms of ease of elimination of stannyl groups from vinyl and particularly allyl compounds (see Chapters 2 and 4-4). In these radical polymerization attempts, the C–Sn bonds thus act as scavengers and prevent further chain propagation. It follows that ionic polymerization should be possible. This has been proved for trimethylallyltin and for its copolymerization with α-olefins (propene) (535), using composite organometallic catalysts (Ziegler catalysts) such as $Et_3Al + TiCl_4$. Other examples of this type are given in the patent.

2-Tributylstannylbutadiene (**3**) (48) occupies a rather special position. It is both a vinyltin and an allyltin compound, but nevertheless is reported to be polymerized smoothly by means of azo-compounds to yield the stannyl-substituted *cis*-1,4-polybutadiene (**4**).

$$n\,H_2C{=}\underset{(\mathbf{3})}{\overset{SnBu_3}{\overset{|}{C}}}{-}CH{=}CH_2 \xrightarrow{\text{Cat.}} \underset{(\mathbf{4})}{\left[-H_2C-\overset{SnBu_3}{\overset{|}{C}}{=}CH-CH_2-\right]_n} \xrightarrow[-n\,Bu_3SnCl]{n\,Cl_2} \underset{(\mathbf{5})}{\left[-H_2C-\overset{Cl}{\overset{|}{C}}{=}CH-CH_2-\right]_n}$$

This can then be converted with chlorine into poly-*cis*-1,4-chloroprene (**5**) (49). The stereospecificity of the reaction is considered by the authors to be due, in part, to the bulky stannyl groups.

Radical polymerization of other stannylated olefins is usually just as easy as that of unsubstituted olefins. Organotin derivatives of styrene have been studied repeatedly. The earliest of the relevant publications (51) (1951) concerns *p*-triethylstannyl-α-methylstyrene, which was later found to polymerize most easily with azobisisobutyronitrile (400). The synthetic scope was enlarged considerably by the introduction of the easily accessible *p*-vinylphenylmagnesium chloride (495) and the corresponding lithium derivative. Many different stannyl-substituted styrenes have now been made by Grignard or modified Grignard syntheses. These include *p*-$H_2C{=}CHC_6H_4SnR_3$ with R = Me, Et (388), Ph (459, 613, 315), C_6H_{11} (390), and (*p*-$H_2C{=}CHC_6H_4)_2SnR_2$ with R = Ph, also *p*-$H_2C{=}C(Me)C_6H_4SnMe_3$ (613), and (*p*-$H_2C{=}CHC_6H_4)_4Sn$ (181). 4-Vinyl-4′-triphenylstannylbiphenyl (390) has been prepared and polymerized as well as copolymerized with other olefins. Thermal bulk polymerization is possible at 140°C, but radical polymerization initiated by azobisisobutyronitrile at 70—90°C is preferable (390, 613).

Comparison of compounds *p*-$H_2C{=}CHC_6H_4MMe_3$ with M = Pb, Si, C, Ge, Sn shows that the rate of polymerization decreases in that order (613). *p*-$H_2C{=}CHC_6H_4SnR_3$ undergoes radical polymerization at 80°C, with the rate practically unaffected by the nature of R (390). Other workers found the triethyl derivative difficult to polymerize, admittedly under different conditions (388). $TiCl_4$ proved comparatively the best catalyst, but results were disappointing even here.

The polymerized products are generally transparent solids, described variously as glasses or translucent films. More precise data are usually lacking.

The triphenylstannyl ester of *p*-vinylbenzoic acid occupies an intermediate position between these styrene derivatives and the stannyl esters to be discussed in the next section. It can be polymerized thermally at 100—150°C, or rapidly with azobisisobutyronitrile at 100°C, to yield colorless to yellow insoluble products (394).

Stannyl esters of acrylic acid and related compounds have been studied very thoroughly. They are mostly obtained smoothly from the free acid and triorganotin hydroxide or oxide (see Chapter 7). Derivatives of acrylic acid $H_2C{=}CHCOOSnR_3$ with R = Me, Et, Bu, Ph (314, 536) and $H_2C{=}CHCOOSnR_2OCOCH{=}CH_2$ (314), as well as analogous derivatives of methacrylic acid (314, 395, 536) and cinnamic acid $PhCH{=}CH\text{-}COOSnR_3$ (789), have been prepared in this way. They could all be polymerized smoothly in bulk, in solution, or in suspension, either thermally or with the aid of radical initiators. Copolymerizations with methyl methacrylate and methyl ω-hydroxyenanthate (389), styrene, acrylonitrile, cyclopentadiene, and divinylbenzene have been described, and the properties of the products studied in detail. In every case, polymerization involved typical vinyl chain propagation without participation by the stannyl substituents. Only in one case was a low molecular weight tetraorganotin eliminated, with attendant cross-linking of the remaining material (389). Otherwise, the higher alkyl groups tended to give elastomers, while ethyl or methyl groups yielded glass-like polymers. Some of the materials show intense absorption of X-rays. Polymeric tributylstannyl methacrylate is a creeping, rubbery substance. Copolymerization with dibutyltin dimethacrylate on the other hand affords a transparent highly elastic material that may then be vulcanized with dibutyltin oxide, sulfur, and dibutyltin dihydride (92, 314).

21-2-2 Other organotin polymers

Organotin-substituted polyethers have been described. They were obtained from tin-containing epoxides made either by epoxidation of unsaturated organotins or by stannylation of epoxides. The tin may also be bonded through oxygen or sulfur (499) (polyethers with tin in the main chain were discussed in Chapter 21-1):

$$2\,\mathrm{EtCH{-}CH(CH_2)_4OH}\ (\text{epoxide, }\backslash O/) + (\mathrm{MeO})_2\mathrm{SnBu_2} \xrightarrow{-2\,\mathrm{MeOH}} \left[\mathrm{EtCH{-}CH(CH_2)_4O}\ (\text{epoxide, }\backslash O/)\right]_2 \mathrm{SnBu_2}$$

Further possibilities of this system appear to have remained unnoticed.

Organotin-substituted silicones are obtained by hydrolysis of suitable monomers, e.g. $Me_3SnCH_2Si(Me)Cl_2$. The place of the halogen can also be taken by –OR or –OCOR (770).

22 Analysis of organotin compounds

Proof of purity and identity of organotin compounds can often be supplied on the basis of classical and, usually, rapidly determined criteria. Because of the good thermal stability of most organotins this is true even for melting and boiling points. For liquids, use of the refractive index is strongly recommended. There is a large collection of data in the tables of this book, with further data in the references. Molecular weight can generally be determined cryoscopically. Difficulties have been encountered only with ditins (315, 416, 583) and with a few aryltins, probably because of formation of solid solutions (564). Thermistor-type vapor-pressure osmometers have proved extremely useful here and in many other instances, including molecular weights above 1000 (564, 583). The isothermal distillation method is useful in certain cases.

Determinations have sometimes, e.g. with organotin hydrides, to be carried out in an inert atmosphere. This creates few difficulties since the necessary equipment is available; some of it was first designed for organoaluminum chemistry (402, 928).

22-1 SEPARATION TECHNIQUES

There is no need here for discussion of customary methods such as fractional distillation or crystallization. These can usually be applied without modification in organotin chemistry.

22-1-1 Paper chromatography

Mixtures of different organotin halides, both aromatic and aliphatic, are often investigated by paper chromatography owing to the simplicity of equipment, ease of operation, and saving of time. Apart from organotin halides, the method can also be applied to compounds which can be converted into halides in the chromatographic solvent, e.g. stannoxanes, hydroxides, stannyl esters, alkoxides, and amines, and even some tetra-organotins. Mixtures of these with tetraorganotins have also been separated. Reverse-phase techniques are usually adopted now, the paper being impregnated with dinonyl phthalate, tritolyl phosphate, 2-phenoxy-ethanol (890), olive oil (604, 701), or α-bromonaphthalene (604). Mobile phases which have proved useful include trimethylpentane–acetic acid or (more frequently) methanol–hydrochloric acid (604, 701, 890). The bands are made visible usually by spraying with a solution of Catechol Violet (604, 890) (see Chapter 22-3).

Tetraorganotins are first cleaved on the chromatogram, before spraying, by UV irradiation (890). Similarly, aryl compounds are first sprayed with aqueous $HgCl_2$ to make them receptive to the dyestuff.

Mixtures of the most diverse species have been analyzed successfully by this technique (128, 209, 229, 604, 701, 890), but quantitative evaluation presents difficulties.

22-1-2 Thin-layer chromatography

Tetra-, tri-, di-, and mono-organotin compounds, even in traces, can be separated and detected satisfactorily on silica gel (113). Similar separations have been achieved with mixtures of different diorganotin compounds (855). Isopropyl ether–1.5% glacial acetic acid is an excellent eluent for routine analysis (546a). This type of chromatography is especially useful in the case of non-volatile or thermally unstable organotin compounds. It has therefore found widespread application.

22-1-3 Gas–liquid chromatography

This is certainly the most versatile method of separation; it is also capable of considerable development. Organotin compounds are thermally sufficiently stable to make decomposition (or comproportionation or disproportionation) on the column a rare occurrence. Good separations have been achieved with mixtures of organotin halides and tetraorganotins (230, 556), aliphatic organotin hydrides and ditins (583, 602), mixtures of tetramethyltin and trimethyltin hydride (664), and with various products of hydrostannation of unsaturated compounds. (For examples see refs. 6a, 206a, 439a, 465a, 468, 555, 555a, 555b, 592, 664, and 812a.)

Apart from special column packings (reviewed in ref. 664), packings of silicone oil, or silicone or Apiezon grease, are widely used. Hydrogen is used as carrier gas in most cases, but helium or nitrogen are used also. The most common detector is the thermal conductivity cell, and also the FID despite trouble due to deposition of SnO_2 in the latter. In many cases a high splitting ratio of the carrier gas stream is necessary.

Rapid and accurate assessment of many reactions, in respect of composition and uniformity of products and reaction mechanism, has made this a highly fruitful technique in many areas of organotin chemistry. As examples one might quote the comproportionation of tetraorganotins (665), and the isomerism and stereochemistry of the addition of organotin hydrides to olefins (6a, 471, 555b, 602) and alkynes (465a, 468) (see Chapter 9 on hydrostannation). The number of successful gas-chromatographic separations is increasing rapidly.

Separations can be achieved on a preparative scale, even with rather sensitive compounds. For example, diethyldimethyltin has been purified on a column of polyethylene glycol 400 on kieselguhr (1 in. × ca. 40 ft.) (304). All the isomers resulting from hydrostannation of 1,3-pentadiene (piperylene), including *cis–trans*-isomers, have been separated in capillary columns (FID, polypropylene glycol, 0.1 in. × ca. 165 ft.), and most of them also on a preparative scale. In the case of Me_3SnH adducts a column with polypropylene glycol ($\frac{3}{8}$ in. × 20 ft.) proved suitable, and in the case of Et_3SnH adducts a column with Reoplex 400 ($\frac{3}{8}$ in. × 20 ft.); Chromosorb A with hydrogen as carrier gas was used in both cases (6a, 555b).

22-2 ELEMENTAL ANALYSIS

C and H analysis of organotin compounds has often presented difficulties. However, tubes charged with copper oxide–lead chromate and, in addition, a platinum gauze cylinder (796), have given satisfactory results (491), as has a silver vanadate catalyst (144). The following equipment is in use in the author's laboratory, where it has been tested extensively and produces good results in routine analysis of tin compounds of the most diverse nature. There is a standard semi-micro combustion train for C,H,O determination, type U/E (Heraeus, Hanau), charged with the normal combustion catalyst and, in addition, CeO_2 (Auer-Remy, Hamburg) and $PbCrO_4$ (granulated, analytical grade; E. Merck, Darmstadt). Procedure and details correspond to the instruction manual for the apparatus. Sample weights are usually about 5—15 mg.

Halogen bonded to tin can be determined simply by potentiometric titration with $AgNO_3$ after alkaline hydrolysis. For halogen bonded to carbon, and for determination of sulfur, decomposition of the sample in a Wurzschmitt micro-bomb can be recommended.

For tin, macro-scale oxidation of the organotin and weighing as SnO_2 is still the best method. It was introduced in 1910 by P. Pfeiffer and improved in 1928 by K. A. Kocheskov; it has been modified several times since (315). The method is not particularly precise, but it is simple and rapid. Several thousand determinations have been carried out quite satisfactorily over a number of years in the author's laboratory. Silica test-tubes of 20 mm. diameter are used, and samples of 150—200 mg. are oxidized with a mixture of equal parts of conc. HNO_3 and conc. H_2SO_4. If oxidation is incomplete, a further portion of acid mixture is added. Highly volatile organotins are first treated dropwise at −15°C with elemental bromine which is then carefully evaporated. Other methods are in use for special cases (315). Alternatively, the sample may first be burnt in a Wurzschmitt bomb and the tin then determined fairly accurately by colorimetry with dithiol (202).

For industrial use, particularly for detection of very small amounts of tin, a flame-photometric method has been developed which utilizes chemiluminescence in the flame and measures it at 2706 Å (747). The method has proved highly satisfactory in routine analysis. There is also a microanalytical method which makes use of the mineralization technique of Schöniger (702). Small amounts of tin can be determined in aqueous solution (after suitable opening out) by means of atomic absorption spectroscopy. The limit of detection is about 0.1 p.p.m. (127a). Some of the functional-group determinations described in the following section provide further methods for estimating small and very small amounts of tin.

22-3 DETERMINATION OF FUNCTIONAL GROUPS

Purity tests and structure determinations, as well as analyses of technical products, frequently depend on functional-group determination. These can elucidate e.g. the nature of the stannyl groups present (SnR_3, SnR_2, or SnR), their ratio in a mixture, or the presence of Sn–Sn or Sn–H groups. The methods available are rather few, but several rapid, accurate, and specific procedures have been developed in the last few years. (Spectroscopic methods are discussed in Chapter 23.)

Mono- and di-organotins form a blue complex with Catechol Violet in alcoholic solution. The complex is destroyed at once by ethylenediaminetetraacetate solution, the color changing from blue to yellow, and can thus be titrated very accurately (111). At pH 1.3—2.1, the titration determines only monoorganotin groups Sn–R (190b). Dibutyltin groups can be assayed colorimetrically as the diphenylcarbazone complex (801). Mono- and tri-alkyl groups do not interfere, and neither does tetrabutyltin. Dithizone may be used to distinguish between dibutyl- and dioctyl-tin groups (132). A potentiometric titration of the sum of SnR_3 and SnR_2

compounds with alkali has been described, as has a selective amperometric estimation of SnR_2 compounds by means of quinolinol (663a). Mixtures of aliphatic triorganotins (R up to C_4) are best analyzed by gas chromatography (Chapter 22-1-3). Triphenyltin groups can be determined polarographically (539), but this may sometimes require preliminary enrichment by two-phase partition (84). With the aid of a chloroform–tartrate buffer the method can achieve quantitative separation from diphenyltin groups (83). This last method has also proved useful in functional-group analysis of aromatic linear and cyclic polytins (564).

Determination of Sn–H groups has been used to follow reaction rates, as first described in refs. 578 and 597. This can be done by IR spectroscopy, but often also rapidly and with little trouble by acidimetric titration. The following is a suitable procedure (584): the sample is introduced (under an inert atmosphere) into one limb of a Zerewitinoff Y-tube, and mixed carefully with an excess of dichloroacetic acid from the other limb. The hydrogen evolved is collected in a gas burette, when 22.4 ml. indicates 1 milliequivalent of Sn–H. A colorimetric determination of organotin hydrides makes use of isatin or ninhydrin (211a).

Sn–Sn groups are titrated easily at 20°C with a standard solution of iodine in benzene (564). Sn–C groups (with the exception of some mentioned in the next paragraph) are usually not affected, but Sn–N, Sn–O, and Sn–H are. The method is important for the analysis of polytins. With aliphatic organotins, it is often helpful to start by degrading any Sn–Sn groups present with chlorine in CH_2Cl_2 at -70°C (579). The excess of chlorine must be destroyed (still at -70°C) by passing in propylene, since it would otherwise attack Sn–C bonds also (above -40°C). Cleavage by halogen may be followed by other group determinations described above.

Sn–C bonds in Sn–C≡C, Sn–CH_2Ph, and Sn–C–C=C groups (596) can be titrated at 20°C with iodine in benzene just like Sn–Sn bonds. Sn–C=C groups do not interfere.

Mixtures of stannoxanes and triorganotin hydroxides may be titrated with Karl Fischer reagent (451).

23 Spectroscopic investigation

23-1 ULTRAVIOLET SPECTROSCOPY

Tetraethyltin shows only one band, of average intensity, at about 220 nm. In the corresponding ditin this band is broader and more intense. The spectrum of the aliphatic-substituted cyclononatin is remarkable: the maximum is still at about 220 nm., but the intensity has increased enormously, most likely because of the eighteen Sn–C bonds per molecule (579). The band extends just into the visible, which accounts for the pale yellow color. The effect is undoubtedly due to delocalized electrons in the ring of tin atoms. Tetraaryltins absorb at about 230 nm., i.e. not far removed from their aliphatic analogs. The molar extinction coefficient ε is however much larger (see Table 23-1). Substituents (Me, Cl) in the *para*-position have a negligible effect (267). Hexaphenylditin absorbs less strongly in this region, but shows a new band at 247 nm. which has been assigned to the Sn–Sn group (182). There are also phenyl bands at 245—270 nm. which are common to nearly all phenyltin compounds, e.g. Ph_3SnH. The corresponding tritin has a further band at 275 nm. (564). The colorless dodecaphenylcyclohexatin is distinguished by a relatively broad band at 280 nm. but it tails off sharply (564) in contrast to that of the aliphatic cyclononatin. It can be assumed that ring mesomerism is inhibited by the twelve electron-withdrawing Ph groups. A similar band (at 277 nm.) is found in the case of the tetrahedral $(Ph_3Sn)_4Sn$, and is assigned to the Sn–Sn groups. The related Ge and Pb compounds show similar bands (182).

Table 23-1 summarizes some of the data. The material available for comparison is still rather fragmentary. Even so, investigation and assignment of fine-structure has been attempted for some aromatic organotins

Table 23-1
Ultraviolet Absorptions of Some Organotins

COMPOUND	λ_{max}(nm.)	ε	SOLVENT[a]	REF.
Et_4Sn	219	1530	C_6H_{12}	579
$Et_3SnSnEt_3$	222	1970	C_6H_{12}	579
$Bu_3SnSnBu_3$	<215		C_6H_{12}	182
$(Et_2Sn)_9$	217	~100,000	C_6H_{12}	579
Et_3SnOPh	~230	10,000	C_6H_{12}	267
	278	1882		
Ph_4Sn	228	77,000	C_6H_{12}	564
	245—270	~15,000		
$(p\text{-}MeC_6H_4)_4Sn$	227	~70,000	C_6H_{12}	267
	255—265	~1740		
$(p\text{-}ClC_6H_4)_4Sn$	~235	~100,000	C_6H_{12}	267
	260—270	~1890		
$Ph_3SnSnPh_3$	222	57,000	C_6H_{12}	273, 564
	247	34,000		
	245—270	weak		
Ph_8Sn_3	220	~100,000	C_6H_{12}	564
	247	52,000		
	275	49,000		
$(Ph_2Sn)_6$	218	~100,000	C_6H_{12}	564
	280	~42,000		
	245—270	~55,000		
$(Ph_3Sn)_4Ge$	276	73,000	$CHCl_3$	182
$(Ph_3Sn)_4Sn$	277	79,000	$CHCl_3$	182
$(Ph_3Sn)_4Pb$	298	595,000	$CHCl_3$	182

[a] Cyclohexane or chloroform.

(267), leading to the conclusion that the Sn–aryl bond has partial double-bond character (see Chapter 2-1).

23-2 INFRARED AND RAMAN SPECTROSCOPY

The use of infrared spectroscopy as an aid to preparative organotin chemistry goes back many years. In this context it is a very important spectroscopic method. Numerous data relating to specific IR bands and their use for purity control, or for tracing the course of a reaction, are found scattered through the literature. Where appropriate, these are also referred to in many places in this book.

There have been several systematic studies and assignments. Only the groups working with H. Kriegsmann (424) and R. A. Cummins (168) will be mentioned here. The latter has concerned itself particularly with the

Table 23-2

Infrared and Raman Absorptions of Organotins[a]

BOND: IN	IR BANDS[b] ($cm.^{-1}$)	RAMAN BANDS ($cm.^{-1}$)	ASSIGNMENT
Sn–C:			
SnMe	1180—1210 (100)	1183 (98)	C–H deformation
	770—790 (100)	764 (190)	rock
	520—570 (421a, 795)	524 (190)	asym. stretch of $SnMe_3$
	506—525 (421a, 795)	504—512 (98, 190)	sym. stretch of $SnMe_3$
	~152	126 (98)	asym. skeletal bend deformation
	~100		sym. skeletal bend
SnEt	1180—1190 (478)		C–H deformation
	660—735 vs		rock
	496—535 s-vs (300)	506 (300)	asym. stretch
	475—495 m-s		sym. stretch
	~132		asym. skeletal bend
	~86		sym. skeletal bend
SnPr	~1185 (116, 168)		C–H deformation
	580—600	588 (163)	asym. stretch
	500—510	500 (163)	sym. stretch
SnBu	585—605 (116, 168, 526a)	582 (163)	asym. stretch
	500—515	500 (163)	sym. stretch

SnPh	1060—1075 s (292)		C–H deformation (aromatic)
	263—283 (670)		asym. stretch
	226—263 (670)		sym. stretch
$SnCH{=}CH_2$	930—960 s (292)		C–H deformation
$SnCH{=}CHPh$	~985		
$Et_3SnCH_2CR{=}CH_2$	870—880 (812)		
$Ph_3SnCH_2CR{=}CH_2$	894 (812)		C–H deformation
$SnC{\equiv}C$	1210 (560)		
	890		
	555		$SnC{\equiv}C$ asym. stretch
	2135—2145		$C{\equiv}C$
Sn–H:			
R_3SnH	1777—1840 vs (346)	1808—1844 (424)	stretch
	506—570 s (424)	506—561 (424)	deformation
$R_2Sn(X)H$ (X = F,Cl,Br,I)	1830—1877 vs		stretch
R_2SnH_2	1820—1858 vs		stretch
$RSnH_3$	1855—1880 vs (346)		stretch
SnH_4	1860, 1910		stretch
Sn–D:			
R_3SnD	1300—1305 (594)		
Ph_3SnD	~1325 (594)		stretch
Sn–O:			
$(R_3Sn)_2O$	737—790 vs (421a, 478)		asym. stretch
	395—415 m (421a)	411	sym. stretch
$(R_2SnO)_n$	563—580 vs (478)		asym. stretch
	400—430		sym. stretch

Table 23-2—continued

BOND: IN	IR BANDS[b] (cm.$^{-1}$)	RAMAN BANDS (cm.$^{-1}$)	ASSIGNMENT
Sn–O:			
OSnO, $R_2Sn(OR')_2$	~600 (526a)		asym. stretch
	~480		sym. stretch
SnOO	~550		stretch
SnOSi	960—980		asym. stretch
SnOH	885—910 s (222, 478)		deformation
SnOD	~662 (222)		deformation
R_3SnOR'	470—520 s (168, 526a)		stretch
SnOC	960—1100 (526a)		stretch
Sn–S:			
SnSC	355—368		
SnSGe	355 (755)		
SnSPb	365 (755)		
SnSSn	366—386 vs (478)	366—386 (421)	asym. stretch
	315—330 s-vs	319—320 (421)	sym. stretch
Sn–Se:			
SnSeSn	~268		asym. stretch
	~224		sym. stretch
Sn–N:			
SnNSn	712—728 (798)		asym. stretch
SnNSi	~879		asym. stretch
R_3SnNR_2	465—530		stretch
	780, 872		
$(R_3Sn)_3N$	712—728		asym. stretch
	<400		sym. stretch

Sn–F:			
R_3SnF	328—350 (96)		stretch
	372 (670)		stretch
R_2SnF_2	550—580 s (478)		asym. stretch
	440—460		sym. stretch
	340—370 vs (478)		
Sn–Cl:			
R_3SnCl	315—335 s (478)	315—331 (419)	stretch
	105—150		deformation
R_2SnCl_2	325—355 s (478)		stretch
$RSnCl_3$	352—385 vs (163, 478, 836)	358, 372—376	stretch
($SnCl_4$	368, 403 (117)	366	stretch)
Sn–Br:			
R_3SnBr	215—225 m (478)	231 (419)	stretch
R_2SnBr_2	240—255 s (24, 478)		stretch
$RSnBr_3$	255—260 s (478)	233, 260 (163)	stretch
($SnBr_4$	220, 280 (117)		stretch)
Sn–I:			
R_3SnI	~175	169—177 (419)	stretch
R_2SnI_2	180—205 (24)		stretch
$RSnI_3$	200—210		stretch
(SnI_4	149, 216 (117)		stretch)
Sn–Sn:			
Me_6Sn_2		190 (98)	

[a] Characteristic bands of substituents (e.g. –OH, –COOR), which are not affected appreciably by the stannyl group, have been omitted.
[b] m medium, s strong, vs very strong.

effect of different C–C bond conformations in alkyl groups of three or more carbons attached to tin (see also Chapter 2-1). The stretching band at 600 cm.$^{-1}$, which is intensified at low temperature or in the solid, is assigned to the *trans*-conformation, and that at 500 cm.$^{-1}$, which becomes weaker under these conditions, to the *gauche*-conformation (see Table 23-2). Similar assignments have been made for the Sn–CH_2 bands at 670 and 700 cm.$^{-1}$ (168).

The characteristic Sn–H band at about 1800 cm.$^{-1}$ is used quite often in the laboratory (476, 551, 553). Its position depends on the number of alkyl groups attached to the tin. ν(Sn–H) also increases with increasing electronegativity of the substituents (424, 591), e.g. in the following series (in 10—30% solution in cyclohexane) (591):

	t-Bu_3SnH	*s*-Bu_3SnH	*i*-Bu_3SnH	*n*-Bu_3SnH	Ph_3SnH	
ν(Sn–H)	1777	1791	1808	1812	1841	cm.$^{-1}$

Polar solvents lower the frequency (424). Data for individual compounds will be found in Table 8-1.

Several other absorption frequencies have been measured with similar precision, and assignments derived (summaries in refs. 165, 478, and 507). The data are summarized in Table 23-2. Unless indicated otherwise by reference, the values are taken from measurements in the author's laboratory or from a table given by D. H. Lohmann (477).

Raman spectroscopy is used especially for confirmation of assignments. Data are included in Table 23-2, with references. Further information will be found in refs. 98, 190, and 723.

23-3 NUCLEAR MAGNETIC RESONANCE SPECTROSCOPY

23-3-1 Proton magnetic resonance spectroscopy

The number of laboratories engaged in NMR studies of organotin compounds is increasing rapidly. The volume of published data is considerable and growing fast (e.g. 71, 135, 136, 179, 225, 346, 442, 443, 480, 507—509, 675, 739, 755, 833, 860a, and 861). A comprehensive review (508) deals with the entire field of NMR spectroscopy of organometallic compounds up to late 1964. It lists 420 references (a few as late as 1965) and contains tabulated data for 54 organotin compounds. This review is here supplemented by reference to later papers.

Three naturally occurring tin nuclides have magnetic moments: ^{115}Sn, ^{117}Sn, and ^{119}Sn, all of them with nuclear spin $I = \frac{1}{2}$. Because of their higher relative abundance, ^{117}Sn and ^{119}Sn are the important ones (^{115}Sn 0.35%, ^{117}Sn 7.67%, ^{119}Sn 8.68%). The spectra are often complicated, since both isotopes couple with α-CH and β-CH as well as with directly bonded H, and since, furthermore, the substituents themselves can couple via the tin. Investigation has therefore been largely confined to methyltin compounds.

Tables 23-3—23-6 give τ-values (p.p.m., $Me_4Si = 10$) and coupling constants J in Hz for some important organotins. For most of these, J(Sn–CH) is given in the form $J(^{117}Sn)/J(^{119}Sn)$, the value for ^{119}Sn being always greater than that for ^{117}Sn.

As a general rule, τ($Sn–CH_3$) decreases with increasing electronegativity of the other substituents, and with their number. Coupling constants usually differ more than τ-values. The increase in $J(Sn–CH_3)$ in the series

$$Me_4Sn < Me_3SnCl < Me_2SnCl_2 < MeSnCl_3 \quad \text{(173a, 195a)}$$

(see Table 23-3), and in a number of other series, has been ascribed to increasing *s*-character of the bonding. Thus, salts of the *sp*-hybridized Me_2Sn^{2+} ion (see Table 23-5) show particularly large coupling constants (approx. 100). Changes in the coordination sphere are also reflected in J(Sn–CH): the values for Me_3SnCl are 58 in CCl_4, 67 in pyridine (complex with penta-coordinate tin!), and 70 in tetramethylene sulfoxide (796a). The dependence of τ(Sn–CH) on the bond character is given as (71a):

$$sp^3 < sp^2 < sp \qquad sp^3 < sp^3d < sp^3d^2$$

Changes in J(Sn–CH) can be used to explain the formation of donor–acceptor complexes (507) and to interpret the bonding in complexes (345a).

Table 23-3 can be used to compare the coupling constants J(Sn–CH) for different substituents. Comparison shows that

$$J(Sn–CHHal_2) < J(Sn–CH_2Hal) < J(Sn–CH_3) < J(Sn–CH_2Ph)$$

$$\approx J\left(Sn–CH{=}C\begin{smallmatrix}/\\ \backslash\end{smallmatrix}\right) < J(Sn–CH_2Me)$$

The weak coupling in halomethyl compounds is noteworthy (534b). In vinyltins, *trans*-J(Sn–CH) coupling is appreciably greater than *cis*-coupling or geminal coupling (see Table 23-3). Increasing substitution of the vinyl hydrogens lowers J(Sn–CH)(*trans*) from 180 to about 80, and J(Sn–CH)(*cis*) from about 90 to about 40. The value of J(Sn–CH)(geminal), approx. 90, is taken as evidence of $d\pi$–$p\pi$ interaction (74b).

An anomalous relationship between β-CH coupling and α-CH coupling is found with organotins just as with some other organometallic compounds. The fact that J(Sn–β-CH) (about 70) is larger than J(Sn–α-CH)

Table 23-3
NMR Data for Some Organotins and Organotin Halides

FORMULA	τ	$J^{117/119}$	REFERENCES
Me_3SnR	τ(Me)	$J(^{117/119}Sn–Me)$	91a, 508, 798a
Me_3SnMe	9.93—9.96	51.5—52.0/54.3—54.4	
Me_3SnEt	9.97—9.95	49.9—50.2/52.2—52.6	
Me_3SnPr	9.96—9.93	50.0—50.2/52.3—52.5	
Me_3SnBu	9.97—9.92	49.9—50.1/52.2—52.5	
Me_3Sn*t*-Bu	9.99—9.97	47.5—47.9/49.7—50.2	
Me_3SnCH_2Ph	9.98	50.8/52.9	
Me_3SnPh	9.73	52.2/54.6	
Me_3Sn-1-Np	9.60	51.6/54.1	
Et_3SnEt	Me 8.8	Sn–Me 68.1/71.2[a]	195a, 508
	CH_2 9.2	$Sn–CH_2$ 65.7/69.2[a]	798a
Pr_3SnPr	—	$Sn–CH_\alpha$ 49.1(?)	195a, 508
		$Sn–CH_\beta$ 67.2	
$Sn(CH_2CHMe_2)_4$	CH_2 9.21	—	715a
	CH 9.06		
	Me 8.08		
⪞$SnCH_2CH_2CH_2Me$	α-CH_2 9.20	—	715a
	β- and γ-CH_2 8.2—8.9		
	Me 9.09		

FORMULA	τ	$J(^{117/119}Sn–Me)$	$J(^{117/119}Sn–Et)$[b]	REF.
Me_4Sn		52.0/54.3	—	798a
Me_3SnEt		50.5/52.8	74.4/77.9	
Me_3SnBu		50.4/52.6	—	
$MeSnEt_3$		47.1/49.3	68.6/72.0	
Et_4Sn		—	65.7/69.2	
$PhSnMe_3$		52.9/55.2	—	
$PhSnEt_3$		—	70.5/74.0	
Ph_3SnEt		—	79.7/83.2	
			$Sn–CH_2Ph$	
$(PhCH_2)_4Sn$		—	56.8/59.0	798a
$(PhCH_2)_3SnMe$		48.7/50.9	57.9/60.2	
$(PhCH_2)_2SnMe_2$		49.9/52.2	59.2/61.5	
$PhCH_2SnMe_3$		51.0/53.4	60.4/62.7	

[a] Different coupling constants 49.1/51.4 and 30.8/32.2 are reported in ref. 508.
[b] This refers to α-CH_2–Sn coupling.

Table 23-3—continued

FORMULA	τ		$J(^{119}Sn\text{–}Me)$	$J(^{119}Sn\text{–}CH)$	REF.
	Me	CH			
Me_3SnCCl_3	9.50	—	56.0	—	
$Me_3SnCHCl_2$	9.66	4.30	57.0	15.0	
Me_3SnCH_2Cl	9.78	6.97	56.4	19.4	
Me_3SnCBr_3	9.50	—	57.2	—	
$Me_3SnCHBr_2$	9.64	4.52	56.2	13.2	
Me_3SnCH_2Br	9.74	7.34	56.4	18.0	175e
$Me_2Sn(Br)CHCl_2$	8.97	4.04	—	16.0	534a
$Me_2Sn(Br)CH_2Cl$	9.10	6.53	—	23.0	
$Me_2Sn(Br)CH_2Br$	9.07	6.82	—	19.8	

FORMULA	τ(H)	$J(^{117/119}Sn\text{–}CH)$	REFERENCES
Me_3SnCl	9.34—9.37	56.0/58.5 in $CHCl_3$ 67.5/70.5 in H_2O	71a, 861a
$(MeCH_2)_3SnCl$	8.64	38.7	
$(ClCH_2)_3SnCl$	6.31	18.3	
Me_3SnBr	9.20	56.0	
$(MeCH_2)_3SnBr$	8.61	37.5	
$(BrCH_2)_3SnBr$	6.67	16.4	
Me_2SnCl_2	8.84	66.6/69.7 in $CHCl_3$ 103/108 in H_2O	
$(ClCH_2)_2SnCl_2$	6.17	18.7	
Me_2SnBr_2	8.70	63.5/66 in $CHCl_3$ 103/108 in H_2O	
$(BrCH_2)_2SnBr_2$	6.47	16.4	
$MeSnCl_3$	8.35	95.3/99.7	
$ClCH_2SnCl_3$	5.92	19.4	
$Sn(CH_G{=}CH_CH_T)_4$ (Sn[C=C with H_G, H_C, H_T]$_4$)	H_C 4.3 H_G 3.7, H_T 3.9	Sn–H_C 90.6 Sn–H_G 98.3, Sn–H_T181.1	74b, 508[b]
$Sn(CH_2\text{-}CH_G{=}CH_CH_T)_4$	CH_2 8.1, H_C 5.2 H_G 4.1, H_T 5.3	Sn–CH_2 63	74b[b]
$Bu_2Sn\text{-}(CH_G{=}CH_CH_T)_2$	H_C 4.33 H_G 3.55, H_T 3.86 in Bu: H_α 9.20 H_β, H_γ 8.2—8.9 H_δ 9.09		715a

[b] For further data on substituted vinyltins and allyltins, see refs. 465d, 468, and 469.

Table 23-4
NMR Data for Organotin Hydrides

FORMULA	τ(Sn–H)		$J(^{117/119}Sn{-}H)$		REFERENCES
Me_3SnH	5.27	(Me 9.82)	1664/1744	(HSnCH 2.4) (SnMe 54/56)	346, 507, 508
Et_3SnH	5.00—5.24		1539/1611		
Pr_3SnH	5.21		1533/1605		
i-Pr_3SnH	4.85		1439/1505		
Bu_3SnH	5.22		1532/1609		
Ph_3SnH	3.03, 3.16		1850/1935		
Me_2SnH_2	5.55 5.24	(Me 9.8)	1717/1797	(HSnCH 2.65) (SnMe 57.6/60.2)	345b, 346, 507
Et_2SnH_2	5.25		1616/1691		
Pr_2SnH_2	5.42		1615/1685		
i-Pr_2SnH_2	5.07		1540/1612		
Bu_2SnH_2	5.43		1619/1690		
t-Bu_2SnH_2	4.7	(Me 8.73)	1484/1554	(SnCCH 68/71)	509
Ph_2SnH_2	3.98		1842/1927		
$MeSnH_3$	5.86	(Me 9.73)	1770/1852	(HSnCH 2.7) (SnMe 62)	346, 507
$EtSnH_3$	5.66		1710/1790		
$PrSnH_3$	5.83		1710/1790		
i-$PrSnH_3$	5.46		1672/1750		
$BuSnH_3$	5.71		1716/1796		
$PhSnH_3$	4.98		1836/1921		
SnH_4	6.11—6.15		1842/1933		

FORMULA	τ(Sn–H)		$J(^{117/119}Sn{-}H)$		REF.
R_2SnHX					345b
	X = Cl	X = Br	X = Cl	X = Br	
R = Me	2.88	3.31	2128/2228	2082/2178	
Et	2.67	3.18	1940/2031	—	
Pr	2.68	3.06	1914/2002	1862/1954	
i-Pr	2.70	3.13	1745/1828	1706/1786	
Bu	2.66	2.91	1890/1983	1875/1964	
	2.90		2108/2208		

(about 50), and is of opposite sign, has been tentatively interpreted in terms of a single electron term between the tin and the H atoms of the methyl group, with some participation by *d*-electrons (145a).

In compounds of the type SnOCH– or SnNCH–, Sn–O–CH coupling or Sn–N–CH coupling has been found in some cases (see Table 23-5), but not in others, such as Me_3SnOMe and Me_3SnNMe_2. The occurrence

Table 23-5
NMR Data for Some Other Organotin Compounds

FORMULA	τ(Sn–Me)	$J(^{117/119}$Sn–Me)	REFERENCES
$Me_3SnOCOCH_2Cl$		56.4/58.9	345a, 796a
$Me_3SnOCOCHCl_2$		56.9/59.6	
$Me_3SnOCOCCl_3$		56.5/59.0	
Me_3Sn oxinate	9.5	54/57	
Me_2Sn(oxinate)$_2$	9.56	67.9/71.2	
$Me_2Sn(acac)_2$	9.51	95.0/99.3	
Me_2Sn(picrate)$_2$	9.40	73.5/77.6	
$Me_2Sn(ClO_4)_2$		102.0/107.0	
$Me_2Sn(NO_3)_2$		104.3/108.7	
$Me_2Sn(OMe)_2$		71.3/74.4	
$Me_3SnSnMe_3$	9.79	47.0/49.6	195a
$Me_3SnOSnMe_3$	9.73		195a
Me_3SnLi	10.4	14	195a
Ph_4Sn	2.3—2.5[a]		508
$\mathit{Me}_3SnN(\mathit{CH}_2Me)_2$	9.82[a], 6.99[a]	56[a], 45[a]	534b
$Ph_3SnN(\mathit{Me})COOO\mathit{t}$-Bu	7.3[a]	28[a]	74c
$Bu_3SnOC\mathit{H}(CCl_3)OMe$	5.05[a]	20.4[a]	176
$Bu_3SnOC\mathit{H}(CCl_3)OSnBu_3$	4.64[a]	10.5[a]	176
$R_3SnOC\mathit{H}_2CCl_3$	5.9[a]	27.5—31.0[a]	465d

[a] Involving atoms and groups italicized in formulae.

of Sn–N–CH coupling in the stannylamines is temperature-dependent. Attempts have been made to explain this by an exchange mechanism (465d, 468, 469).

Organotin hydrides have been explored in great detail (see Table 23-4). They show a remarkably large Sn–H coupling constant (about 1600—2200).

The following rules can be derived from the data available to-date:

$\tau(R_3SnH)$ (R =) *i*-Pr < *n*-Bu ≈ *n*-Pr ≈ Et < Me (346)

τ(SnH) $SnH_4 > RSnH_3 > R_2SnH_2 > R_3SnH$

J(Sn–H) $SnH_4 > RSnH_3 > R_2SnH_2 > R_3SnH$

In each case,

J(Sn–H) (R =) Ph > Me > Et ≈ *n*-Pr ≈ *n*-Bu > *i*-Pr

For compounds R_2SnHX (345b):

J(Sn–H) (X =) H < I < Br < Cl

τ(Sn–H) (X =) H < I < Br < Cl

J(Sn–H) (R =) *i*-Pr < *n*-Bu < *n*-Pr < Et < Me

23-3-2 ^{119}Sn magnetic resonance spectroscopy

As mentioned earlier, tin has magnetically active isotopes, all with $I = \frac{1}{2}$. Resonance spectroscopy is therefore possible with these nuclei as well as with protons. The ^{119}Sn isotope (15.87 MHz at 10^4 Gauss) is more suitable than ^{117}Sn. τ and J are both strongly dependent on the solvent (195a). Again, many ^{119}Sn NMR spectra show complex spin-coupling patterns in confirmation of proton NMR results. To-date there have been few ^{119}Sn NMR studies (173a, 195a, 534b). A selection of the available data is given in Table 23-6.

Table 23-6

^{119}Sn Chemical Shifts (p.p.m. relative to Me_4Sn = 0) (173a, 175f, 534b)

Compound	Shift	Compound	Shift
$Me_3SnSnMe_3$	+113	$Me_3SnCHBr_2$	−42
Me_3SnBu	+1.8	Me_3SnCBr_3	−101.2
Bu_3SnCN	+48	Me_3SnNEt_2	−58.2
Me_4Sn	0	$Me_3SnOSnBu_3$	−82
Me_3SnCH_2Cl	−4	Me_3SnOCH_2Ph	−100
$Me_3SnCHCl_2$	−33	Me_3SnCl	−166
Me_3SnCCl_3	−84	Me_2SnCl_2	−137
Me_3SnCH_2Br	−36	n-Bu_4Sn	+12
		n-Bu_3SnCl	−143

23-4 MÖSSBAUER SPECTROSCOPY

Mössbauer spectroscopy (general reviews, refs. 259a and 294a) using the naturally occurring isotope ^{119}Sn has already rendered valuable service in organotin chemistry. It is particularly valuable for study of the geometry of the electronic environment, i.e. the coordination number of the tin. The technique has also been of great value in showing that bonds formed by tin are invariably less ionic than had been assumed previously.

A selection of data is given in Table 23-7. In some cases (e.g. $SnHal_4$) it has been possible to establish a linear relationship between the chemical shift δ and the electronegativity of substituents bonded to the tin. More information is usually obtained from the quadrupole splitting Δ (mm./sec.) which depends on the electrostatic field gradient at the tin nucleus, i.e. on distortion of the electronic environment by bonded atoms. Quadrupole splitting is thus expected in unsymmetrical species. However, none is observed in compounds of type $\geq$Sn–M (M = H, Li, Na, Ge, Sn, Pb) or in tetraorganotins. Compounds with Sn–F, Sn–Cl, Sn–Br, Sn–I, Sn–O,

Sn–S, and Sn–N bonds do show quadrupole splitting, which suggests that such splitting is observed only when the bonded atom has an unshared electron pair which can participate in $p\pi$–$p\pi$ bonding with the tin. There seem to be exceptions to this rule; compound types $R_3Sn{-}CF_3$ and $R_3Sn{-}C_6F_5$ both exhibit quadrupole splitting.

Δ also depends on the polar character of the compound:

$$\Delta[Bu_2Sn(OCOMe)_2] < \Delta[Bu_2Sn(OCOCH_2Cl)_2] < \Delta[Bu_2Sn(OCOCCl_3)_2]$$

Attempts have been made to establish a relationship between Δ and the electronegativity of the substituents.

In the compounds R_3SnX, Δ has the following values (642a):

(a) $X = Me, Ph, CH{=}CH_2, SnMe_3, Li$ $\quad\Delta = 0$

(b) $X = CF_3, C_6F_5, C_6Cl_5, C{\equiv}CPh$ $\quad\Delta = 1.0$—1.4

(c) $X = F, Cl, Br, I$ $\quad\Delta = 2.2$—3.9

Using Δ as an indication of ligand symmetry, and hence also of coordination, the high values in group (c) can be seen as a consequence of pentacoordination rather than of differences in electronegativity. Pentacoordination has also been established for Me_3SnClO_4, but not for $(neo\text{-}C_5H_{11})_3SnX (X = F, I, N_3)$ in which it is prevented by steric hindrance.

Stannyl esters show wide variations in their Δ-values (see Table 23-7) and must therefore have different coordination. In tin chelates, values of Δ provide stereochemical information: Δ is ca. 2 for *cis*-configurations and ca. 4 for *trans*-configurations. Mössbauer spectroscopy is also a useful tool for kinetic studies.

Finally, reference should be made to some comprehensive reviews (8, 235a, 259a, 294, 294a, 671a) and some recent papers (106a, 153a, 206b, 207a, 642a).

23-5 MASS SPECTROMETRY

In the last few years, mass spectrometry too has been applied to the structural elucidation and theoretical interpretation of organotins (review in ref. 129b).

Bond dissociation energies for tin (see Chapter 2-1) are relatively low, and certainly well below those for Ge and Si. It is therefore not surprising that organotin molecules, particularly large ones, suffer considerable fragmentation in the mass spectrometer; e.g. while cyclic $Ph_{12}Si_6$ and even

Table 23-7
Mössbauer Parameters for Organotin Compounds

COMPOUND	δ (mm./sec.) rel. to SnO_2	Δ (mm./sec.)	REFERENCES
Me_4Sn	1.22, 1.29	0	153a, 642a
Me_3SnPh	1.16	0	642a
$Me_3SnCH{=}CH_2$	1.30	0	235a, 642a
Ph_4Sn	1.15, 1.21	0	
	1.27, 1.30[a]		
$Ph_3SnCH{=}CH_2$	1.28	0	642a
$Me_3SnC{\equiv}CPh$	1.23	1.17	
Me_3SnCF_3	1.31	1.38	
$Me_3SnC_6F_5$	1.27	1.31	
$(Me_3Sn)_2$	1.46	0	
$(Ph_3Sn)_2$	1.30	0.3	235a
$(Ph_3Sn)_4Sn$	1.33	0.3	
$(Ph_3Sn)_4Ge$	1.13	0.3	
$(Ph_3Sn)_4Pb$	1.39	0.3	
$(Ph_3Sn)_2SnPh_2$	1.06	0.2	
Ph_3SnLi	1.40	0	
Me_3SnNa	1.38[a]	0	153a
Me_3SnF	1.22[a], 1.28	3.86	153a, 642a
Me_3SnCl	1.42, 1.44[a]	3.09, 3.41	106a,153a,642a
	1.45[a]	3.50	
Me_3SnBr	1.30, 1.45[a]	3.0, 3.25	
	1.49	3.40	
Me_3SnI	1.48	3.05	
Ph_3SnF	1.25, 1.29[a]	3.53, 3.90	642a
Ph_3SnCl	1.31, 1.45	2.5, 2.56	
Ph_3SnBr	1.20, 1.37	2.40, 2.48	106a, 642a
	1.40	2.51	
Ph_3SnI	1.20	2.25	642a
n-Bu_3SnCl	1.36, 1.58	2.80, 3.40	106a
Et_3SnBr	1.38, 1.62	3.45, 2.82	
Me_2SnCl_2	1.58, 1.61[a]	3.55	153a
$Me_3SnCl{\cdot}Py$	1.30, 1.45	3.18, 3.44	
Me_3SnOH	1.19	2.91	
Me_3SnNO_3	1.44	4.14	
Bu_2SnO	0.93, 1.03	1.95, 2.08	235a
	1.15	2.2	
SnH_4	1.27	0	293a
$MeSnH_3$	1.24	0	
Me_2SnH_2	1.22—1.23	0	
Me_3SnH	1.15, 1.24	0	153a, 293a
n-Bu_3SnH	1.44	0	293a
n-Bu_2SnH_2	1.42, 1.45	0	235a, 293a
n-Bu_3SnH	1.41	0	293a

COMPOUND	δ (mm./sec.) rel. to SnO_2	Δ (mm./sec.)	REFERENCES
$PhSnH_3$	1.40	0	
Ph_2SnH_2	1.38	0	
Ph_3SnH	1.39	0	
n-Bu_2SnClH	1.56	3.34	
Me_3SnOAc	1.35	3.68	207a
Ph_3SnOAc	1.27	3.40	
$Ph_3SnOCO(CH_2)_8CH{=}CH_2$	0.57	2.31	
$Ph_3Sn(CH_2)_{16}Me$	0.56	2.32	
$Ph_3SnOCOCH(Et)Bu$	1.21	2.26	
$Me_2Sn(oxinate)_2$	0.77	1.98 (*cis*)	206b
$(PyH)_2[Me_2SnCl_4]$	1.59	4.32 (*trans*)	
$Me_2Sn(acac)_2$	1.18	3.93	

[a] Re-calculated from the original values (relative to α-Sn) by adding 2.1.

cyclic $Ph_{12}Ge_6$ give molecular ion peaks and peaks corresponding to a stepwise fragmentation, the largest ion from cyclic $Ph_{12}Sn_6$ is that of a fragment with two tin atoms (427b).

However, smaller organotin molecules often show the molecular peak as well as a characteristic series of fragmentation peaks. Fragmentation rules have been derived (625b), particularly for trialkyltin halides (236a). Appearance potentials for a number of organotin cations, especially R_3Sn^+, are known (625b, 899a). However, calculation of badly needed dissociation energies (e.g. of the Sn–H bond) meets with the difficulty that the required thermochemical data are either not available or so mutually contradictory that a choice must be largely arbitrary (see Chapter 2-1).

As tin has several naturally occurring isotopes, each ion appears in the mass spectrum as a series of peaks close to each other. This can interfere with interpretation, particularly when two fragments differ by only one or two mass units ($-H$, $-2H$).

Mass spectrometry of negative ions could also become important for organotins. Certainly, Et_4Sn gives appreciable $M-1$ fragments, and this has been ascribed to $p\pi$–$d\pi$ stabilization (402a).

24 Toxicity of organotin compounds

Acute toxicities of organotin compounds for warm-blooded animals and single oral ingestion are listed in Table 24-1. However, lists of LD_{50} values often fail to present a clear picture to the chemist. In particular, they rarely enable him to make comparisons with materials normally encountered in a chemical laboratory. Table 24-1 therefore contains an additional list of acute oral toxicities of some common reagents and chemicals by way of a "comparison scale." It will be seen that not a single organotin compound is as highly toxic as, say fluoroacetic acid, sodium selenate, or some digitalis components. The most poisonous of the organotins investigated to date are the triethyl and trimethyl derivatives. These are comparable with e.g. methyl isocyanate, 1,3-propanediol, or sodium selenate and selenite, and approach the toxicity of hydrocyanic acid. Next in order are some diethyltins and tripropyltins. The majority of these compounds, as well as all butyltins, higher alkyltins, and phenyltins tested so far, belong to the categories (817) "moderately toxic" (LD_{50} 50—500 mg./kg. for oral ingestion), "slightly toxic" (LD_{50} 500—5000 mg./kg.), or "practically non-toxic" (see Table 24-1).

All values were obtained in animal experiments using small mammals. Some examples from Table 24-1 show that the values depend on the species; the rat is usually taken as the standard test animal, and the above definitions (817) of degrees of toxicity are based on this.

Different behavior of several species is clearly shown by using triethyl-[^{113}Sn]tin chloride. Distribution of $SnEt_3$ residues in the rat, the guinea pig, and the hamster is non-uniform, the highest concentration being in rat blood and in the liver of all three species. Sub-cellular fractionation of rat liver, brain, and kidney shows that $SnEt_3$ residues are attached to all fractions, but to different extents. Rat hemoglobin is responsible for the high uptake of $SnEt_3$ residues by rat blood (2 moles/mole of hemoglobin) (709a).

Acute toxicity values must not be applied directly to chronic toxicity; values for the latter are usually much lower, not only for organotins. Table 24-1 also contains some data from chronic feeding studies on experimental animals. At present there are few reliable data on acute toxicity by inhalation. The current American table of "Threshold Limit Values for 1965" gives an MAC value (MAC = maximum allowable concentration) for all organotin compounds of 0.1 mg./m.^{3} (378).

Several organotins, especially those of the series R_3SnX such as triethyltin cyanide, are powerful sternutators or lachrymators either as vapor or as dust, and some are skin irritants. Practically all organotin compounds of appreciable volatility induce headaches if allowed to contaminate the laboratory atmosphere. However, most of them have a strong and unpleasant odor which acts as a warning signal. Wartime studies in Anglo-Saxon countries have established the connection between structure and toxicity of several organotins. None of them was sufficiently toxic for military use (literature summary in ref. 315).

Biochemical experiments indicate that diethyltin compounds interfere with the action of α-keto-acid dehydrogenases and that the effect can be prevented by BAL but not by glutathione. Triethyltin compounds have no effect other than interference with oxidative phosphorylation (7) and mitochondrial functions. Tissue examination shows an increased water content in the central nervous system, more marked in the spinal chord than in the brain. However, the results must not be taken as proof of processes in the living experimental animal (59, 315, 379). The effect of these substances on different species of animal varies remarkably (846). For rats and rabbits, intravenous injection gives the same results with both dialkyltins and trialkyltins. Oral administration gives different results for the two series of compounds: it appears that the dialkyltin compounds are unable to penetrate some barrier in the animal organism; they therefore show little or no effect on the central nervous system (45, 380). The action of R_3Sn-compounds is said to be similar to that of 2,4-dinitrophenol. In animal experiments, triphenyltin acetate in sub-acute amounts acts mainly as a stimulant, and in higher doses as a depressant. The ultimate cause of death is respiratory paralysis. In some instances, the experimenters were able to restore spontaneous respiration by intravenous injection of Cardiazol (845). For the same alkyl group, maximum toxicity is found in the compound of type R_3SnX, followed by R_2SnX_2 and R_4Sn. $RSnX_3$ is the least toxic or non-toxic. This sequence is not always obvious, since alkyl groups may be split off in the animal organism (especially the liver) and R_4Sn thus converted gradually into the much more toxic R_3SnX (161).

Table 24-1

Acute Oral Toxicities (LD_{50})[a] *of Organotin Compounds*

(*Right-hand Side: Toxicities of Well Known Chemicals and Reagents, for Comparison*)

ORGANOTIN COMPOUND	LD_{50} (mg./kg.)	TEST ANIMAL[c]	REF.	REAGENTS AND LABORATORY CHEMICALS[b]	LD_{50} (mg./kg.)	TEST ANIMAL[c]
Me_3Sn acetate	9.1	R	58	Sodium fluoroacetate	0.22	R
Et_3Sn acetate	4.0	R	58	Digitoxin	0.25[g]	C
Et_3SnCl	>6[d]	R	846	Potassium cyanide	1.6	D
$(Et_3Sn)_2SO_4$	8.5	R	846		10—15	R
Et_2SnCl_2	21[d]	R	846	Hydrogen cyanide	4	Rb
tolerated dose	16[d]	R	846			
Et_2Sn dilaurate	210	R	375	Sodium selenate	4[h]	Rb
Pr_3Sn acetate	118	R	58	Vitamin D_2	4—5[g]	D
$Pr_3SnCH_2CH_2COONa$	700	M	359	1,3-Propanediol	6	M
i-Pr_3Sn acetate	44[e]	R	58	*N,N*-Diethylsuccinamide propyl ester	6.3	M
i-Pr_3SnOH	44.3	R	377			
Bu_3Sn acetate	133	R	377	Methyl isothiocyanate	8.5[g]	Rb
	380	R	58	Ethyl isothiocyanate	10[i]	Rb

[a] Single dose (in mg./kg. body-weight) causing death within 4 days in 50% of the test animals. More values for LD_{50} and percutaneous toxicity are given by O. R. Klimmer (379a).

[b] From *Handbook of Toxicology* (817).

[c] C = cat, D = dog, G = guinea pig, M = mouse, R = rat, Rb = rabbit.

[d] Intravenous injection.

[e] Other authors find *i*-Pr_3SnCl less toxic than the *n*-propyl derivative (45).

[f] Depending on the solvent or emulsifying agent.

[g] "Lethal dose", not otherwise defined.

[h] LD_{100}.

[i] Minimal LD.

Table 24-1—continued

ORGANOTIN COMPOUND	LD_{50} (mg./kg.)	TEST ANIMAL[c]	REF.	REAGENTS AND LABORATORY CHEMICALS[b]	LD_{50} (mg./kg.)	TEST ANIMAL[c]
$(Bu_3Sn)_2O$	112—194[f]	R	192, 379a	Methyldi-(β-chloroethyl)amine·HCl	10	R
percutaneous	605	R	377			
percutaneous	11,700	R	192	Sodium arsenite	14—30[g]	Rb
30 days continuous feeding (mg./kg. feed)	>100	R	192			
Bu_3Sn salicylate	137	R	377	Methacrylonitrile	15[g]	M
Bu_3Sn oleate	194—225[f]	R	377	Ethylenimine	15	R
Bu_3Sn undecylate	205	R	377	*N*,*N*-Diethylaminoethyl chloride	17	R
Bu_2SnCl_2	100	R	380			
$(Bu_2SnO)_n$	600—800[f]	R	377	2,4-Dinitrophenol	20—30	D
Bu_2Sn di-monobutylmaleate	120—170	R	377	*p*-Dinitrobenzene	29.4	Rb
				Alkylmercury chloride	30	R
Bu_2Sn di-2-ethylhexylthioglycollate	510	R	377	Acrolein	46	R
Bu_2Sn dilaurate	175	R	380	Acrylonitrile	50	G
	>500	R	38			
$(BuSnS_{1.5})_n$	> 20,000	R	379a	Nicotine	50	R
$(C_6H_{13})_3Sn$ acetate	1000	R	58	Allyl alcohol	52	Rb
$(C_8H_{17})_3Sn$ acetate	>1000	R	58	γ-Hexachlorocyclohexane	60	Rb
$[(C_8H_{17})_3Sn]_2SO_4$	>6000	R	375	Hydroquinone	70	C
$(C_8H_{17})_3Sn$ laurate	>6000	R	375	Pentachlorophenol	78	R
$[(C_8H_{17})_2SnO]_n$	2500	R	377	Barium chloride	90[g]	D
$(C_8H_{17})_2Sn$ β-mercaptopropionate	2100	R	377	Ethylene chlorohydrin	95	R
$(C_8H_{17})_2Sn$ maleate	4500	R	377	Tri-*o*-cresyl phosphate	100[g]	Rb
$(C_8H_{17})_2Sn$ di-monobutylmaleate	2030	R	377	α- and β-Naphthol	100—150[g]	C
				Quinone	130	R
				Methyl iodide	150—220	R

Table 24-1—continued

ORGANOTIN COMPOUND	LD_{50} (mg./kg.)	TEST ANIMAL[c]	REF.	REAGENTS AND LABORATORY CHEMICALS[b]	LD_{50} (mg./kg.)	TEST ANIMAL[c]
$(C_8H_{17})_2Sn$ di-isooctylthioglycollate	1975	R	377	DDT	180	M
				Isoquinoline	350	R
percutaneous	2250	R	377	Dimethyl sulfate	440	R
				Quinoline	460	R
$(C_8H_{17})_2Sn$ dilaurate	>6000	R	380	Triethylamine	460	R
$(n\text{-}C_{16}H_{33})_2SnCl_2$	>10,000	R	379a			
Ph_3Sn acetate	136	R	376			
	21	G	376			
	30—50	Rb	376			
percutaneous	~450	R	376	Phenol	530	R
170 days continuous feeding (mg./kg.)	>25	R	376	Propanol	1870	R
$(Ph_3Sn)_2S$	≥700	M	359	Cyclohexanol	2000	R
$(Ph_3Sn)_2S_2$	>700	M	359			
$Ph_3Sn(CH_2)_3COOH$	>10,000	M	359			

The nature of X exerts only slight, and usually indirect, influence, e.g. on solubility or volatility. In the triethyl and tripropyl series, the hydrides are less toxic than the corresponding chlorides, but more toxic than the tetraorganotins (129).

It appears that on only one occasion have organotin compounds caused major damage to human life, but that occasion amounted to a catastrophe. In 1954 there was introduced in France under the protected trademark Stalinon a therapeutic product containing diethyltin diiodide as the active ingredient. Each capsule was supposed to contain 15 mg. of the substance, the recommended dose being 6 capsules per day for a course of treatment lasting eight days. The composition of the product was however not uniform, contrary to intentions, and it contained some monoethyl- and up to 10% of triethyl-tin compounds, the latter about ten times as toxic to mammals as the diethyltin compound. A comprehensive report of the whole affair is given by Barnes and Stoner (59). Symptoms observed 2—4 days after oral ingestion of this preparation included violent headaches, vomiting, dizziness, abdominal pains, evidence of paralysis, edema of the papilla at the back of the eye (being evidence of cerebral edema), and ultimately cramp, loss of consciousness, and respiratory paralysis. Where poisoning did not prove fatal, the after-effects included headaches, debility, impaired vision, and other neurological conditions. O. R. Klimmer (379) thinks that poisoning by other organotin compounds would produce similar symptoms. (The statement was made in relation to triphenyltin acetate which is used as a fungicide. However, no case of human poisoning by this material has been reported.) He also suggests the following therapy (379): "Administration of BAL is pointless: therefore only symptomatic treatment by gastric lavage, aperient salts, oxygen, and artificial respiration need be considered. There may be cerebral edema to combat."

Skin damage represents an industrial hazard of some importance in the manufacture and processing of tributyltin compounds, and to a lesser extent in that of dibutyltin compounds. Process workers may contract contact dermatitis or suffer from sensitization. Sorption through the skin is also possible, particularly when the skin comes into contact with organotins dissolved in oils, fats, or other lipid solvents (379a).

There have been no indications of carcinogenic activity of organotin compounds either from chronic studies on animals or from biochemical investigation of cell metabolism. Barnes considers that, on the basis of their biochemical mode of action as understood at present, these materials are completely free from suspicion in this direction (quoted in ref. 377).

Butylthiostannonic acid, a substance in technical use, is described as non-poisonous (377). *n*-Octyl-organotins are so little toxic either acutely or chronically (380) that they are incorporated even in food wrappers and containers (plastic foil and bottles) (41). Hexabutylstannoxane (tributyltin oxide), in common technical use e.g. as wood preservative, has been found in oral ingestion trials on monkeys to be less than half as toxic as the widely used wood preservative pentachlorophenol (703). Apart from slight temporary skin "burns" there are no reports of toxic symptoms having arisen from handling of these materials. Triphenyltin acetate is used on a large scale as a fungicide, yet no case of human poisoning has been reported (376, 379). Extensive experiments (on beet leaves) have shown that the residues of this fungicide and its degradation products, which may occur for a limited time in a limited number of places, are entirely unobjectionable for the animal, for the quality of the animal products, or for the health of the consumer of these products (104).

For further items of information and detailed toxicological discussion the reader should consult a literature summary up to 1959 (315), and critical reviews by Stoner, Barnes, et al. (58, 59, 828a), by Tauberger and Klimmer (846), and by Ascher and Nissim (45). Klimmer, in a more recent survey, gives many new results and data from his own work (379a). Another recent paper by Klimmer reviews the toxicological aspects of organotin applications in agriculture (379b).

Thus, most organotin compounds are no more toxic than other common laboratory or technical chemicals. There has not been any report of fatal poisoning or even permanent damage arising from handling of organotin compounds in chemical laboratories or factories, although the number of laboratories and chemical works dealing with organotins is considerable, as shown by the list of references.

Naturally, tidiness and a responsible approach to experimentation are prerequisites for organotin work in a chemical laboratory. This includes dismantling and cleaning of used equipment. Preliminary rinsing with alkalis or alkaline cleansers has proved useful, since this transforms the more highly volatile organotin compounds R_3SnX and R_2SnX_2 into less volatile stannoxanes or non-volatile diorganotin oxides. Oxidation by rinsing with bromine or chlorine in water or hydrocarbon solvents serves a similar purpose. This leads to elimination of alkyl groups and to transformation at least into the less toxic or non-toxic diorgano- and mono-organo-tins. The wearing of rubber gloves for these operations is recommended, and good fume hoods are essential. The extensive experience gained in the author's laboratory and in other laboratories (491) over many years proves that the handling of organotin compounds under these conditions presents no particular difficulty. The worst that has befallen any of the experimenters has been occasional headaches of varying severity, mostly of short duration but sometimes persisting for a few days, and the nuisance of offensive odors. Both of these could have been avoided in every case.

25 Technology and commercial applications

A whole range of organotin compounds has found practical application and is therefore prepared on an industrial scale. Reliable estimates put world production in 1948 at a few hundred kilograms only. By 1956 this had grown to 300 tons, by 1960 to 1500 tons, and by 1962 to ca. 3000 tons (488) of which 2100 tons were produced in Europe. Consumption in 1963 in the United States alone was 1600 tons (40). Production in 1965 has been estimated at 5000—6000 tons,* and for 1968 at 10,000 tons.† Future development is viewed with optimism, and new factories are planned or in course of construction (40). Two reviews describe the state of things in 1965 (170a, 712a), and one in 1969 (351a).

It is impossible to give here a complete picture of the development or even the present state of organotin technology. An appreciable part of the required information is inaccessible, for obvious reasons. Also, the publications of firms and individuals with commercial interests in this field are not always designed solely to convey objective information. Some of the developments can be traced only from the patent literature. In this connection one might quote a comment by H. V. Smith (805), with which the present author finds himself in complete agreement: "It has been necessary to omit a number of the author's private conclusions regarding the validity and originality of some of the patents described. Similarly it was found unwise to attempt to trace the tortuous course of the complex system of licence agreements which has been erected around these patents." This chapter must therefore remain fragmentary. Certainly the

* Estimate by the Tin Research Institute, Greenford, Middlesex, England; production in Russia and other Eastern countries is not included.

† Dr. E. S. Hedges, Tin Research Institute, during the Organotin Symposium at Frankfurt/Main, November 1968.

patent literature cannot be reviewed in its entirety. An attempt will be made to give the reader some insight into the subject and to enable him to consult appropriate sources for further information.

25-1 INDUSTRIAL PREPARATION

Until 1962 organotin compounds on a commercial scale were produced exclusively by Grignard methods (329, 608) (for total production see previous section). It is remarkable that this inherently complicated method (see Chapter 3-1-2), which requires large quantities of solvents, was ever developed successfully into a technical process. One factory equipped to produce propyl-, butyl-, octyl-, and phenyl-tin compounds uses glass-lined reaction vessels each of 2250 liter capacity (329). Another firm makes tetrabutyltin using toluene [activated by an ether (683)] as solvent. Excellent yields are reported (258) with magnesium, $SnCl_4$, and butyl chloride, all prepared and purified by the firm's own processes. Because of the inhibiting action of trace metals, glass vessels are again required. A description and flow-sheet of the plant have been published (258). Yet another process starts from alkyl halide, tin metal and sodium metal (see also Chapter 3-1-1) (289). The most important step appears to be the reaction of sodium with R_2SnCl_2 and 2 moles of RCl to give R_4Sn. Only part of the tetraorganotin appears to be taken off as product, the remainder being converted back into R_2SnCl_2 (see below) and re-cycled.

Tetraalkyltins are also made by alkylation of $SnCl_4$ with organoaluminums, which are nowadays available as major industrial chemicals in any desired amount. Alkali halides (324) or ethers and amines (605) are added to complex the resulting $AlCl_3$ (see Chapter 3-1-2). Solvents are not required, and the process appears to be of technical importance (608). A similar process has been patented for aryltins (892).

This process achieves complete alkylation of the tin halide in a single stage and without an excess of R_3Al. In principle this can also be done with Grignard compounds, but alkylation tends to be incomplete in practice, leading to mixtures of R_4Sn with R_3SnCl and R_2SnCl_2. Methods of separation, e.g. by means of methanol (in which R_4Sn is insoluble) (718) or alkalis (257), have been patented. In many cases the organotin halides need not be separated, e.g. when the next stage involves comproportionation with tin tetrahalide (see below). The organotin halide content of the crude tetraorganotin is determined, and allowance is made for it when calculating the amount of tin tetrahalide required (258).

The next stage usually involves transformation of the tetraorganotin, obtained by one of the above methods, into organotin halides. This is generally done by comproportionation (700) (see Chapter 5-1-3).

Compounds of type R_3SnCl and R_2SnCl_2 are thus accessible on the technical scale; improvements are claimed from time to time (625a). Recently, organotin trihalides have become available also (527, 556a, 603).

Organotin halides can also be made by partial alkylation of $SnCl_4$ (324, 605). However, preparation direct from tin metal without the intervention of some other organometallic alkylating agent may be more desirable. For R_2SnX_2 compounds, E. Frankland (212) had already shown the way in 1849, since he obtained the very first organotin compound by this method! (see Chapter 5-1-1). Technical application had to wait until the present decade (33). It appears that reaction of alkyl iodides with tin and a magnesium catalyst (899), and catalytic preparation of dibutyltin dibromide from butyl bromide and tin, have both attained technical importance. In the latter, hydrolysis to dibutyltin oxide yields NaBr, which is then treated with butanol and acid to regenerate butyl bromide (40).

The organotin halides prepared as above are mostly converted with alkali into stannoxanes, e.g. into the much used hexabutylstannoxane (tributyltin oxide) (258) or into diorganotin oxides. Further syntheses can then be performed with these two groups of compounds (see Chapters 17-1 and 17-2). Polyhydric alcohols such as ethylene glycol, glycerol, and sorbitol have been recommended as stabilizers for stannoxanes (901).

Several processes have been patented (822) for condensing these organotin oxides with reactive OH or SH groups with elimination of water, e.g. with phenols (312) and other hydroxy-compounds (311, 398, 529), but especially with mercaptans (374, 398, 502) such as mercaptocarboxylic esters (500), mercapto-heterocyclics (82), and phenylenedimethylenethiols (501). Trimercaptides (823) can be obtained in this way from stannonic acids and 3 moles of RSH. One can also start from organotin halides and treat these with e.g. NaOR, NaSR, or Na_2S_2 (157, 502). Sulfides $(R_2SnS)_3$ are accessible by the same method (157).

25-2 ORGANOTINS AS POLYMER STABILIZERS

Certain organotin compounds, especially the R_2SnX_2 type, are excellent stabilizers for polyvinyl chloride (PVC), neoprene, and other polymers against degradation by light and oxygen, and decomposition during hot fabrication. This was the first major application of organotins* and appears still to be the main outlet. Although the number of known PVC stabilizers is quite large, the organotins are superior to all others for colorless trans-

* The very first technical application appears to have been that of tetraphenyltin as a transformer-oil additive (146) (see Chapter 25-4).

parent foils and discs, including ones made from rigid PVC. They are best suited to the stringent requirements, e.g. of extrusion pressing of discs (40, 488). At present their application is restricted by their relatively high cost, but this should improve fairly soon. The first successful compounds were dibutyltin dilaurate (900) and maleate (681) (1943). There followed a flood of patent applications (reviewed up to 1959 in ref. 315). A detailed description of organotin compounds in use by 1958, and attendant technical problems, is given by H. V. Smith (805). More recently, there have been added hydroxy-compounds such as dibutyltin bis(propylene glycol)maleate (506) and sulfur compounds, e.g. thioglycollic esters $Bu_2Sn(SCH_2COOC_8H_{17})_2$ and $R_3SnSC(S)NR'_2$ (R = Bu, C_6H_{13}; R' = Et, H) (201). Among these newer stabilizers are dioctyl compounds (329, 495) used particularly for PVC bottles and wrapping foils for the foodstuffs industry (41).* Non-toxic* sulfur-containing monobutyltin compounds are used for the same purpose (351).

Organotin compounds are also suitable as stabilizers for chlorinated polyethylenes (201), vinyl copolymers (201, 381), silicones (122), and polyamides (306).

The mode of action of organotin stabilizers, especially for PVC, is the subject of much conjecture. Several mechanisms are involved in the degradation of PVC by light, heat, and oxygen. One of these involves formation of conjugated poly-enes by a polar mechanism (with elimination of HCl) and subsequent oxidation giving rise to carbonyl groups (498, 890a). In presence of organotin stabilizers, no C=O absorptions are observed (207). The organotins are therefore described as "antioxidants," but they are also regarded as "HCl-acceptors" (288, 498). How they might combine with HCl is again far from clear. Experiments with dibutyltin diacetate labeled with ^{14}C in the alkyl groups have certainly established that the Sn–C links are the real active agent. The acetyl groups play no part in the stabilization (350). It is assumed that the Sn–C bond undergoes radical cleavage (see also Chapter 4-4). Radical sites on the polymer chain would then be saturated by butyl groups, thus inhibiting crosslinking, reaction with oxygen, and start of radical-chain reactions. It is certain that this is only part of the answer to the problem of stabilizer action (148, 315).

The remarkably long induction period (before commencement of HCl elimination) associated with dibutyltins and dioctyltins deserves special

* The relevant legal requirements vary from country to country (351a). E.g., in the U.S.A. the Food and Drug Administration permits di-*n*-octyltin *S,S'*-bis(isooctylmercaptoacetate) and polymeric di-*n*-octyltin maleate as stabilizers for PVC for package of foodstuffs. The specifications are very detailed, and strict limits are imposed on the concentration of mono- and tri-*n*-octyl, isooctyl, and other residues (42d). For requirements in the Federal Republic of Germany, reference should be made to *Mitteilungen des Bundesgesundheitsamts*: *Gesundheitliche Beurteilung von Kunststoffen im Rahmen des Lebensmittel-Gesetzes.*

comment. It appears that no chemical reaction occurs during this period, and this part of the protective action is therefore ascribed to a physical action, i.e. to quenching. The organotin compounds are assumed to transform irradiation energy directly into heat energy, thus preventing breakage of chemical links in the polymer over a distance of as much as 25 Å. A very strong absorption band, in the near infrared, found in active organotin stabilizers, is important in this connection (890a).

25-3 ORGANOTINS AS INSECTICIDES AND FUNGICIDES

Inorganic tin compounds are non-toxic or only very slightly toxic towards mammals, insects, and fungi. As early as 1929 it was realized that organotins might be active against moths (282). However, their biocidal activity was not investigated on a broader front until 1950. The pioneer work was done by the group in Utrecht (G. J. M. van der Kerk, J. G. A. Luijten, and others) and independently in the research laboratories of Farbwerke Hoechst A.G. [K. Härtel (281) et al.]. Since then, several other groups have made contributions to knowledge in this rapidly expanding field of application of organotin compounds. The literature up to 1959 is surveyed in reference 315. The situation at the end of 1964 (with the exceptions of fungicidal applications in agriculture) is described by K. R. S. Ascher and S. Nissim (45) with a large number of references. For later reviews (1967, 1969), see refs. 129a and 351a.

In several of the applications in this field—but by no means in all—the organotin compound in question may come into contact with human beings or with food intended for human consumption (plants, domestic animals). Harmful effects must therefore be avoided (see Chapter 24 on toxicity) and the legal requirements of the various countries observed carefully. The need thus arose for extensive studies in the physiology of nutrition, in analysis, and in toxicology (e.g. ref. 103). New microanalytical methods had to be devised and the limits of others extended (111, 113, 114) (see also Chapter 22). Some biological applications, in use at present or in course of testing, are given below. As explained already, the scope and objects of this book permit only a brief survey. This should provide those interested with the essential information and with references for further study.

High fungicidal activity is found in trialkyltins and triaryltins R_3SnX, much less in compounds of type R_2SnX_2, and only occasionally in tetraorganotins (491) (where it may be due to elimination of an alkyl group). The nature of X is of little consequence provided that it bestows adequate solubility and is capable of exchange in aqueous media. The activity of aliphatic R_3SnX compounds is greatest when the sum of the alkyl group carbons is between 9 and 12, irrespective of whether the three alkyl groups are the same or different (491). High toxicity towards fungi is thus the exception rather than the rule in this series (359). This specific activity (in conjunction with the notable lack of toxicity of inorganic tin compounds) differentiates the R_3SnX compounds sharply from other metal-containing fungicides (e.g. Hg, Cu, Cd, Zn) the metal of which is toxic either in all combined forms or as the free ion. The activity of numerous aliphatic and aromatic organotins against various micro-organisms has been summarized in extensive tables (354, 359, 491, 492, 621). As a rule, aromatic organotins are less phytotoxic than aliphatic ones (912). Triphenyltin acetate (Brestan of Farbwerke Hoechst A.G.) is of importance in combating fungus diseases in large-scale agriculture, especially in root crops like potatoes* and sugar-beet (281). It is highly effective against diseases of either leaves or roots and tubers (303). Against these, as also against late blight of celery and anthracnose of beans, it is superior to other fungicides. It has also been used successfully against diseases of cocoa trees and ground-nut plants, and in many other cases. Its "biological half-life" is only about 3—4 days in the open and 7—8 days under glass, since water and light degrade it to inorganic tin compounds. (For analyses of residues and evaluation of these see references 102 and 104.) It has been combined with the thiocarbamate Maneb to improve plant tolerance (281). Triphenyltin hydroxide (Du-Ter of N. V. Philips-Duphar) is marketed for the same purposes.

Fungicidal organotin compounds, e.g. triphenyltin chloride and tributyltin oxide [TBTO or BioMET of Metal and Thermit Corporation (912)], are added to paints, and used to impregnate potato sacks (jute), sisal and manila ropes, textiles, leather, tent canvas, sail cloth, and nets against attack by molds and fungi, especially in moist tropical climates. They are also used to prevent formation of algae, and bacterial and mold slimes, in the wood pulp (649) and paper industries, and in cooling towers (329, 912). Triethyltin hydroxide, tributyltin acetate, and tributyltin oxide [e.g. the commercial product OZ of the Osmose Wood Preserv. Co., which contains Butinox (tributyltin oxide) dissolved in mineral oil (37)] are used or recommended as preservatives for wood (298, 703) including pit props and wooden electricity distribution poles. Detailed test reports are given in references 35, 200, and 703. Certain active organotins can be introduced as self-emulsifying additives (528), and others can be applied as stable dispersions (652). Antiseptic properties on non-metallic surfaces are claimed for tripropyl- and tributyl-tin fluorides (525).

* The Food and Drug Administration of the U.S.A. has agreed to establish "a tolerance of 0.05 parts per million for residues of the fungicide triphenyltin hydroxide in or on the raw agricultural commodity potatoes" (42b).

Tributyltin oxide (696) and a "polyethoxy-amine of tributyltin chloride" (522) are recommended as bactericides and biostatics inside buildings, including cases of endemic hospital diseases caused by strains of bacteria that have become resistant to other materials. They are also recommended as antibacterials for laundry purposes. Tributyltin benzoate, sometimes in combination with formaldehyde (Incidin made by Desowag-Chemie GmbH), is listed as yet another disinfectant (269), while some further compounds are given as bactericidal seed dressings. *para*-Substituted triethyltin phenoxides have been proposed as nematocides for agricultural use (316).

Dibutyltin dilaurate and maleate and other stannyl esters are used in veterinary medicine. Tetraisobutyltin is used as an anthelmintic against worm and protozoal infections of poultry, both as prophylactic and as remedy (315).

Compounds used to protect canal locks, embankments, and ships hulls of wood (863) or metal* against overgrowths, ship-worms (703), and barnacles, especially while ships are in harbour (860), and also for protection of other underwater coatings, include tributyltin oxide (703) and sulfide (534) and triphenyltin chloride. Absorbed into porous materials they have been found to protect measuring instruments during extended oceanographic investigations for periods of months and even years against overgrowths (533).

Triphenyltin compounds are used against snails (carriers of bilharzia in eastern and African countries), and also against pests of paddy rice.

Tributyl- and triphenyl-tin chloride and triphenyltin acetate are employed for long-term protection against rodents (912), e.g. by storage of foodstuffs in impregnated sacks.

Insecticidal properties of organotin compounds were noted as early as 1929 (282). Surprisingly, these observations were not followed up until recent years (309, 354), since when there has been rapid development (81) (reviewed in ref. 45). Mono-, di-, and tetra-organotins are non-toxic or barely toxic to insects (81, 386). In contrast, triorganotin compounds, whether aliphatic or aromatic, are mostly highly toxic. Toxicity towards insects is nearly the same for derivatives from the methyl to about the pentyl (386), in contrast to fungicidal activity which goes through a maximum (see above). Practical application is of course restricted to compounds that cannot become a danger to man. In field trials, good insecticidal activity has been achieved with triphenyltin acetate, chloride, and hydroxide against citrus rust mite (*Phyllocoptruta oleivorus*), tobacco

* The U.S. Aluminum Association has recommended that organotin-based paints be used exclusively as anti-foulants on aluminum-hulled boats (42a).

hormworm (*Protoparce sexta*), and cabbage-moth caterpillars (*Mamestra brassicae*) (45).

An interesting mode of attack on insect pests stems from the observation that triphenyltin acetate and hydroxide (and other R_3SnX compounds) strongly inhibit feeding even of those species against which they show only slight insecticidal activity. To this category belong the larvae of the clothes moth (*Tineola bisselliella*), of the house fly (44c), and of the carpet beetle (309), and termites against which wood can be protected for a whole year by a single application of tributyltin oxide (200). Other larvae include those of the potato tuber moth (*Gnorimoschema operculella*) and the striped maize borer (*Chilo agamemnon*). This type of protection of sugar-beet, tobacco (42c), and other commercial crops by triphenyltins has become important especially in Middle Eastern countries, where it is used e.g. against the climbing cutworm or cotton-leaf worm (*Prodenia litura*) and the black cutworm (*Agrotis ipsilon*). Even larvae starved beforehand show no inclination to feed on treated leaves. The cause lies in inhibition of the feeding process, not in some form of repellent action (45—47).

Considerable chemosterilant activity against the male housefly is observed for triphenyltin acetate and hydroxide (44b), whereas trialkyltins seem to be inactive in this respect.

25-4 VARIOUS APPLICATIONS

Organotin compounds, e.g. dibutyltin dilaurate or tetraphenyltin, are added to chlorinated lubricating oils (6), transformer oils (315, 329, 475), and solvents (148) as inhibitors against corrosion by HCl. Probably the mechanism again involves capture of reactive radicals by cleavage of the Sn–C bond (see also Chapters 4-4 and 25-1). The same mechanism may account for the protective action of organotins against degradation of rubber and similar polymers (148) (summary up to 1959 in ref. 315). The substances exhibiting this activity are identical with, or very similar to, those mentioned in Chapter 25-1. Catalytic activity is claimed for organotins in vulcanization (57), and in the production of silicones, polyesters, and polyolefins (127e, 315, 329, 837). Dibutyltin dilaurate, stannous octoate, and some other tin compounds are used in the production of polyurethanes. They catalyze the addition of the hydroxy-compound to the isocyanate [for possible mechanisms (185, 684) see also the work of A. G. Davies et al. in Chapter 19-2-2], and enable the rate of addition to be kept in step with the rate of CO_2 evolution. This was difficult with earlier catalysts, e.g. triethylamine, and meant that the process had often to be carried out in two stages. Some of the organotin compounds are about 10^4 times as active, and thus render a one stage process feasible. They show synergistic action with the amine-type catalysts (39, 307, 329). The kinetics of the doubly catalyzed reaction of isocyanates with alcohols (684) and ureas (185) have been investigated in greater detail and mechanisms proposed. In the latter reaction, butyltin trichloride is 28 times as effective as triethyltin chloride.

Organotin compounds are added to lubricating oils not only for removal of HCl but also to improve the lubricating properties as such, especially for high-pressure applications. Dibutyltin sulfide, which was in the first instance picked more or less by chance, has proved superior to other additives both in mineral oil and in silicone oil lubricants. The term "non-corrosive film formation" is used (44). In this connection, an unspecified organotin compound has proved an excellent anti-wear additive for high-performance two-stroke motors, e.g. for chain saws, welding plants, and outboard motors (784). Several patents are concerned with this particular application (315).

Organotin compounds have also been proposed for deposition of a conducting film on glass (electroluminescence), prevention of static charge, improved adhesion of glass fibers to synthetic resins, and protection of silicones intended for the textile and paper industries.

References

1. Abel, E. W., and D. B. Brady, *J. Chem. Soc.*, 1192 (1965).
1a. Abel, E. W., and D. A. Armitage, *Adv. Organomet. Chem.*, **5**, 1 (1967).
2. Abel, E. W., D. Brady, and B. R. Lerwill, *Chem. and Ind.*, 1333 (1962).
2a. Abel, E. W., and C. R. Jenkins, *J. Chem. Soc.*, *A*, 1344 (1967).
3. Addison, C. C., W. B. Simpson, and A. Walker, *J. Chem. Soc.*, 2360 (1964).
4. Adrlanov, K. A., *J. Polymer Sci.*, **52**, 257 (1961).
5. Adrova, N. A., M. M. Koton, and V. A. Klages, *Vysokomol. Soed.*, **3**, 1041 (1961).
6. Airs, R. S., and H. C. Evans (Shell Res. Ltd.), Brit. Pat. 833,873 (May 4, 1960), *C.A.*, **54**, 21745 (1960): J. P. McDermott (Esso Research and Engng. Co.), U.S. Pat. 2,786,814 (March 26, 1957), *C.A.*, **51**, 10892 (1957).
6a. Albert, H.-J., Dissertation, University of Giessen, 1969; H.-J. Albert, W. P. Neumann, W. Kaiser, and H.-P. Ritter, *Chem. Ber.*, in the press.
7. Aldridge, W. N., and J. E. Cremer, *Biochem. J.*, **61**, 406 (1955).
7a. Aleksandrov, Yu. A., V. N. Glushakova, and B. A. Radbil, *Tr. Khim. Tekhnol.*, 69 (1967).
8. Aleksandrov, Yu. A., K. P. Mitrofanov, O. Yu. Okhlobystin, L. S. Polak, and V. S. Shpinel, *Doklady Akad. Nauk SSSR*, **153**, 370 (1963).
8a. Aleksandrov, Yu. A., and N. G. Sheyanov, *Zh. Obshch. Khim.*, **36**, 953 (1966).
8b. Aleksandrov, Yu. A., N. G. Sheyanov, and V. A. Shushunov, *Zh. Obshch. Khim.*, **38**, 1352 (1968).
9. Aleksandrov, Yu. A., and V. A. Shushunov, *Doklady Akad. Nauk. SSSR*, **140**, 595 (1961).
10. Aleksandrov, Yu. A., and V. A. Shushunov, *Trudy po Khim. i Khim. Tekhnol.*, **4**, 644 (1962).
11. Aleksandrov, Yu. A., and V. A. Shushunov, *Zh. Obshch. Khim.*, **35**, 115 (1965); Aleksandrov, Yu. A., and B. A. Radbil, *ibid.*, **36**, 543 (1966).
12. Aleksandrov, Yu. A., and N. N. Vyshinskii, *Trudy po Khim. i Khim. Tekhnol.*, **4**, 656 (1962).
13. Alleston, D. L., and A. G. Davies, *Chem. and Ind.*, 551 (1961).
14. Alleston, D. L., and A. G. Davies, *Chem. and Ind.*, 949 (1961).
15. Alleston, D. L., and A. G. Davies, *J. Chem. Soc.*, 2050 (1962).
16. Alleston, D. L., and A. G. Davies, *J. Chem. Soc.*, 2465 (1962).
17. Alleston, D. L., A. G. Davies, and B. N. Figgis, *Proc. Chem. Soc.* 457 (1961); *Angew. Chem.*, **73**, 683 (1961).
18. Alleston, D. L., A. G. Davies, and M. Hancock, *J. Chem. Soc.*, 5744 (1964).
19. Alleston, D. L., A. G. Davies, M. Hancock, and R. F. M. White, *J. Chem. Soc.* 5469 (1963).
20. Allred, A. L., and E. G. Rochow, *J. Inorg. Nucl. Chem.*, **5**, 269 (1958); **17**, 215 (1961); **20**, 167 (1961).
21. Amberger, E., H. P. Fritz, C. G. Kreiter, and M. R. Kula, *Chem. Ber.*, **96**, 3270 (1963).
22. Amberger, E., and M. R. Kula, *Chem. Ber.*, **96**, 2560 (1963).
23. Amberger, E., M. R. Kula, and J. Lorberth, *Angew. Chem.*, **76**, 145 (1964).
24. Cf. ref. 117.
25. Anderson, D. G., M. A. M. Bradney, B. A. Loveland, and D. E. Webster, *Chem. and Ind.*, 505 (1964).
25a. Anderson, D. G., J. R. Chipperfield, and D. E. Webster, *J. Organomet. Chem.*, **12**, 323 (1968).
26. Anderson, H. H., *J. Org. Chem.*, **19**, 1766 (1954).

27. Anderson, H. H., *J. Amer. Chem. Soc.*, **79**, 4913 (1957).
28. Anderson, H. H., *J. Org. Chem.*, **22**, 147 (1957).
29. Anderson, H. H., *Inorg. Chem.*, **1**, 647 (1962).
30. Anderson, H. H., *Inorg. Chem.* **3**, 108 (1964).
31. Anderson, H. H., *Inorg. Chem.*, **3**, 912 (1964).
32. Anderson, H. H., and J. A. Vasta, *J. Org. Chem.*, **19**, 1300 (1954).
33. Andreas, H., O. Klump, and I. Menzel (Deutsche Advance Prod. GmbH.), Brit. Pat. 902,360 (August 1, 1962), *C.A.*, **58**, 550 (1963).
34. Andrews, T. M., F. A. Bower, B. R. La Liberte, and J. C. Montermoso, *J. Amer. Chem. Soc.*, **80**, 4102 (1958).
35. Anon., *Tin and its Uses*, **47**, 11 (1959).
36. Anon., *Tin and its Uses*, **48**, 14 (1959).
37. Anon., *Tin and its Uses*, **50**, 9 (1960). Further references there.
38. Anon., Huntington Research Centre Rep., 1952–1961, pp. 27, 52.
39. Anon., *Tin and its Uses*, **55**, 7 (1962).
40. Anon., *Chemical Age*, **91**, 229 (1964); *Chem. and Engng. News*, **42**, No. 3, 25 (1964).
41. Anon., *Tin and its Uses*, **67**, 3 (1965).
42. Anon., Report of IUPAC Symposium on Natural Products, Kyoto, 1964, *Nachr. aus Chem. und Technik*, **13**, 27 (1965).
42a. Anon., *Tin and its Uses*, **78**, 4 (1968).
42b. Anon., *Tin and its Uses*, **78**, 1 (1968), cited from U.S. Federal Register, Washington, D.C., **33**, No. 78, p. 1, April 20, 1968.
42c. Anon., *Tin and its Uses*, **78**, 2 (1968), cited from *Indian J. Entomology*, **29**, 18 (1967).
42d. Anon., *Tin and its Uses*, **77**, 14 (1968), gives a review.
43. D'Ans, J., and H. Gold., *Chem. Ber.*, **92**, 3076 (1959).
44. Antler, M., *Ind. Engng. Chem.*, **51**, 753 (1959).
44a. Arnold, D. R., and V. Y. Abraitys, *Chem. Comm.*, 1053 (1967).
44b. Ascher, K. R. S., J. Meisner, and S. Nissim, *World Rev. Pest Control*, 7, 84 (1968).
44c. Ascher, K. R. S., J. Moscowitz, and S. Nissim, cited in *Tin and Its Uses*, **73**, 8 (1967).
45. Ascher, K. R. S., and S. Nissim, *World Rev. Pest Control*, **3**, 188 (1964); survey giving 151 references.
46. Ascher, K. R. S., and G. Rones, *Internat. Pest Contr.*, **6** (4), 4 (1964).
47. Ascher, K. R. S., and S. Nissim, *Internat. Pest. Contr.*, **7** (July), 21 (1965); cited in *Tin and Its Uses*, **68**, 12 (1965).
48. Aufdermarsh, C. A., *J. Org. Chem.*, **29**, 1994 (1964).
49. Aufdermarsh, C. A., and R. C. Ferguson, Lecture abstract, *Chem. and Engng. News*, **41**, 42 (1963).
50. Austin, R. P., *J. Amer. Chem. Soc.*, **54**, 3726 (1932).
51. Bachman, G. B., C. L. Carlson, and M. Robinson, *J. Amer. Chem. Soc.*, **73**, 1964 (1951).
52. Backer, H. J., and J. Kramer, *Rec. Trav. Chim.*, **53**, 1101 (1934).
53. Bähr, G., *Z. Anorg. Allg. Chem.*, **256**, 107 (1948).
54. Bähr, G., and R. Gelius, *Chem. Ber.*, **91**, 812 (1958).
55. Bähr, G., and R. Gelius, *Chem. Ber.*, **91**, 818 (1958).
56. Bähr, G., and G. Zoche, *Chem. Ber.*, **88**, 1450 (1955).
56a. Balasova, L. D., A. B. Bruker, and L. Z. Soborovskii, *Zh. Obshch. Khim.*, **35**, 2207 (1965).
57. Baranovskaya, N. B., A. A. Berlin, M. Z. Zakharova, A. I. Mizikin, and E. N. Zil'berman, S.S.S.R. Pat. 126,115 (February 10, 1960), *C.A.*, **54**, 16008 (1960).
58. Barnes, J. M., and H. B. Stoner, *Brit. J. Industr. Med.*, **15**, 15 (1958).

59. Barnes, J. M., and H. B. Stoner, *Pharmacol. Rev.*, **11**, 211 (1959); review with references.
60. Bartlett, P. D., and T. Funahashi, *J. Amer. Chem. Soc.*, **84**, 2596 (1962).
61. Baukov, J., J. Belavin, and J. F. Lucenko, *Zh. Obshch. Khim.*, **35**, 1092 (1965).
62. Baum, G. A., and W. J. Considine, *J. Polymer. Sci.*, *B*, **1**, 517 (1963).
63. Baum, G. A., and W. J. Considine, *J. Org. Chem.*, **29**, 1267 (1964).
64. Bajer, F. J., and H. W. Post, *J. Org. Chem.*, **27**, 1422 (1962).
65. Beattie, I. R., *Quart. Rev.*, **17**, 382 (1963).
66. Beattie, I. R., and T. Gilson, *J. Chem. Soc.*, 2585 (1961).
67. Beattie, I. R., and G. P. McQuillan, *J. Chem. Soc.*, 1519 (1963).
68. Beattie, I. R., G. P. McQuillan, and R. Hulme, *Chem. and Ind.*, 1429 (1962).
69. Beck, W., and E. Schuierer, *Chem. Ber.*, **97**, 3517 (1964).
70. Beg, M. A. A., and H. C. Clark, *Chem. and Ind.*, 140 (1962).
71. Berghe, E. V. van den, and G. P. van der Kelen, *Bull Soc. Chim. Belges*, **74**, 479 (1965).
71a. Berghe, E. V. van den, and G. P. van der Kelen, *J. Organomet. Chem.*, **6**, 522 (1966); **11**, 479 (1968).
72. Beumel, O. F., Jr., Dissertation, Univ. of Durham, New Hampshire, 1960; cited in ref. 429.
72a. Birnbaum, E. R., and P. H. Javora, *J. Organomet. Chem.*, **9**, 379 (1967).
73. Blake, D., G. E. Coates, and J. M. Tate, *J. Chem. Soc.* 618 (1961).
74. Blake, D., G. E. Coates, and J. M. Tate, *J. Chem. Soc.* 756 (1961).
74a. Blaukat, U., Diploma Thesis, Univ. of Giessen, 1968; W. P. Neumann and U. Blaukat, *Angew. Chem.*, **81**, 625 (1969).
74b. Blears, D. J., S. S. Danyluk, and S. Cawley, *J. Organomet. Chem.*, **6**, 284 (1966).
74c. Bloodworth, A. J., *J. Chem. Soc.,C*, 2380 (1968).
75. Bloodworth, A. J., and A. G. Davies, *Proc. Chem. Soc.* 264 (1963).
76. Bloodworth, A. J., and A. G. Davies, *Proc. Chem. Soc.* 315 (1963).
77. Bloodworth, A. J., and A. G. Davies, *Chem. Comm.* 24 (1965).
78. Bloodworth, A. J., and A. G. Davies, *J. Chem. Soc.* 5238 (1965).
79. Bloodworth, A. J., and A. G. Davies, *J. Chem. Soc.*, 6245, 6858 (1965).
80. Bloodworth, A. J., and A. G. Davies, *J. Chem. Soc.*, *C*, 299 (1966). Further references there.
80a. Bloodworth, A. J., A. G. Davies, and I. F. Graham, *J. Organomet. Chem.*, **13**, 351 (1968).
80b. Bloodworth, A. J., A. G. Davies, and S. C. Vasishta, *J. Chem. Soc.*, *C*, 1309, 2640 (1967).
81. Blum, M. S., and F. A. Bower, *J. Econ. Ent.*, **50**, 84 (1957); M. S. Blum, J. J. Pratt, Jr., *ibid.*, **53**, 445 (1960).
82. Boboli, E., J. Kamionska, E. Kremky, and A. Pazgan, Pol. Pat. 47,960 (January 25, 1964), *C.A.*, **61**, 14710 (1964).
83. Bock, R., S. Gorbach, and H. Oeser, *Angew. Chem.*, **70**, 272 (1958).
84. Bock, R., H. T. Niederauer, and K. Behrends, *Z. Anal. Chem.*, **190**, 33 (1962).
85. Boeseken, J., and J. J. Rutgers, *Rec. Trav. Chim.*, **42**, 1017 (1923).
86. Borisov, A. E., and N. V. Novikova, *Izvest. Akad. Nauk SSSR* 1670 (1959).
87. Bost, R. W., and P. Borgstrom, *J. Amer. Chem. Soc.*, **51**, 1922 (1929).
88. Bott, R. W., C. Eaborn, and T. W. Swaddle, *J. Chem. Soc.* 2342 (1963).
89. Bott, R. W., C. Eaborn, and D. R. M. Walton, *J. Organomet. Chem.*, **2**, 154 (1964).
90. Bott, R. W., C. Eaborn, and J. A. Waters, *J. Chem. Soc.* 681 (1963).
91. Boué, S., M. Gielen, and J. Nasielski, *Bull. Soc. Chim. Belges*, **73**, 864 (1964).
91a. Boué, S., M. Gielen, and J. Nasielski, *Bull. Soc. Chim. Belges*, **76**, 559 (1967).
91b. Boué, S., M. Gielen, and J. Nasielski, *Tetrahedron Letters*, 1047 (1968).

92. Braun, D., *Angew Chem.*, **73**, 197 (1961); review giving further references.
93. Braye, E. H., W. Hübel, and I. Caplier, *J. Amer. Chem. Soc.*, **83**, 4406 (1961).
94. Brinckman, F. E., and F. G. A. Stone, *J. Inorg. Nucl. Chem.*, **11**, 24 (1959).
95. Brockway, L. O., and H. O. Jenkins, *J. Amer. Chem. Soc.*, **58**, 2036 (1936).
96. Brown, D. H., A. Mohammed, and D. W. A. Sharp, *Spectrochim. Acta*, **21**, 1013 (1965).
97. Brown, H. C., *Hydroboration*, W. A. Benjamin, Inc., New York, 1962.
98. Brown, M. P., E. Cartmell, and G. W. A. Fowles, *J. Chem. Soc.*, 506 (1960).
99. Brown, M. P., and G. W. A. Fowles, *J. Chem. Soc.*, 2811 (1958).
100. Brown, M. P., R. Okawara, and E. G. Rochow, *Spectrochim. Acta*, **16**, 595 (1960).
101. Brown, T. L., and G. L. Morgan, *Inorg. Chem.* **2**, 736 (1963).
102. Brüggemann, J., K. Barth, and K. H. Niesar, *Zentralbl. Vet. Med.,A*, **11**, 20 (1964).
103. Brüggemann, J., and O. R. Klimmer, *Zentralbl. Vet. Med.,A*, **11**, 1 (1964), and four following papers.
104. Brüggemann, J., O. R. Klimmer, and K. H. Niesar, *Zentralbl. Vet. Med., A*, **11**, 40 (1964).
105. Bruker, A. B., L. D. Balashova, and L. Z. Soborovskij, *Doklady. Akad. Nauk SSSR*, **135**, 843 (1960).
106. Bryan, R. F., *Proc. Chem. Soc.* 232 (1964).
106a. Bryochova, E. V., G. K. Semin, V. I. Goldanskii, and V. V. Khrapov, *Chem. Comm.*, 491 (1968).
107. Buchman, O., M. Grosjean, J. Nasielski, and B. Wilmet-Devos, *Helv. Chim. Acta*, **47**, 1688 (1964). Further references there.
108. Buchman, O., M. Grosjean, and J. Nasielski, *Helv. Chim. Acta*, **47**, 1695 (1964). Further references there.
109. Buisson, R., M. Lesbre, and J. G. A. Luijten, *Bull. Soc. Chim. France*, 754 (1956).
109a. Bulten, E. J., and J. G. Noltes, *Tetrahedron Letters*, 4389 (1966).
110. Burg, A. B., and J. R. Spielman, *J. Amer. Chem. Soc.*, **83**, 2667 (1961).
111. Bürger, K., *Z. Lebensmittel-Unters. u. -Forschung*, **114**, 1, 10 (1961).
112. Bürger, K., *Z. Anal. Chem.* **183**, 438 (1961).
113. Bürger, K., *Z. Anal. Chem.* **192**, 280 (1963).
114. Bürger, K., *Tin and its Uses*, **61**, 7 (1964).
115. Burt, S. L. (Union Carbide and Carbon Corp.), U.S. Pat. 2,583,084 (January 22, 1952), *C.A.* **47**, 146 (1953).
116. Butcher, F. K., W. Gerrard, E. F. Mooney, R. G. Rees, and H. A. Willis, *Spectrochim. Acta*, **20**, 51 (1964).
117. Butcher, F. K., W. Gerrard, E. F. Mooney, R. G. Rees, H. A. Willis, A. Anderson, and H. A. Gebbie, *J. Organomet. Chem.*, **1**, 431 (1964).
118. Bychkov, V. T., and N. S. Vyazankin, *Zh. Obshch. Khim.*, **35**, 687 (1965).
119. Cahours, A., *Liebigs Ann. Chem.*, **114**, 354 (1860).
120. Cahours, A., and A. Riche, *Compt. Rend.*, **35**, 91 (1852).
121. Calas, R., J. Valade, and J.-C. Pommier, *Compt. Rend.* **255**, 1450 (1962).
122. Caldwell, J. R. (Eastman Kodak Co.), U.S. Pat. 2,720,507 (October 11, 1955), *C.A.*, **50**, 2205 (1956).
123. Calingaert, G., H. A. Beatty, and H. R. Neal, *J. Amer. Chem. Soc.*, **61**, 2755 (1939).
124. Calingaert, G., H. Soroos, and V. Hnizda, *J. Amer. Chem. Soc.*, **62**, 1107 (1940). Further references there.
125. Calingaert, G., H. Soroos, and G. W. Thomson, *J. Amer. Chem. Soc.*, **62**, 1542 (1940). Further references there.
126. Campbell, I. G. M., G. W. A. Fowles, and L. A. Nixon, *J. Chem. Soc.*, 3026 (1964).
127. Campbell, I. G. M., G. W. A. Fowles, and L. A. Nixon, *J. Chem. Soc.*, 1389 (1964).

127a. Capacho-Delgado, L., and D. C. Manning, *Spectrochim. Acta*, **22**, 1505 (1966).
127b. Carberry, E., and R. West, *J. Organomet. Chem.*, **6**, 583 (1966).
127c. Cardin, D. J., and M. F. Lappert, *Chem. Comm.*, 1034 (1967).
127d. Carlsson, D. J., and K. U. Ingold, *J. Amer. Chem. Soc.*, **90**, 1055 (1968).
127e. Carrick, W. L., and R. W. Kluiber (Union Carbide Corp.), Brit. Pat. 886,920 (January 10, 1962), *C.A.*, **57**, 7464 (1962).
128. Cassol, A., L. Magon, and R. Barbier, *J. Chromatog.*, **19**, 57 (1965).
129. Caujolle, F., M. Lesbre, D. Meynier, and G. de Saqui-Sannes, *Compt. Rend.*, **243**, 987 (1956).
129a. Chalmers, L., *Mfg. Chem. Aerosol News*, **38**, 37 (1967).
129b. Chambers, D. B., F. Glockling, and J. R. C. Light, *Quart. Rev.*, **22**, 317 (1968).
130. Chambers, R. D., and T. Chivers, *Proc. Chem. Soc.*, 208 (1963).
131. Chambers, R. F., and P. C. Scherer, *J. Amer. Chem. Soc.*, **48**, 1054 (1926).
131a. Chao-lun, T., Z. Ren-xi, and M. Shiun-tsun, *Kexue Tongbao* (Peking), **17**, No. 2, 71 (1966).
132. Chapman, A. H., M. W. Duckworth, and J. W. Price, *Brit. Plastics*, 78 (1959),
133. Chatt, J., and A. A. Williams, *J. Chem. Soc.*, 4403 (1954).
134. Clark, H. C., S. G. Furnival, and J. T. Kwon, *Canad. J. Chem.*, **41**, 2889 (1963); C. Barnetson, H. C. Clark, and J. T. Kwon, *Chem. and Ind.*, 458 (1964).
134a. Clark, H. C. and R. G. Goel, *Inorg. Chem.* **4**, 1428 (1965).
135. Clark, H. C., J. T. Kwon, L. W. Reeves, and E. J. Wells, *Inorg. Chem.*, **3**, 907 (1964).
136. Clark, H. C., J. T. Kwon, L. W. Reeves, and E. J. Wells, *Canad. J. Chem.*, **42**, 941 (1964).
137. Clark, H. C., and J. T. Kwon, *Canad. J. Chem.*, **42**, 1288 (1964).
138. Clark, H. C., and R. J. O'Brien, *Proc. Chem. Soc.*, 113 (1963); *Inorg. Chem.*, **2**, 1020 (1963).
139. Clark, H. C., R. J. O'Brien, and A. L. Pickard, *Inorg. Chem.*, **2**, 740 (1963). Some of the results later amended in *J. Organomet. Chem.*, **4**, 43 (1965).
140. Clark, H. C., R. J. O'Brien, and J. Trotter, *Proc. Chem. Soc.*, 85 (1963); *J. Chem. Soc.*, 2332 (1964).
141. Clark, H. C., and J. H. Tsai, *Chem. Comm.*, 111 (1965).
142. Clark, H. C., and C. J. Willis, *J. Amer. Chem. Soc.*, **82**, 1888 (1960).
143. Coates, G. E., and K. Wade, *Organometallic Compounds*, 3rd Edn., Methuen & Co., London, 1968, Vol. I.
144. Colaitis, D., and M. Lesbre, *Bull. Soc. Chim. France*, 1069 (1952).
145. McCombie, H., and B. C. Saunders, *Nature*, **159**, 491 (1947).
145a. Considine, W. J., *J. Organomet. Chem.*, **5**, 263 (1966).
146. Considine, W. J., personal communication.
147. Considine, W. J., G. A. Baum, and R. C. Jones, *J. Organomet. Chem.*, **3**, 308 (1965).
148. Considine, W. J., A. Ross, and R. J. Daum, lecture given at conference on Industrial Syntheses and Applications of Organometallics, New York, June 1964.
149. Considine, W. J., and J. J. Ventura, *Chem. and Ind.*, 1683 (1962).
150. Considine, W. J., and J. J. Ventura, *J. Org. Chem.*, **28**, 221 (1963).
151. Considine, W. J., J. J. Ventura, A. J. Gibbons, Jr., and A. Ross, *Canad. J. Chem.*, **41**, 1239 (1963).
152. Considine, W. J., J. J. Ventura, B. G. Kushlefsky, and A. Ross, *J. Organomet. Chem.*, **1**, 299 (1964).
153. Cooke, D. J., G. Nickless, and F. H. Pollard, *Chem. and Ind.*, 1493 (1963).
153a. Cordey-Hayes, M., R. D. Peacock, and M. Vucelic, *J. Inorg. Nucl. Chem.*, **29**, 1177 (1967).

154. Cottrell T. L., *The Strengths of Chemical Bonds*, 2nd Edn., Butterworths, London, 1958.
155. Cragg, R. H., and M. F. Lappert, *J. Chem. Soc.*, *A*, 82 (1966).
156. Crain, R. D., and P. E. Koenig, lecture abstract 1959, cited in ref. 314.
157. Crauland, M. (S. A. des Manuf., St. Gobain), D.B. Pat. 1,046,053 (December 11, 1958), *C.A.*, **54**, 24401 (1960).
157a. Creemers, H. M. J. C., A. J. Leusink, J. G. Noltes, and G. J. M. van der Kerk, *Tetrahedron Letters*, 3167 (1966).
158. Creemers, H. M. J. C., and J. G. Noltes, *Rec. Trav. Chim.*, **84**, 382 (1965). Further references there.
159. Creemers, H. M. J. C., and J. G. Noltes, *Rec. Trav. Chim.*, **84**, 590 (1965).
159a. Creemers, H. M. J. C., and J. G. Noltes, *Rec. Trav. Chim.*, **84**, 1589 (1965).
159b. Creemers, H. M. J. C., and J. G. Noltes, *J. Organomet. Chem.*, **7**, 237 (1967).
160. Creemers, H. M. J. C., J. G. Noltes, and G. J. M. van der Kerk, *Rec. Trav. Chim.*, **83**, 1284 (1964).
161. Cremer, J. E., *Biochem. J.*, **67**, 28P (1957); **68**, 685 (1958).
162. Cullen, W. R., D. S. Dawson, and G. E. Styan, *J. Organomet. Chem.*, **3**, 406 (1965).
163. Cummins, R. A., *Austral. J. Chem.*, **16**, 985 (1963).
164. Cummins, R. A., *Austral. J. Chem.*, **17**, 594 (1964).
165. Cummins, R. A., and P. Dunn, *The Infrared Spectra of Organotin Compounds* (Bibliography, 198 Spectra), Dept. of Supply, Austral. Defence Sci. Serv., Report 266 (1963).
166. Cummins, R. A., and P. Dunn, *Austral. J. Chem.*, **17**, 185 (1964).
167. Cummins, R. A., and P. Dunn, *Austral. J. Chem.*, **17**, 411 (1964).
168. Cummins, R. A., and J. V. Evans, *Spectrochim. Acta*, **21**, 1016 (1965). Further references there.
168a. Curtis, M. D., and A. L. Allred, *J. Amer. Chem. Soc.*, **87**, 2554 (1965).
169. Daněk, O., *Coll. Czech. Chem. Comm.*, **26**, 2035 (1961).
170. Dannley, R. L., and W. A. Aue, *J. Org. Chem.*, **30**, 3845 (1965).
170a. Daum, R. J., *Ann. New York Acad. Sci.*, **125**, Art. 1, 229 (1965).
171. Dave, L. D., D. F. Evans, and G. Wilkinson, *J. Chem. Soc.*, 3684 (1959).
172. Davidson, W. E., K. Hills, and M. C. Henry, *J. Organomet. Chem.*, **3**, 285 (1965).
173. Davies, A. G., *Trans. New York Acad. Sci.*, *Ser. II*, **26**, 923 (1964). Review.
173a. Davies, A. G., *Chemistry in Britain*, **4**, 403 (1968).
173b. Davies, A. G., *Synthesis*, 56 (1969).
174. Davies A. G., and I. F. Graham, *Chem. and Ind.*, 1622 (1963).
175. Davies, A. G., and M. Hancock, unpublished. Cited in ref. 82.
175a. Davies, A. G., and P. G. Harrison, *J. Chem. Soc.*, *C*, 298, 1313 (1967).
175b. Davies, A. G., and T. N. Mitchell, *J. Organomet. Chem.*, **6**, 568 (1966).
175c. Davies, A. G., T. N. Mitchell, and W. R. Symes, *J. Chem. Soc.*, *C*, 1311 (1966).
175d. Davies, A. G., P. R. Palan, and S. C. Vasishtha, *Chem. and Ind.*, 229 (1967).
175e. Davies, A. G., and T. N. Mitchell, *J. Chem. Soc.*, *C*, 1896 (1969).
175f. Davies, A. G., P. G. Harrison, J. D. Kennedy, T. N. Mitchell, R. J. Puddephatt, and W. McFarlane, *J. Chem. Soc.*, *C*, 1136 (1969).
176. Davies, A. G., and W. R. Symes, *Chem. Comm.*, 25 (1965); *J. Organomet. Chem.*, **5**, 394 (1966); *J. Chem. Soc.*, *C*, 1009 (1967).
176a. Davies, J. V., A. E. Pope, and H. A. Skinner, *Trans. Faraday Soc.*, **59**, 2237 (1963).
176b. Des Tombe, F. J. A., G. J. M. van der Kerk, H. M. J. C. Creemers, and J. G. Noltes, *Chem. Comm.*, 914 (1966); *J. Organomet. Chem.*, **13**, P9 (1968).
177. Dillard, C. R., E. H. McNeill, D. E. Simmons, and J. B. Yeldell, *J. Amer. Chem. Soc.*, **80**, 3607 (1958).
177a. Dörfelt, C., A. Janek, D. Kobelt, E. F. Paulus, and H. Scherer, *J. Organomet. Chem.*, **14**, P22 (1968).

178. Dörfelt, Chr., E. Reindl, and K. Härtel (Farbwerke Hoechst), D.B. Pat. 1,079,329 (April 7, 1960), *C.A.*, **55**, 14382 (1961).
179. Dreeskamp, H., *Z. Naturforsch.*, **19b,** 139 (1964), and private communications.
180. Drefahl, G., and D. Lorenz, *J. Prakt. Chem.*, **24**, 106 (1964).
181. Drefahl, G., G. Plötner, and D. Lorenz, *Angew. Chem.*, **72**, 454 (1960).
182. Drenth, W., M. J. Janssen, G. J. M. van der Kerk, and J. A. Vliegenthart, *J. Organomet. Chem.*, **2**, 265 (1964).
183. Dub, M., *Organometallic Compounds*, Vol. II, Springer, Berlin, 1961.
184. Dunn, P., and T. Norris, Austral. Dept. of Supply, Rept. 269, p. 21 (1964).
185. Dyer, E., and R. B. Pinkerton, *J. Appl. Polymer Sci.*, **9**, 1713 (1965).
186. Eaborn, C., and K. C. Pande, *J. Chem. Soc.*, 1566 (1960).
186a. Eaborn, C., H. L. Hornfeld, and D. R. M. Walton, *J. Chem. Soc.*, *B*, 1036 (1967).
186b. Eaborn, C., A. R. Thompson, and D. R. M. Walton, *Chem. Comm.*, 1051 (1968).
187. Eaborn, C., and J. A. Waters, *J. Chem. Soc.*, 1131 (1962).
188. Eberly, K. C. (Firestone Tire and Rubber Co.), U.S. Pat. 2,560,034 (July 10, 1951), *C.A.*, **46**, 7583 (1952).
189. Edgell, W. F., and C. H. Ward, *J. Amer. Chem. Soc.*, **76**, 1169 (1954).
190. Edgell, W. F., and C. H. Ward, *J. Mol. Spectr.*, **8**, 343 (1962).
190a. Edmondson, R. C., and M. J. Newlands, *Chem. Comm.*, 1219 (1968).
190b. Efer, J., D. Quaas, and W. Spichale, *Z. Chem.*, **5**, 390 (1965).
190c. Egger, R., Dissertation, Univ. of Munich, 1961.
191. Egmond, J. C. van, M. J. Janssen, J. G. A. Luijten, G. J. M. van der Kerk, and G. M. van der Want, *J. Appl. Chem.*, **12**, 17 (1962).
192. Elsea, J. R., and O. E. Paynter, *Am. Med. Ass. Arch. Industr. Health*, **18**, 214 (1958).
193. Emeléus, H. J., and S. F. A. Kettle, *J. Chem. Soc.*, 2444 (1958).
194. Emeléus, H. J., and J. J. Zuckerman, *J. Organomet. Chem.*, **1**, 328 (1964).
195. Emmert, B., and W. Eller, *Ber.*, **44**, 2328 (1911).
195a. Emsley, J. W., J. Feeney, and L. H. Sutcliffe, *High Resolution Nuclear Magnetic Resonance Spectroscopy*, Pergamon Press, Oxford, 1968.
196. Eskin, I. T., A. N. Nesmeyanov, and K. A. Kocheshkov, *Zh. Obshch. Khim.*, **8**, 35 (1938), *C.A.*, **32**, 5386 (1938).
197. Evans, F. W., R. J. Fox, and M. Szwarc, *J. Amer. Chem. Soc.*, **82**, 6414 (1960).
198. Evers, E. C., lecture abstract, 1959; cited in ref. 314.
199. Evieux, E. (Soc. Usines Chim. Rhône-Poulenc), Fr. Pat. 1,153,234 (March 4, 1958), and supplements.
200. Fahlstrom, G. B., *Proc. Amer. Wood Preserv. Ass.*, **54**, 178 (1958); cited in *Tin and its Uses*, **46**, 4 (1959).
201. Farbwerke Hoechst AG., Brit. Pat. 841,151 (July 13, 1960), *C.A.*, **54**, 26007 (1960).
202. Farnsworth, M., and J. Pekola, *Anal. Chem.*, **31**, 410 (1959).
203. Farrar, W. V., and H. A. Skinner, *J. Organomet. Chem.*, **1**, 434 (1964).
204. Fergusson, J. E., W. R. Roper, and C. J. Wilkins, J. Chem. Soc., 3716 (1965).
205. Fineman, M. A., and R. Daignault, *J. Inorg. Nucl. Chem.*, **10**, 205 (1959).
206. Finholt, A. E., A. C. Bond, Jr., K. E. Wilzbach, and H. I. Schlesinger, *J. Amer. Chem. Soc.*, **69**, 2692 (1947).
206a. Fish, R. H., H. G. Kuivila, and I. J. Tyminski, *J. Amer. Chem. Soc.*, **89**, 5861 (1967).
206b. Fitzsimmons, B. W., N. J. Seeley, and A. W. Smith, *Chem. Comm.*, 390 (1968).
207. Flitcroft, N., and H. D. Kaesz, *J. Amer. Chem. Soc.*, **85**, 1377 (1963).
207a. Ford, B. F. E., B. V. Lienge, and J. R. Sams, *Chem. Comm.*, 1333 (1968).
208. Fox, V. W., J. G. Hendricks, and H. J. Ratti, *Ind. Engng. Chem.*, **41**, 1774 (1949).
209. Franc, J., M. Wurst, and V. Moudrý, *Coll. Czech. Chem. Comm.*, **26**, 1313 (1961).
210. Del Franco, G. J., P. Resnick, and C. R. Dillard, *J. Organomet. Chem.*, **4**, 57 (1965).
211. Frankel, M., D. Gertner, D. Wagner, and A. Zilkha, *J. Org. Chem.*, **30**, 1596 (1965).

211a. Frankel, M., D. Wagner, D. Gertner, and A. Zilkha, *Israel J. Chem.*, **4**, 183 (1966).
212. Frankland, E., *Liebigs Ann. Chem.*, **71**, 171, esp. 212 (1849).
213. Frankland, E., *J. Chem. Soc.*, **2**, 267 (1850).
214. Frankland, E., *Liebigs Ann. Chem.*, **85**, 329 (1853).
215. Freedman, H. H., *J. Amer. Chem. Soc.*, **83**, 2194 (1961).
216. Freedman, H. H., *J. Amer. Chem. Soc.*, **83**, 2195 (1961).
217. Freedman, H. H., *J. Org. Chem.*, **27**, 2298 (1962).
218. Freedman, H. H., and R. S. Gohlke, *Proc. Chem. Soc.*, 249 (1963).
219. Freedman, H. H., and D. R. Petersen, *J. Amer. Chem. Soc.*, **84**, 2837 (1962).
220. Freeman, J. P., *J. Amer. Chem. Soc.*, **80**, 5954 (1958).
221. Freidlina, R. Ch., and E. Z. Tchukovskaya, *Doklady Akad. Nauk SSSR*, **150**, 1055 (1963) (for example).
222. Friebe, E., and H. Kelker, *Z. Anal. Chem.*, **192**, 267 (1963).
223. Fritz, G., and H. Scheer, *Z. Naturforsch.*, **19b**, 537 (1964).
224. Fritz, G., and H. Scheer, *Z. Anorg. Allg. Chem.*, **338**, 1 (1965).
225. Fritz, H. P., and C. G. Kreiter, *J. Organomet. Chem.*, **4**, 313 (1965).
226. Fuchs, O., and H. W. Post, *Rec. Trav. Chim.*, **78**, 566 (1959).
227. Fuchs, R., and H. Gilman, *J. Org. Chem.*, **22**, 1009 (1957).
228. Fulton, R. F., Dissertation, Purdue Univ., Lafayette, Indiana, 1960.
229. Gasparič, J., and A. Cee, *J. Chromatog.*, **8**, 393 (1962).
230. Geissler, H., and H. Kriegsmann, *Z. Chem.*, **5**, 423 (1965).
231. Gelius, R., *Chem. Ber.*, **93**, 1759 (1960); *Angew. Chem.*, **72**, 322 (1960).
232. George, T. A., K. Jones, and M. F. Lappert, *J. Chem. Soc.*, 2157 (1965).
232a. George, T. A., and M. F. Lappert, *Chem. Comm.*, 463 (1966).
233. George, T. A., and M. F. Lappert, unpublished; cited in ref. 336.
233a. George, T. A., and M. F. Lappert, *J. Organomet. Chem.*, **14**, 327 (1968).
234. Gerrard, W., E. F. Mooney, and R. G. Rees, *J. Chem. Soc.*, 740 (1964).
235. Gerrard, W., and R. G. Rees, *J. Chem. Soc.*, 3510 (1964).
235a. Gibb, T. C., and N. N. Greenwood, *J. Chem. Soc.*, *A*, 43 (1966).
236. Gibbons, A. J., A. K. Sawyer, and A. Ross, *J. Org. Chem.*, **26**, 2304 (1961).
236a. Gielen, M., and G. Mayence, *J. Organomet. Chem.*, **12**, 363 (1968).
237. Gielen, M., and J. Nasielski, *J. Organomet. Chem.*, **1**, 173 (1963).
238. Gilman, H., and C. E. Arntzen, *J. Amer. Chem. Soc.*, **72**, 3823 (1950).
238a. Gilman, H., W. H. Atwell, and F. K. Cartledge, *Adv. Organomet. Chem.*, **4**, 1 (1966).
239. Gilman, H., and F. K. Cartledge, *Chem. and Ind.*, 1231 (1964); *J. Organomet. Chem.*, **5**, 48 (1966).
240. Gilman, H., F. K. Cartledge, and S.-Y. Sim, *J. Organomet. Chem.*, **4**, 332 (1965).
241. Gilman, H., and J. Eisch, *J. Org. Chem.*, **20**, 763 (1955).
242. Gilman, H., and C. W. Gerow, *J. Amer. Chem. Soc.*, **78**, 5823 (1956); *J. Org. Chem.*, **22**, 334 (1957).
243. Gilman, H., and L. A. Gist, Jr., *J. Org. Chem.*, **22**, 368 (1957).
244. Gilman, H., and R. W. Leeper, *J. Org. Chem.*, **16**, 466 (1951).
245. Gilman, H., O. L. Marrs, and S.-Y. Sim, *J. Org. Chem.*, **27**, 4232 (1962).
246. Gilman, H., O. L. Marrs, W. J. Trepka, and J. W. Diehl, *J. Org. Chem.*, **27**, 1260 (1962).
247. Gilman, H., and H. W. Melvin, Jr., *J. Amer. Chem. Soc.*, **71**, 4050 (1949).
248. Gilman, H., and S. D. Rosenberg, *J. Amer. Chem. Soc.*, **74**, 531 (1952).
249. Gilman, H., and S. D. Rosenberg, *J. Amer. Chem. Soc.*, **74**, 5580 (1952).
250. Gilman, H., and S. D. Rosenberg, *J. Amer. Chem. Soc.*, **75**, 2507 (1953).
251. Gilman, H., and S. D. Rosenberg, *J. Amer. Chem. Soc.*, **75**, 3592 (1953).

252. Gilman, H., and S. D. Rosenberg, *J. Org. Chem.*, **18**, 1554 (1953). Further references there.
253. Gilman, H., and S. D. Rosenberg, *J. Org. Chem.*, **24**, 2063 (1959).
254. Gilman, H., and G. L. Schwebke, *J. Amer. Chem. Soc.*, **85**, 1016 (1963); E. Hengge, H. Reuter, and R. Petzold, *Angew. Chem.*, **75**, 677 (1963).
254a. Gilman, H., and G. L. Schwebke, *Adv. Organomet. Chem.*, **1**, 1 (1966).
255. Gilman, H., and T. C. Wu, *J. Amer. Chem. Soc.*, **77**, 3228 (1955).
256. Glockling, F., *Quart. Rev.*, **20**, 45 (1966); numerous references.
257. Gloskey, C. R. (Metal and Thermit Corp.), U.S. Pat. 2,718,522 (September 20, 1955), *C.A.*, **50**, 8709 (1956).
258. Gloskey, C. R., *Chem. Engng. Progr.*, **58**, No. 9, 71 (1962).
259. Goddard, D., and A. E. Goddard, *J. Chem. Soc.*, **121**, 256 (1922).
259a. Goldanskii, V. J., *The Mössbauer Effect and its Applications in Chemistry*, Consultants Bureau, New York, 1964 (Van Nostrand, Princeton, 1966).
260. Goldanskii, V. J., E. F. Makarov, R. A. Stukan, V. A. Trukhtanov, and V. V. Khrapov, *Doklady Akad. Nauk SSSR*, **151**, 357 (1963).
261. Gol'dstein, I. P., E. N. Gur'yanova, and E. D. Delinskaya, K. A. Kocheshkov, *Doklady Akad. Nauk SSSR*, **136**, 1079 (1961).
261a. Gormley, J. J., and R. G. Rees, *J. Organomet. Chem.*, **5**, 291 (1966).
261b. Götze, H.-J., Dr. rer. nat. Dissertation, Univ. of Bonn, 1968.
262. Gould, E. S., *Mechanism and Structure in Organic Chemistry*, Holt, Rinehart & Winston, 1959. German 2nd Edition, Verlag Chemie, Weinheim, 1964. Further references there.
262a. Grady, G. L., and H. G. Kuivila, *J. Org. Chem.*, **34**, 2014 (1969).
263. McGrady, M. M., and R. St. Tobias, *J. Amer. Chem. Soc.*, **87**, 1909 (1965).
264. Grant, D., and J. R. van Wazer, *J. Organomet. Chem.*, **4**, 229 (1965).
265. Green, B. S., D. B. Sowerby, and K. J. Wihksne, *Chem. and Ind.*, 1306 (1960).
266. Griffiths, V. S., and G. A. W. Derwish, *J. Mol. Spectroscopy*, **3**, 165 (1959).
267. Griffiths, V. S., and G. A. W. Derwish, *J. Mol. Spectroscopy*, **7**, 233 (1961). Further references there.
268. Grohn, H., and R. Paudert, *Z. Chem.*, **3**, 89 (1963); *Angew. Chem.*, **73**, 716 (1961).
269. Grün, L., and H. H. Fricker, *Tin and its Uses*, **61**, 8 (1964). On adsorption and biological activity see H. Schulz-Utermöhl and H. Weissenstein, *Tin and its Uses*, **61**, 7 (1964).
270. Grüttner, G., *Ber.*, **50**, 1808 (1917).
271. Grüttner, G., and E. Krause, *Ber.*, **49**, 2666 (1916).
272. Grüttner, G., E. Krause, and M. Wiernik, *Ber.*, **50**, 1549 (1917).
273. Hague, D. N., and R. H. Prince, *Proc. Chem. Soc.*, 300 (1962).
274. Haltmeier, A., Diploma Thesis, Univ. of Giessen, 1963.
275. O'Hara, M., R. Okawara, and Y. Nakamura, *Bull. Chem. Soc. Japan*, **38**, 1379 (1965).
276. Harada, T., *Bull. Chem. Soc. Japan*, **14**, 472 (1939).
277. Harada, T., *Sci. Papers, Inst. Phys. Chem. Res.* (*Tokyo*), **36**, 497 (1939).
278. Harada, T., *Sci. Papers, Inst. Phys. Chem. Res.*, **38**, 115 (1940).
279. Harada, T., *Bull. Chem. Soc. Japan*, **17**, 281, 283 (1942).
280. Harada, T., and T. Okuba, *Sci. Papers, Inst. Phys. Chem. Res.*, **42**, 59 (1947).
281. Härtel, K., *Angew. Chem.*, **70**, 135 (1958). Lecture given at the Organotin Symposium, Frankfurt/Main, November 1963; cited in *Tin and its Uses*, **61**, 7 (1964), and earlier papers.
282. Hartman, E., P. Kümmel, and M. Hardtmann (I. G. Farben), Ger. Pat. 485,646 (November 8, 1929), *C.A.*, **24**, 1230 (1930); also other patents of I. G. Farben.
283. Hartmann, H., *Naturwiss.*, **52**, 303 (1965) (with K. Meyer); *ibid.*, **52**, 304 (1965) (with M. K. El A'ssar).

283a. Hartmann, H., *Liebigs Ann. Chem.*, **714**, 1 (1968).
284. Hartmann, H., E. Dietz, K. Komorniczyk, and W. Reiss, *Naturwiss.*, **48**, 570 (1961); H. Hartmann, B. Karbstein, and W. Reiss, *ibid.*, **52**, 59 (1965).
285. Hartmann, H., B. Karbstein, P. Schaper, and W. Reiss, *Naturwiss.*, **50**, 373 (1963).
286. Hartmann, H., H. Wagner, B. Karbstein, M. K. El A'ssar, and W. Reiss, *Naturwiss.*, **51**, 215 (1964). Further references there.
287. Cf. ref. 629.
288. Haward, R. N., *Chem. and Ind.*, 1442 (1964).
288a. Hayashi, T., K. Iyoda, and I. Shiihara, *J. Organomet. Chem.*, **10**, 81 (1967).
288b. Hayashi, T., J. Uchimura, S. Matsuda, and S. Kikkawa, *Kogyo Kagaku Zasshi*, **70**, 714 (1967).
289. Hechenbleikner, I., and K. R. Molt (Carlisle Chem. Works), U.S. Pat. 3,059,012 (October 16, 1962). Further references there. *C.A.*, **58**, 6860 (1963).
290. Henderson, A., and A. K. Holliday, *J. Organomet. Chem.*, **4**, 377 (1965).
290a. Hengge, E., and U. Brychcy, *Monatsh.*, **97**, 1309 (1966).
291. Hengge, E., and K. Pretzer, *Chem. Ber.*, **96**, 470 (1963).
292. Henry, M. C., and J. G. Noltes, *J. Amer. Chem. Soc.*, **82**, 555 (1960).
293. Henry, M. C., and J. G. Noltes, *J. Amer. Chem. Soc.*, **82**, 561 (1960).
293a. Herber, R. H., and G. I. Parisi, *Inorg. Chem.*, **5**, 769 (1966).
294. Herber, R. H., and H. A. Stöckler, *Trans. N.Y. Acad. Sci., Ser. II*, **26**, 929 (1964).
294a. Herber, R. H., "Chemical Applications of Mössbauer Spectroscopy," *Progr. Inorg. Chem.*, **8**, 1—41 (1968).
295. Heymann, E., Diploma Thesis, Univ. of Giessen, 1963.
296. Heymann, E., Dissertation, Univ. of Giessen, 1965.
297. Hieber, W., and R. Breu, *Chem. Ber.*, **90**, 1270 (1957).
298. Hof, T., and J. G. A. Luijten, *Timb. Techn.*, No. 2236 (1959); cited in *Tin and its Uses*, **64**, 5 (1964).
299. Hoffmann, E. G., *Liebigs Ann. Chem.*, **629**, 104 (1960); E. G. Hoffmann, and W. Tornau, *Angew. Chem.*, **73**, 578 (1961).
300. Hoffmann, E. G., *Z. Electrochem.*, **64**, 616 (1960).
300a. Hollaender, J., unpublished, Univ. of Giessen, 1968; reviewed in *Angew. Chem.*, **81**, 296 (1969).
301. Holmes, J. R., and H. D. Kaesz, *J. Amer. Chem. Soc.*, **83**, 3903 (1961).
302. Holmes, J. M., R. D. Peacock, and J. C. Tatlow, *Proc. Chem. Soc.*, 108 (1963).
303. Holmes, T. D., and I. F. Storey, *Plant Pathology*, **11**, 139 (1962).
304. Höppner, K., U. Prösch, and H. Wiegleb, *Z. Chem.*, **4**, 31 (1964).
305. Höppner, K., and D. Walkiewitz, *Z. Chem.*, **2**, 23 (1962).
306. Horn, C. F., and H. Vineyard (Union Carbide Corp.), Brit. Pat. 899,896 (June 27, 1962), *C.A.*, **57**, 11391 (1962).
307. Hostettler, F., and E. F. Cox, *Ind. Engng. Chem.*, **52**, 609 (1960).
308. Huang, H. H., and K. M. Hui, *J. Organomet. Chem.*, **2**, 288 (1964).
309. Hueck, H. J., and J. G. A. Luijten, *J. Soc. Dyers and Colorists*, **74**, 476 (1958).
310. Hulme, R., *J. Chem. Soc.*, 1524 (1963).
311. Hulse, R., and H. J. Twitchett (Imperial Chem. Ind.), Brit. Pat. 899,948 (June 27, 1962), *C.A.*, **58**, 9137 (1963).
312. Hulse, R., and H. J. Twitchett (Imperial Chem. Ind.), Brit. Pat. 957,841 (May, 13, 1964), *C.A.*, **61**, 9526 (1964).
313. Ibekwe, S. D., and M. J. Newlands, *Chem. Comm.*, 114 (1965).
313a. Ibekwe, S. D., and M. J. Newlands, *J. Chem. Soc.*, 4608 (1965).

314. Ingham, R. K., and H. Gilman, "Organopolymers of Group IV Elements," in *Inorganic Polymers* (Ed. F. G. A. Stone and W. A. G. Graham), Academic Press, New York, 1962, p. 321. Gives numerous further references.
315. Ingham, R. K., S. D. Rosenberg, and H. Gilman, *Chem. Rev.*, **60**, 459 (1960); comprehensive literature review.
315a. Ingold, C. K., *Helv. Chim. Acta*, **47**, 1191 (1964).
315b. Itoi, K., and S. Kumano, *Kogyo Kagaku Zasshi*, **70**, 82 (1967); Itoi, K., Fr. Pat. 1,368,522; *C.A.*, **62**, 2794 (1965).
316. Jacobi, E., S. Lust, O. Zima, and A. van Schoor (E. Merck AG), D.A.S., 1,045,716 (December 4, 1958), *C.A.*, **54**, 25546 (1960).
317. Janssen, M. J., and J. G. A. Luijten, *Rec. Trav. Chim.*, **82**, 1008 (1963).
318. Janssen, M. J., J. G. A. Luijten, and G. J. M. van der Kerk, *Rec. Trav. Chim.*, **82**, 90 (1963).
319. Janssen, M. J., J. G. A. Luijten, and G. J. M. van der Kerk, *J. Organomet. Chem.*, **1**, 286 (1964).
320. Jenkner, H. (Kali-Chemie A.G.), D.B. Pat. 1,106,325 (May 10, 1961), *C.A.*, **56**, 2472 (1962).
321. Jenkner, H. (Kali-Chemie A.G.), Brit. Pat. 871,642 (June 23, 1961), *C.A.*, **55**, 27809 (1961).
322. Jenkner, H., *Chem.-Ztg.*, **86**, 527, 563 (1962).
323. Jenkner, H. (Kali-Chemie A.G.), Brit. Pat. 900,132 (July 4, 1962), *C.A.*, **57**, 15151 (1962).
324. Jenkner, H., and H. W. Schmidt (Kali-Chemie A.G.), D.A.S. 1,048,275 (January 8, 1959), *C.A.*, **54**, 19486 (1960).
325. Johnson, O. H., *Chem. Rev.*, **48**, 259 (1951).
326. Johnson, O. H., *J. Org. Chem.*, **25**, 2262 (1960).
327. Johnson, O. H., and H. E. Fritz, *J. Org. Chem.*, **19**, 74 (1954).
328. Johnson, O. H., H. E. Fritz, D. O. Halvorson, and R. L. Evans, *J. Amer. Chem. Soc.*, **77**, 5857 (1955).
329. Johnson, W. A., *Tin and its Uses*, **54**, 5 (1962).
330. Johnson, W. K., *J. Org. Chem.*, **25**, 2253 (1960).
331. Jolly, W. L., *Angew. Chem.*, **72**, 268 (1960).
332. Jones, K., and M. F. Lappert, *Proc. Chem. Soc.*, 358 (1962).
333. Jones, K., and M. F. Lappert, *Proc. Chem. Soc.*, 22 (1964).
334. Jones, K., and M. F. Lappert, *Proc. Chem. Soc.*, 1944 (1965).
335. Jones, K., and M. F. Lappert, *J. Organomet. Chem.*, **3**, 295 (1965); review giving further references.
336. Jones, K., and M. F. Lappert, *Organomet. Chem. Rev.*, **1**, 67 (1966); review giving further references.
337. Jones, W. J., W. C. Davies, S. T. Bowden, C. Edwards, V. E. Davis, and L. H. Thomas, *J. Chem. Soc.*, 1446 (1947).
338. Jones, W. J., D. P. Evans, T. Gulwell, and D. C. Griffiths, *J. Chem. Soc.*, 39 (1935).
339. Joshi, K. K., and P. A. H. Wyatt, *J. Chem. Soc.*, 3825 (1959).
340. Kaabak, L. V., and A. P. Tomilov, *Zh. Obshch. Khim.*, **33**, 2808 (1963). Further references there.
341. Kaesz, H. D., *J. Amer. Chem. Soc.*, **83**, 1514 (1961).
342. Kaesz, H. D., J. R. Phillips, and F. G. A. Stone, *J. Amer. Chem. Soc.*, **82**, 6228 (1960).
343. Kaesz, H. D., and F. G. A. Stone, "Vinylmetallics," in *Organometallic Chemistry* (Ed. H. Zeiss), Reinhold Publ. Corp., New York, 1960.

344. Karantassis, T., and C. Vassiliades, *Compt. Rend.*, **205**, 460 (1937).
345. Kasai, N., K. Yasuda, and R. Okawara, *J. Organomet. Chem.*, **3**, 172 (1965).
345a. Kawakami, K. T., and R. Okawara, *J. Organomet. Chem.*, **6**, 249 (1966).
345b. Kawakami, K., T. Saito, and R. Okawara, *J. Organomet. Chem.*, **8**, 377 (1967).
346. Kawasaki, Y., K. Kawakami, and T. Tanaka, *Bull. Chem. Soc. Japan*, **38**, 1102 (1965).
347. Kawasaki, Y., T. Tanaka, and R. Okawara, *Bull. Chem. Soc. Japan*, **37**, 903 (1964).
348. Kekulé, A., *Liebigs Ann. Chem.*, **109**, 190 (1861).
349. Kelen, G. P. van der, L. Verdonck, and D. van de Vondel, *Bull. Soc. Chim. Belges*, **73**, 733 (1964).
349a. Kelso, R. G., K. W. Greenlee, J. M. Derfer, and C. E. Boord, *J. Amer. Chem. Soc.*, **77**, 1751 (1955).
350. Kenyon, A. S., U.S. Nat. Bur. Standards Circ. 525, p. 81 (1953), *C.A.*, **48**, 7338 (1954).
351. Kerk, G. J. M. van der, lecture given at the Organotin Symposium Frankfurt/Main, November 1963; cited in *Tin and its Uses*, **61**, 5 (1964).
351a. Kerk, G. J. M. van der, J. G. A. Luijten, J. G. Noltes, and H. M. J. C. Creemers, *Chimia*, **23**, 313 (1969).
352. Kerk, G. J. M. van der, and J. G. A. Luijten, *J. Appl. Chem.*, **4**, 301 (1954).
353. Kerk, G. J. M. van der, and J. G. A. Luijten, *J. Appl. Chem.*, **4**, 307 (1954).
354. Kerk, G. J. M. van der, and J. G. A. Luijten, *J. Appl. Chem.*, **4**, 314 (1954).
355. Kerk, G. J. M. van der, and J. G. A. Luijten, *J. Appl. Chem.*, **6**, 49 (1956).
356. Kerk, G. J. M. van der, and J. G. A. Luijten, *J. Appl. Chem.*, **6**, 56 (1956).
357. Kerk, G. J. M. van der, and J. G. A. Luijten, *J. Appl. Chem.*, **6**, 93 (1956).
357a. Kerk, G. J. M. van der, and J. G. A. Luijten, *Org. Synth.*, **36**, 86 (1956).
358. Kerk, G. J. M. van der, and J. G. A. Luijten, *J. Appl. Chem.*, **7**, 369 (1957).
359. Kerk, G. J. M. van der, J. G. A. Luijten, J. C. van Egmond, and J. G. Noltes, *Chimia*, **16**, 36 (1962). Further references there.
360. Kerk, G. J. M. van der, J. G. A. Luijten, and M. J. Janssen, *Chimia*, **16**, 10 (1962).
361. Kerk, G. J. M. van der, J. G. A. Luijten, and J. G. Noltes, *Chem. and Ind.*, 352 (1956).
362. Kerk, G. J. M. van der, J. G. A. Luijten, and J. G. Noltes, *Angew. Chem.*, **70**, 298 (1958).
363. Kerk, G. J. M. van der, and J. G. Noltes, *J. Appl. Chem.*, **9**, 106 (1959).
364. Kerk, G. J. M. van der, and J. G. Noltes, *J. Appl. Chem.*, **9**, 176 (1959).
365. Kerk, G. J. M. van der, and J. G. Noltes, *J. Appl. Chem.*, **9**, 179 (1959).
366. Kerk, G. J. M. van der, J. G. Noltes, and J. G. A. Luijten, *J. Appl. Chem.*, **7**, 356 (1957).
367. Kerk, G. J. M. van der, J. G. Noltes, and J. G. A. Luijten, *J. Appl. Chem.*, **7**, 366 (1957).
368. Kerk, G. J. M. van der, J. G. Noltes, and J. G. A. Luijten, *Rec. Trav. Chim.*, **81**, 853 (1962).
369. Kettle, S. F. A., *J. Chem. Soc.*, 2936 (1959).
370. Kharasch, M. S., and A. L. Flenner, *J. Amer. Chem. Soc.*, **54**, 674 (1932).
370a. Khoo, L. E., and H. H. Lee, *Tetrahedron Letters*, 4351 (1968).
371. Kipping, F. B., *J. Chem. Soc.*, **131**, 2365 (1928).
372. Kipping, F. B., *Proc. Roy. Soc., A*, **159**, 139 (1937).
373. Kleiner, F. G., Diploma Thesis, Univ. of Giessen, 1963.
373a. Kleiner, F. G., and W. P. Neumann, *Liebigs Ann. Chem.*, **716**, 19 (1968).
374. Klemchuk, P. P. (Geigy A.G.), Fr. Pat. 1,355,999 (April 26, 1963), *C.A.*, **61**, 3148 (1964).
375. Klimmer, O. R., lecture given at the Organotin Symposium Frankfurt/Main, November 1957; private communication.

376. Klimmer, O. R., *Arzneimittel-Forsch.*, **13**, 432 (1963); *Zentralbl. Vet. Med.,A*, **11**, 29 (1964).
377. Klimmer, O. R., lecture given at the Organotin Symposium Frankfurt/Main, November 1963; cited in *Tin and its Uses*, **61**, 6 (1964).
378. Klimmer, O. R., private communication.
379. Klimmer, O. R., *Pflanzenschutz- und Schädlingsbekämpfungsmittel. Abriss einer Toxikologie und Therapie von Vergiftungen*, Hundt-Verlag, Hattingen (Ruhr), 1964.
379a. Klimmer, O. R., lecture given at the Organotin Symposium Frankfurt/Main, November, 1968, *Arzneimittel-Forsch.*, **19**, 934 (1969).
379b. Klimmer, O. R., *Pflanzenschutz-Berichte*, **37**, 57 (1968).
380. Klimmer, O. R., and I. U. Nebel, *Arzneimittel-Forsch.*, **10**, 44 (1960).
381. Knowles, D. G., *Plastics Technol.*, **6**, 35, 42 (1960).
382. Kocheshkov, K. A., E. M. Panov, and N. N. Zemlyanskii, *Izvest. Akad. Nauk SSSR*, 2255 (1961).
382a. Kochkin, D. A., and I. N. Azerbaev, *Tin and Lead Organic Monomers and Polymers* (Russian), Nauka (Science Publishing), Alma-Ata, S.S.S.R., 1968.
383. Kochkin, D. A., and J. N. Cirgadze, *Zh. Obshch. Khim.*, **32**, 4007 (1962).
384. Kochkin, D. A., L. V. Luk'yanova, and E. B. Reznikova, *Zh. Obshch. Khim.*, **33**, 1945 (1963).
385. Kochkin, D. A., and I. B. Tsekmareva, *Zh. Obshch. Khim.*, **31**, 3010 (1961).
386. Kochkin, D. A., V. I. Vashkov, and V. P. Dremova, *Zh. Obshch. Khim.*, **34**, 325 (1964).
387. Kochkin, D. A., and S. G. Yerenikina, *Doklady Akad. Nauk SSSR*, **139**, 1375 (1961).
387a. Kockel, B., and N. Grün, personal communication.
388. Kolesnikov, G. S., and S. L. Davydova, *Zh. Obshch. Khim.*, **29**, 2042 (1959).
389. Kolesnikov, G. S., and G. T. Gurgenidze, *Izvest. Akad. Nauk SSSR*, 1275 (1962).
390. Koton, M. M., and L. F. Dokukina, *Doklady Akad. Nauk SSSR*, **152**, 1357 (1963).
391. Koton, M. M., and T. M. Kiseleva, *Zh. Obshch. Khim.*, **27**, 2553 (1957).
392. Koton, M. M., and T. M. Kiseleva, *Zh. Obshch. Khim.*, **32**, 7136 (1958). Gives references to older work.
393. Koton, M. M., and T. M. Kiseleva, *Doklady Akad. Nauk SSSR*, **130**, 86 (1960).
394. Koton, M. M., and T. M. Kiseleva, *Izvest. Akad. Nauk SSSR*, 1783 (1961).
395. Koton, M. M., T. M. Kiseleva, and F. S. Florinsky, *J. Polymer. Sci.*, 52, 237 (1961); *Angew. Chem.*, **72**, 712 (1960).
396. Koton, M. M., T. M. Kiseleva, and V. A. Paribok, *Doklady Akad. Nauk. SSSR*, **125**, 1263 (1959).
397. Koton, M. M., T. M. Kiseleva, and N. P. Zapevalova, *Zh. Obshch. Khim.*, **30**, 186 (1960).
398. Koninklijke Industr. Mij., Fr. Pat. 1,320,473 (March 8, 1963), *C.A.*, **59**, 8788 (1963).
399. Korshak, V. V., A. M. Polyakova, V. F. Mironov, and A. D. Petrov, *Izvest. Akad. Nauk SSSR*, 178 (1959).
400. Korshak, V. V., A. M. Polyakova, and E. S. Tambovtseva, *Izvest. Akad. Nauk SSSR*, 742 (1959).
401. Korshak, V. V., S. V. Rogozhin, and T. A. Makarova, *Vysokomol. Soed.*, **4**, 1297 (1962).
402. Köster, R., and P. Binger, *Adv. Inorg. Chem.*, **7**, 263 (1965).
402a. Kostyanovsky, R. G., *Tetrahedron Letters*, 2721 (1968).
402b. Kostyanovsky, R. G., and A. K. Prokof'ev, *Izvest. Akad. Nauk SSSR*, 274 (1968).
Kozeschkow, K. A., see also Kocheskhov, K. A.
403. Kozeschkow, K. A., *Ber.*, **62**, 996 (1929).
404. Kozeschkow, K. A., *Ber.*, **66**, 1661 (1933).

405. Kozeschkow, K. A., M. M. Nadj, and A. P. Alexandrow, *Ber.*, **67**, 1348 (1934). Further references there.
406. Kraus, C. A., and R. H. Bullard, *J. Amer. Chem. Soc.*, **48**, 2131 (1926).
407. Kraus, C. A., and R. H. Bullard, *J. Amer. Chem. Soc.*, **51**, 3605 (1929).
408. Kraus, C. A., and C. C. Callis, *J. Amer. Chem. Soc.*, **45**, 2624 (1923).
409. Kraus, C. A., and W. N. Greer, *J. Amer. Chem. Soc.*, **44**, 2629 (1922).
410. Kraus, C. A., and W. N. Greer, *J. Amer. Chem. Soc.*, **45**, 3078 (1923).
411. Kraus, C. A., and W. N. Greer, *J. Amer. Chem. Soc.*, **47**, 2568 (1925).
412. Kraus, C. A., and A. M. Neal, *J. Amer. Chem. Soc.*, **51**, 2403 (1929).
413. Kraus, C. A., and W. V. Sessions, *J. Amer. Chem. Soc.*, **47**, 2361 (1925).
414. Krause, E., *Ber.*, **51**, 1447 (1918).
415. Krause, E., and R. Becker, *Ber.*, **53**, 173 (1920).
416. Krause, E., and A. von Grosse, *Die Chemie der Metallorganischen Verbindungen*, Verlag Borntraeger, Berlin, 1937; Reprint, Verlag Sändig, Wiesbaden 1965.
417. Krause, E., and R. Pohland, *Ber.*, **57**, 532 (1924).
418. Krause, E., and K. Weinberg, *Ber.*, **63**, 381 (1930).
419. Kriegsmann, H., and H. Geissler, *Z. Anorg. Allg. Chem.*, **323**, 170 (1963). Further references there.
420. Kriegsmann, H., and H. Hoffmann, *Z. Anorg. Allg. Chem.*, **321**, 224 (1963).
421. Kriegsmann, H., and H. Hoffmann, *Z. Chem.*, **3**, 268 (1963). Further references there.
421a. Kriegsmann, H., H. Hoffmann, and S. Pischtschan, *Z. Anorg. Allg. Chem.*, **308**, 212 (1961).
422. Kriegsmann, H., H. Hoffmann, and S. Pischtschan, *Z. Anorg. Allg. Chem.*, **315**, 283 (1962).
423. Kriegsmann, H., and K. Ulbricht, *Z. Chem.*, **3**, 67 (1963).
424. Kriegsmann, H., and K. Ulbricht, *Z. Anorg. Allg. Chem.*, **328**, 90 (1964). Further references there.
425. Kuchen, W., and H. Buchwald, *Chem. Ber.*, **92**, 227 (1959).
426a. Kühlein, K., unpublished work.
427. Kühlein, K., Dissertation, Univ. of Giessen, 1966; W. P. Neumann, and K. Kühlein, *Liebigs Ann. Chem.*, **702,** 13 (1967).
427a. Kühlein, K., W. P. Neumann, and H. Mohring, *Angew. Chem.*, **80**, 438 (1968).
427b. Kühlein, K., and W. P. Neumann, *J. Organomet. Chem.*, **14**, 317 (1968).
428. Kuivila, H. G., *J. Org. Chem.*, **25**, 284 (1960).
429. Kuivila, H. G., *Adv. Organomet. Chem.*, **1**, 47 (1964).
430. Kuivila, H. G., and O. F. Beumel, Jr., *J. Amer. Chem. Soc.*, **80**, 3250 (1958).
431. Kuivila, H. G., and O. F. Beumel, Jr., *J. Amer. Chem. Soc.*, **80**, 3798 (1958).
432. Kuivila, H. G., and O. F. Buemel, Jr., *J. Amer. Chem. Soc.*, **83**, 1246 (1961).
433. Kuivila, H. G., and O. F. Beumel, Jr., (Research Corp.), U.S. Pat. 2,997,485 (August 22, 1961), *C.A.*, **57**, 866 (1962).
434. Kuivila, H. G., and E. R. Jakusik, *J. Org. Chem.*, **26**, 1430 (1961).
435. Kuivila, H. G., and P. L. Levins, *J. Amer. Chem. Soc.*, **86**, 23 (1964).
436. Kuivila, H. G., and L. W. Menapace, *J. Org. Chem.*, **28**, 2165 (1963).
437. Kuivila, H. G., L. W. Menapace, and C. R. Warner, *J. Amer. Chem. Soc.*, **84**, 3584 (1962).
438. Kuivila, H. G., W. Rahman, and R. H. Fish, *J. Amer. Chem. Soc.*, **87**, 2835 (1965).
439. Kuivila, H. G., A. K. Sawyer, and A. G. Armour, *J. Org. Chem.*, **26**, 1426 (1961).
439a. Kuivila, H. G., and R. Sommer, *J. Amer. Chem. Soc.*, **89**, 5616 (1967).
439b. Kuivila, H. G., R. Sommer, and D. C. Green, *J. Org. Chem.*, **33**, 1119 (1968).
440. Kuivila, H. G., and J. A. Verdone, *Tetrahedron Letters*, 119 (1964).

441. Kuivila, H. G., and E. J. Walsh, Jr., *J. Amer. Chem. Soc.*, **88**, 571, 576 (1966).
442. Kula, M. R., E. Amberger, and K. K. Mayer, *Chem. Ber.*, **98**, 634 (1965).
443. Kula, M. R., E. Amberger, and H. Rupprecht, *Chem. Ber.*, **98**, 629 (1965).
444. Kula, M. R., C. G. Kreiter, and J. Lorberth, *Chem. Ber.*, **97**, 1294 (1964).
445. Kula, M. R., J. Lorberth, and E. Amberger, *Chem. Ber.*, **97**, 2087 (1964).
446. Kupchik, E. J., and P. J. Calabretta, *Inorg. Chem.*, **3**, 905 (1964).
447. Kupchik, E. J., and R. E. Connolly, *J. Org. Chem.*, **26**, 4747 (1961).
448. Kupchik, E. J., and R. J. Kiesel, *J. Org. Chem.*, **29**, 764 (1964).
449. Kupchik, E. J., and R. J. Kiesel, *J. Org. Chem.*, **29**, 3690 (1964).
450. Kupchik, E. J., and T. Lanigan, *J. Org. Chem.*, **27**, 3661 (1962).
451. Kushlefsky, B., and A. Ross, *Anal. Chem.*, **34**, 1966 (1962); Kushlefsky, B., J. Simmons, and A. Ross, *Inorg. Chem.*, **2**, 187 (1963).
452. Ladenburg, A., *Ber.*, **3**, 353 (1870).
453. Ladenburg, A., *Liebigs Ann. Chem., Suppl.*, **8**, 55 (1872).
454. Lambourne, H., *J. Chem. Soc.*, **121**, 2533 (1922); **125**, 2013 (1924).
454a. Lappert, M. F., and J. Lorberth, *Chem. Comm.*, 836 (1967).
454b. Lappert, M. F., and B. Prokai, *Adv. Organomet. Chem.*, **5**, 225 (1967).
454c. Lappert, M. F., J. Simpson, and T. R. Spalding, *J. Organomet. Chem.*, **17**, P1 (1969).
455. Larsson, E., *Trans. Chalmers Univ. Technol.* (Gothenburg), **94**, 15 (1950), *C.A.*, **45**, 1494 (1951).
455a. Lautsch, W. F., A. Tröber, et. al., *Z. Chem.*, **3**, 415 (1963).
456. Leavitt, F. C., T. A. Manuel, and F. Johnson, *J. Amer. Chem. Soc.*, **81**, 3163 (1959).
457. Leavitt, F. C., T. A. Manuel, F. Johnson, L. U. Matternas, and D. S. Lehman, *J. Amer. Chem. Soc.*, **82**, 5099 (1960).
458. Leavitt, F. C., and L. U. Matternas, *J. Polymer Sci.*, **62**, S68 (1962).
459. Leebrick, J. R., and H. E. Ramsden, *J. Org. Chem.*, **23**, 935 (1958).
460. Leeper, R. W., L. Summers, and H. Gilman, *Chem. Rev.*, **54**, 101 (1954) (Survey).
461. Lehn, W. L., *J. Amer. Chem. Soc.*, **86**, 305 (1964).
462. Lesbre, M., and R. Buisson, *Bull. Soc. Chim. France*, 1204 (1957).
463. Lesbre, M., and R. Dupont, *Compt. Rend.*, **237**, 1700 (1953).
464. Lesbre, M., and G. Glotz, *Compt. Rend.*, **198**, 1426 (1934).
465. Lesbre, M., and I. Serée de Roch, *Bull. Soc. Chim. France*, 754 (1956).
465a. Leusink, A. J., Dissertation, Univ. of Utrecht, 1966.
465b. Leusink, A. J., H. A. Budding, and W. Drenth, *J. Organomet. Chem.*, **13**, 163 (1968).
465c. Leusink, A. J., H. A. Budding, and W. Drenth, *J. Organomet. Chem.*, **11**, 541 (1968).
465d. Leusink, A. J., H. A. Budding, and J. W. Marsman, *J. Organomet. Chem.*, **9**, 285 (1967); **13**, 155 (1968).
466. Leusink, A. J., H. A. Budding, and J. G. Noltes, *Rec. Trav. Chim.*, **85**, 151 (1966).
466a. Leusink, A. J., W. Drenth, J. G. Noltes, and G. J. M. van der Kerk, *Tetrahedron Letters*, 1263 (1967).
467. Leusink, A. J., and J. W. Marsman, *Rec. Trav. Chim.*, **84**, 1123 (1965).
468. Leusink, A. J., J. W. Marsman, and H. A. Budding, *Rec. Trav. Chim.*, **84**, 689 (1965).
469. Leusink, A. J., J. W. Marsman, H. A. Budding, J. G. Noltes, and G. J. M. van der Kerk, *Rec. Trav. Chim.*, **84**, 567 (1965).
470. Leusink, A. J., and J. G. Noltes, *Rec. Trav. Chim.*, **84**, 585 (1965).
471. Leusink, A. J., and J. G. Noltes, *Tetrahedron Letters*, 335 (1966).
471a. Leusink, A. J., and J. G. Noltes, *Tetrahedron Letters*, 2221 (1966).
472. Leusink, A. J., J. G. Noltes, H. A. Budding, and G. J. M. van der Kerk, *Rec. Trav. Chim.*, **83**, 609 (1964).

473. Leusink, A. J., J. G. Noltes, H. A. Budding, and G. J. M. van der Kerk, *Rec. Trav. Chim.*, **83**, 1036 (1964).
473a. Leusink, A. J., and J. G. Noltes, *J. Organomet. Chem.*, **16**, 91 (1969).
473b. For details see ref. 473a.
474. Lewis, F. B. (Associated Lead Manuf. Ltd.), Brit. Pat. 854,776 (November 23, 1960), *C.A.*, **55**, 11362 (1961).
475. Lewis, C. W., *Ind. Engng. Chem.*, **46**, 366 (1954).
476. Lide, D. R., Jr., *J. Chem. Phys.*, **19**, 1605 (1951); Mathis, R., M. Lesbre, I. S. de Roch, *Compt. Rend.*, **243**, 257 (1956).
477. Lohmann, D. H., Tin Research Institute, Greenford, England, 1965.
478. Lohmann, D. H., *J. Organomet. Chem.*, **4**, 382 (1965).
479. Lorberth, J., *Chem. Ber.*, **98**, 1201 (1965).
479a. Lorberth, J., H. Krapf, and H. Nöth, *Chem. Ber.*, **100**, 3511 (1967).
480. Lorberth, J., and M. R. Kula, *Chem. Ber.*, **97**, 3444 (1964).
481. Lorberth, J., and M. R. Kula, *Chem. Ber.*, **98**, 520 (1965).
482. Lorberth, J., and H. Nöth, *Chem. Ber.*, **98**, 969 (1965).
483. Lorenz, D. H., and E. I. Becker, *J. Org. Chem.*, **27**, 3370 (1962).
484. Lorenz, D. H., and E. I. Becker, *J. Org. Chem.*, **28**, 1707 (1963).
485. Lorenz, D. H., Ph. Shapiro, A. Stern, and E. I. Becker, *J. Org. Chem.*, **28**, 2332 (1963).
486. Löwig, C., *Liebigs Ann. Chem.*, **84**, 308 (1852).
487. Luijten, J. G. A., *Rec. Trav. Chim.*, **82**, 1179 (1963).
488. Luijten, J. G. A., *Metall*, **18**, 814 (1964).
489. Luijten, J. G. A., *Tin and Its Uses*, **67**, 1 (1965).
489a. Luijten, J. G. A., *Rec. Trav. Chim.*, **85**, 873 (1966).
490. Luijten, J. G. A., M. J. Janssen, and G. J. M. van der Kerk, *Rec. Trav. Chim.*, **81**, 202 (1962).
491. Luijten, J. G. A., and G. J. M. van der Kerk, *Investigations in the Field of Organotin Chemistry*, Tin Research Institute, Greenford, 1955.
492. Luijten, J. G. A., and G. J. M. van der Kerk, *J. Appl. Chem.*, **11**, 35 (1961).
493. Luijten, J. G. A., and G. J. M. van der Kerk, *Rec. Trav. Chim.*, **82**, 1181 (1963).
494. Luijten, J. G. A., and G. J. M. van der Kerk, *Rec. Trav. Chim.*, **83**, 295 (1964).
495. Luijten, J. G. A., and S. Pezarro, *British Plastics*, 183 (1957).
495a. Luijten, J. G. A., and F. Rijkens, *Rec. Trav. Chim.*, **83**, 857 (1964).
495b. Lukevits, E. Y., and M. G. Voronkov, *Organic Insertion Reactions of Group IV Elements*, Consultants Bureau, New York, 1966.
496. Lutsenko, I. F., Yu. I. Baukov, and B. N. Chasanov, *Zh. Obshch. Khim.*, **33**, 2791 (1963).
497. Lutsenko, I. F., S. V. Ponomarev, and O. P. Petrii, *Zh. Obshch. Khim.*, **32**, 896 (1962).
498. Mack, G. P., *Kunststoffe*, **43**, No. 3, 94 (1953).
499. Mack, G. P. (Metal and Thermit Corp.), Brit. Pat. 862,430 (March 8, 1961), *C.A.*, **55**, 17649 (1961).
500. Mack, G. P. (Metal and Thermit Corp.), Brit. Pat. 903,068 (August 9, 1962), *C.A.*, **57**, 15152 (1962).
501. Mack, G. P. (Metal and Thermit Chemicals Inc.), Fr. Pat. 1,369,815 (August 14, 1964), *C.A.*, **62**, 586 (1965).
502. Mack, G. P., and J. Heiths (Carlisle Chem. Works), U.S. Pat. 2,745,820 (May 15, 1956), *C.A.*, **51**, 6219 (1957); (Deutsche Advance), Belg. Pat. 578,730 (May 30, 1959), A. Hartmann (Deutsche Advance), D.B. Pat. 1,073,496 (January 21, 1960), *C.A.*, **55**, 10319 (1961).
503. Mack, G. P., and E. Parker (Advance Solvents and Chem. Corp.), U.S. Pat. 2,592,926 (April 15, 1952), *C.A.*, **46**, 11767 (1952).

504. Mack, G. P., and E. Parker (Advance Solvents and Chem. Corp.), U.S. Pat. 2,628,211 (February 10, 1953), *C.A.*, **47**, 5165 (1953).
505. Mack, G. P., and E. Parker (Advance Solvents and Chem. Corp.), U.S. Pat. 2,700,675 (January 25, 1955), *C.A.*, **50**, 397 (1956).
506. Mack, G. P., and E. Parker (Carlisle Chem. Works), U.S. Pat. 2,938,013 (May 24, 1960), *C.A.*, **54**, 20334 (1960).
507. Maddox, M. L., N. Flitcroft, and H. D. Kaesz, *J. Organomet. Chem.*, **4**, 50 (1965).
508. Maddox, M. L., S. L. Stafford, and H. D. Kaesz, *Adv. Organomet. Chem.*, **3**, 1 (1965).
509. Maire, J. C., private communication; J. C. Maire and J. Dutermont, *J. Organomet. Chem.*, **10**, 369 (1967).
510. Maire, J. C., *Compt. Rend.*, **249**, 1359 (1959).
511. Maire, J. C., *Compt. Rend.*, **251**, 1291 (1960).
512. Manulkin, Z. M., *Zh. Obshch. Khim.*, **18**, 299 (1948).
513. Maselli, J. M., Dissertation, Pennsylvania State Univ., 1961, *Diss. Abstr.*, **23**, 836 (1962).
514. Matlach, H., Examination Thesis, Univ. of Giessen, 1966.
515. Matsuda, H., J. Hayashi, and S. Matsuda, *Kogyo Kagaku Zasshi*, **64**, 1951 (1961).
516. Matsuda, H., and S. Matsuda, *Kogyo Kagaku Zasshi*, **63**, 1958 (1960).
517. Matsuda, H., M. Nakamura, and S. Matsuda, *Kogyo Kagaku Zasshi*, **64**, 1948 (1961).
518. Matsuda, H., H. Taniguchi, and S. Matsuda, *Kogyo Kagaku Zasshi*, **64**, 541 (1961).
519. Matsuda, S., and H. Matsuda, *Kogyo Kagaku Zasshi*, **63**, 114 (1960).
520. Matsuda, S., and H. Matsuda, *Bull. Chem. Soc. Japan* **35**, 208 (1962).
521. Matwiyoff, N. A., and R. S. Drago, *Inorg. Chem.*, **3**, 337 (1964).
522. Mazur, P. A. (S. a. M. Chemicals, Ltd.), U.S. Pat. 3,037,039 (May 29, 1962); *C.A.*, **57**, 12536 (1962) and firm's own publication; cited in *Tin and its Uses*, **67**, 13 (1965).
523. Meals, R. N., *J. Org. Chem.*, **9**, 211 (1944).
524. Meixner, M., Examination Thesis, Univ. of Giessen, 1964.
525. Mel'nikov, N. N., N. N. Ivanova, and I. L. Vladimirova, Russ. Pat. 125325 (January 8, 1960), *C.A.*, **54**, 12472 (1960).
526. Menapace, L. W., and H. G. Kuivila, *J. Amer. Chem. Soc.*, **86**, 3047 (1964).
526a. Mendelsohn, J., A. Marchand, and J. Valade, *Bull. Soc. Chim. France*, 2696 (1965); *J. Organomet. Chem.*, **6**, 25 (1966).
526b. Mendelsohn, J., J-Cl. Pommier, and J. Valade, *Bull. Soc. Chim. France*, 745 (1967).
527. Metal and Thermit Corp., Brit. Pat. 739,883 (November 2, 1955), *C.A.*, **50**, 13986 (1956).
528. Metalorgana Ets., Brit. Pat. 921,057 (March 13, 1963), *C.A.*, **59**, 14022 (1963).
529. Metalorgana Ets., Brit. Pat. 938,961 (October 9, 1963), *C.A.*, **60**, 3007 (1964).
530. Meyer, G., *Ber.*, **16**, 1439 (1883); later uses and improvements cited in ref. 315.
531. Meyer, H., Svensk Kemisk Tidskr., **76**, 8 (1964).
532. Meyer, K., Dissertation, Techn. Hochschule, Aachen, 1951.
532a. Migdal, Sh., D. Gertner, and A. Zilkha, *Canad. J. Chem.*, **46**, 2409 (1968).
533. Miller, S. M., *Tin and Its Uses*, **65**, 1 (1964).
534. Miller, S. M., *Ind. Engng. Chem.*, **56**, 226 (1964), cited in *Tin and its Uses*, **67**, 9 (1965).
534a. Mirskov, R. G., and V. M. Vlasov, *Zh. Obshch. Khim.*, **36**, 562 (1966).
534b. Mitchell, T. N., Ph.D. Thesis, Univ. of London, 1967.
534c. Moedritzer, K., *Organomet. Chem. Rev.*, **1**, 179 (1966).
534d. Moedritzer, K., *Adv. Organomet. Chem.*, **6**, 171 (1968).
535. Montecatini s. p. a., Ital. Pat. 589,299 (March 4, 1959), *C.A.*, **55**, 5034 (1961).

536. Montermoso, J. C., T. M. Andrews, and L. P. Marinelli, *J. Polymer Sci.*, **32**, 523 (1958).
537. Morris, H., and P. W. Selwood, *J. Amer. Chem. Soc.*, **63**, 2509 (1941); **64**, 1727 (1942).
537a. Müller, E., Ph.D. Thesis, University of Giessen, 1968.
537b. Müller, E., R. Sommer, and W. P. Neumann, *Liebigs Ann. Chem.*, **718**, 1 (1968).
538. Müller, H. (Bad. Anilin- und Soda-Fabrik), Brit. Pat. 840,619 (July 6, 1960), *C.A.*, **55**, 4363 (1961).
539. Nangniot, P., and P. H. Martens, *Anal. Chim. Acta*, **24**, 276 (1961); cited in *Z. Anal. Chem.* **189**, 388 (1962); S. Gorbach and R. Bock, *Z. Anal. Chem.*, **163**, 429 (1958).
540. Nasarova, L. M., *Zh. Obshch. Khim.*, **31**, 1119 (1961).
540a. Nefedow, O. M., and M. N. Manakow, *Angew. Chem.*, **78**, 1039 (1966).
541. Nefedov, O. M., M. N. Manakov, and A. D. Petrov, *Doklady Akad. Nauk SSSR*, **147**, 1376 (1962).
542. Nesmejanow, A. N., and K. A. Kozeschkow, *Ber.*, **63**, 2496 (1930).
543. Nesmeyanov, A. N., K. N. Anisimov, N. E. Kolobova, and M. J. Zacharova, *Doklady Akad. Nauk SSSR*, **156**, 612 (1964).
544. Nesmeyanov, A. N., A. E. Borisov, and N. V. Novikova, *Izvest. Akad. Nauk SSSR*, 1216 (1959).
545. Nesmeyanov, A. N., A. E. Borisov, N. V. Novikova, and M. A. Osipova, *Izvest Akad. Nauk SSSR*, 263 (1959).
546. Nesmeyanov, A. N., J. F. Lutsenko, and S. V. Ponomarev, *Doklady Akad. Nauk SSSR*, **124**, 1073 (1959).
546a. Neubert, G., *Z. Anal. Chem.*, **203**, 265 (1964).
547. Neumann, W. P., Habilitationsschrift, Univ. of Giessen, 1959.
548. Neumann, W. P., *Angew. Chem.*, **74**, 122 (1962).
549. Neumann, W. P., *Liebigs Ann. Chem.*, **653**, 157 (1962).
550. Neumann, W. P., XIXth IUPAC Congress, London, 1963. Abstr. A, 169.
551. Neumann, W. P., *Angew. Chem.*, **75**, 225 (1963); Gives many other references.
552. Neumann, W. P., *Liebigs Ann. Chem.*, **667**, 1 (1963). Further references there.
553. Neumann, W. P., *Angew. Chem.*, **76**, 849 (1964). Gives many other references.
554. Neumann, W. P., 2nd Internat. Sympos. on Organomet. Chem. Madison, Wisc., U.S.A., September 1965 (Discussion).
554a. Neumann, W. P., *Ann. New York Acad. Sci.*, **159**, 56 (1969).
554b. Neumann, W. P., *Angew. Chem.*, **80**, 48 (1968).
555. Neumann, W. P., and H. J. Albert, unpublished.
555a. Neumann, W. P., H. J. Albert, and W. Kaiser, *Tetrahedron Letters*, 2041 (1967).
555b. Neumann, W. P., H. J. Albert, W. Kaiser, and P. Ritter, *Liebigs Ann. Chem.*, in the press.
556. Neumann, W. P., and G. Burkhardt, *Liebigs Ann. Chem.*, **663**, 11 (1963).
556a. Neumann, W. P., and G. Burkhardt (Studienges. Kohle), D.A.S. 1161893 (January 30, 1964), U.S. Pat. 3,248411 (April 26, 1966).
557. Neumann, W. P., and E. Heymann, *Angew. Chem.*, **75**, 166 (1963).
558. Neumann, W. P., and E. Heymann, *Liebigs Ann. Chem.*, **683**, 11 (1965).
559. Neumann, W. P., and E. Heymann, *Liebigs Ann. Chem.*, **683**, 24 (1965).
560. Neumann, W. P., and F. G. Kleiner, unpublished.
561. Neumann, W. P., and F. G. Kleiner, *Tetrahedron Letters*, 3779 (1964).
562. Neumann, W. P., and K. König, *Tetrahedron Letters*, 495 (1967).
563. Neumann, W. P., and K. König, *Angew. Chem.*, **74**, 215 (1962).
564. Neumann, W. P., and K. König, *Liebigs Ann. Chem.*, **677**, 1 (1964).

565. Neumann, W. P., and K. König, *Liebigs Ann. Chem.*, **677**, 12 (1964).
566. Neumann, W. P., and K. König, *Angew. Chem.*, **76**, 892 (1964).
567. Neumann, W. P., K. König, and G. Burkhardt, *Liebigs Ann. Chem.*, **677**, 18 (1964).
568. Neumann, W. P., K. König, and B. Schneider, unpublished.
569. Neumann, W. P., and K. Kühlein, *Tetrahedron Letters*, 1541 (1963).
570. Neumann, W. P., and K. Kühlein, *Liebigs Ann. Chem.*, **683**, 1 (1965). Further references there.
571. Neumann, W. P., and K. Kühlein, *Angew. Chem.*, **77**, 808 (1965); see also *Chem. and Engng. News*, **43**, No. 38, p. 49 (1965).
571a. Neumann, W. P., and H. Lind, *Angew. Chem.*, **79**, 52 (1967); *Chem. Ber.*, **101**, 2837, 2845 (1968).
572. Neumann, W. P., and H. Matlach, unpublished; see also ref. 514.
573. Neumann, W. P., and M. Meixner, unpublished.
573a. Neumann, W. P., and H. Mohring, unpublished.
574. Neumann, W. P., and E. Müller, unpublished; E. Müller, Diploma Thesis, Univ. of Giessen, 1966.
575. Neumann, W. P., and H. Niermann, *Liebigs Ann. Chem.*, **653**, 164 (1962).
576. Neumann, W. P., H. Niermann, and B. Schneider, *Angew. Chem.*, **75**, 790 (1963); *Liebigs Ann. Chem.*, **707**, 15 (1967).
577. Neumann, W. P., H. Niermann, and R. Sommer, *Angew. Chem.*, **73**, 768 (1961).
578. Neumann, W. P., H. Niermann, and R. Sommer, *Liebigs Ann. Chem.*, **659**, 27 (1962).
579. Neumann, W. P., and J. Pedain, *Liebigs Ann. Chem.*, **672**, 34 (1964).
580. Neumann, W. P., and J. Pedain, *Tetrahedron Letters*, 2461 (1964).
581. Neumann, W. P., and J. Pedain, unpublished; in part, cited in ref. 648.
582. Neumann, W. P., J. Pedain, and R. Sommer, *Liebigs Ann. Chem.*, **694**, 9 (1966).
583. Neumann, W. P., E. Petersen, and R. Sommer, *Angew. Chem.*, **77**, 622 (1965).
584. Neumann, W. P., and K. Rübsamen, *Chem. Ber.*, **100**, 1621 (1967); K. Rübsamen, Diploma Thesis, Univ. of Giessen, 1965.
585. Neumann, W. P., K. Rübsamen, and R. Sommer, *Angew. Chem.*, **77**, 733 (1965).
585a. Neumann, W. P., K. Rübsamen, and R. Sommer, *Chem. Ber.*, **100**, 1063 (1967).
586. Neumann, W. P., and R. Schick, unpublished.
587. Neumann, W. P., R. Schick, and R. Köster, *Angew. Chem.*, **76**, 380 (1964).
588. Neumann, W. P., and B. Schneider, *Angew. Chem.*, **76**, 891 (1964).
589. Neumann, W. P., and B. Schneider, unpublished; B. Schneider, Dissertation, Univ. of Giessen, 1966.
590. Neumann, W. P., and B. Schneider, *Liebigs Ann. Chem.*, **707**, 7 (1967).
591. Neumann, W. P., B. Schneider, and R. Sommer, *Liebigs Ann. Chem.*, **692**, 1 (1966).
592. Neumann, W. P., and R. Sommer, unpublished.
593. Neumann, W. P., and R. Sommer, unpublished; in part, cited in ref. 591.
594. Neumann, W. P., and R. Sommer, *Angew. Chem.*, **75**, 788 (1963).
595. Neumann, W. P., and R. Sommer, unpublished; R. Sommer, Dissertation, Univ. of Giessen, 1964.
596. Neumann, W. P., and R. Sommer, *Angew. Chem.*, **76**, 52 (1964); *Liebigs Ann. Chem.*, **701**, 28 (1967).
597. Neumann, W. P., and R. Sommer, *Liebigs Ann. Chem.*, **675**, 10 (1964).
598. Neumann, W. P., R. Sommer, and H. Lind, *Angew. Chem.*, **76**, 597 (1964).
599. Neumann, W. P., R. Sommer, and H. Lind, *Liebigs Ann. Chem.*, **688**, 14 (1965).
600. Neumann, W. P., R. Sommer, and E. Müller, unpublished.
601. Neumann, W. P., R. Sommer, and E. Müller, *Angew. Chem.*, **78**, 545 (1966).
602. Neumann, W. P., R. Sommer, H. Schwemmler, and M. Scholz, unpublished.

603. Neumann, W. P. (Studienges. Kohle), Fr. Pat. 1,318,310 (February 15, 1963), *C.A.*, **59**, 2858 (1963), D.B. Pat. 1,177,158 (May 6, 1965), *C.A.*, **61**, 14711 (1964).
604. Neumann, W. P., and H. Weller, unpublished; H. Weller, Examination Thesis, Univ. of Giessen, 1965.
605. Neumann, W. P. (K. Ziegler), D.A.S. 1,157,617 (November 21, 1963), *C.A.*, **60**, 3008 (1964); D.A.S. 1,164,407 (March 5, 1964), *C.A.*, **60**, 15910 (1964).
606. Nevett, B. A., and R. S. Tobias, *Chem. and Ind.*, 40 (1963).
607. Niebergall, H., D.B. Pat. 1,087,810 (August 25, 1960), *C.A.*, **55**, 15998 (1961).
608. Niemann, H., lecture given at the Organotin Symposium, Frankfurt/Main, November 28, 1963; cited in *Tin and its Uses*, **61**, 6 (1964).
609. Niermann, H., Dissertation, Univ. of Giessen, 1961.
610. Nitzsche, S., and R. Riedle (Wacker-Chemie GmbH.), U.S. Pat. 2,868,820 (January 13, 1959), *C.A.*, **53**, 13057 (1959).
611. Noltes, J. G., *Rec. Trav. Chim.*, **83**, 515 (1964).
612. Noltes, J. G., *Rec. Trav. Chim.*, **84**, 799 (1965).
613. Noltes, J. G., H. A. Budding, and G. J. M. van der Kerk, *Rec. Trav. Chim.*, **79**, 408, 1076 (1960).
614. Noltes, J. G., and M. J. Janssen, *Rec. Trav. Chim.*, **82**, 1055 (1963).
615. Noltes, J. G., and M. J. Janssen, *J. Organomet. Chem.*, **1**, 346 (1964).
616. Noltes, J. G., and G. J. M. van der Kerk, *Functionally Substituted Organotin Compounds*, Tin Research Institute, Greenford, England, 1958.
617. Noltes, J. G., and G. J. M. van der Kerk, *Chem. and Ind.*, 294 (1959).
618. Noltes, J. G., and G. J. M. van der Kerk, *Rec. Trav. Chim.*, **80**, 623 (1961).
619. Noltes, J. G., and G. J. M. van der Kerk, *Rec. Trav. Chim.*, **81**, 41 (1962).
620. Noltes, J. G., and G. J. M. van der Kerk, *Chimia*, **16**, 122 (1962).
621. Noltes, J. G., J. G. A. Luijten, and G. J. M. van der Kerk, *J. Appl. Chem.*, **11**, 38 (1961).
621a. Nomura, M., S. Matsuda, and S. Kikkawa, *Kogyo Kagaku Zasshi*, **70**, 710 (1967).
622. Normant, H., *Compt. Rend.*, **239**, 1510 (1954).
623. Nosek, J., *Coll. Czech. Chem. Comm.*, **29**, 597 (1964).
624. Nöth, H., and K. H. Hermannsdörfer, *Angew. Chem.*, **76**, 377 (1964).
625. Oakes, V., and R. E. Hutton, *J. Organomet. Chem.*, **3**, 472 (1965).
625a. Oakes, V., and C. Jankowski (Pure Chem. Ltd.), D.A.S. 1,222,503 (August 11, 1966).
625b. Occolowitz, J. L., *Tetrahedron Letters*, 5291 (1966).
626. Ohara, M., and R. Okawara, *J. Organomet. Chem.*, **3**, 484 (1965).
627. Okawara, R., *Proc. Chem. Soc.*, 383 (1961); *Angew. Chem.*, **73**, 683 (1961).
628. Okawara, R., XIXth. IUPAC Congress, London, 1963, Abstr. A, 168.
629. Okawara, R., B. J. Hathaway, and D. E. Webster, *Proc. Chem. Soc.*, 13 (1963).
630. Okawara, R., and M. Ohara, *Bull. Chem. Soc. Japan*, **36**, 623 (1963).
631. Okawara, R., and M. Ohara, *J. Organomet. Chem.*, **1**, 360 (1964).
632. Okawara, R., and E. G. Rochow, Organotin Polymers. Techn. Report, Harvard Univ., Cambridge, Mass., 1960.
633. Okawara, R., and E. G. Rochow, *J. Amer. Chem. Soc.*, **82**, 3285 (1960).
634. Okawara, R., and K. Sugita, *J. Amer. Chem. Soc.*, **83**, 4480 (1961).
635. Okawara, R., and M. Wada, *J. Organomet. Chem.*, **1**, 81 (1963).
635a. Okawara, R., and M. Wada, *Adv. Organomet. Chem.*, **5**, 137 (1967).
636. Okawara, R., D. E. Webster, and E. G. Rochow, *J. Amer. Chem. Soc.*, **82**, 3287 (1960).
637. Okawara, R., D. G. White, K. Fyjitani, and H. Sato, *J. Amer. Chem. Soc.*, **83**, 1342 (1961).

638. Okawara, R., and K. Yasuda, *J. Organomet. Chem.*, **1**, 356 (1964).
639. Olson, D. H., and R. E. Rundle, *Inorg. Chem.*, **2**, 1310 (1963).
639a. Omae, I., S. Matsuda, S. Kikkawa, and R. Sato, *Kogyo Kagaku Zasshi*, **70**, 705 (1967).
640. Oswald, A. A., K. Griesbaum, W. A. Thaler, and B. E. Hudson, Jr., *J. Amer. Chem. Soc.*, **84**, 3897 (1962); W. A. Thaler, A. A. Oswald, and B. E. Hudson, Jr., *ibid.*, **87**, 311 (1965).
641. Paneth, F., and K. Fürth, *Ber.*, **52**, 2020 (1919).
642. Pang, M., and E. I. Becker, *J. Org. Chem.*, **29**, 1948 (1964).
642a. Parish, R. V., and R. H. Platt, *Chem. Comm.*, 1118 (1968).
643. Patil, H. R. H., and W. A. G. Graham, *J. Amer. Chem. Soc.*, **87**, 673 (1965).
644. Patterson, A. M., *J. Amer. Chem. Soc.*, **55**, 3905 (1933).
645. Patterson, A. M., L. T. Capell, and M. A. Magill, *C.A.*, **39**, 5935 (1945).
646. Pauling, L., *The Nature of the Chemical Bond*, 3rd Edn., Cornell Univ. Press, Ithaca, N.Y., 1960, p. 93.
647. Pearce, E. M., *J. Polymer Sci.*, **40**, 273 (1959).
648. Pedain, J., Dissertation, Univ. of Giessen, 1965.
648a. Peddle, G. J. D., and G. Redl, *Chem. Comm.*, 626 (1968).
648b. Pedley, J. B., H. A. Skinner, and C. L. Chernick, *Trans. Faraday Soc.*, **53**, 1612 (1957).
649. Pelissier, M., lecture given at the Organotin Symposium, Paris, March 1964; cited in *Tin and its Uses*, **63**, 11 (1964).
649a. Pereyre, M., Dissertation, Univ. of Bordeaux, 1965.
649b. Pereyre, M., G. Colin, and J. Valade, *Compt. Rend.*, **264**, 1204 (1967).
650. Pereyre, M., and J. Valade, *Bull. Soc. Chim. France*, 2420 (1965). Further references there.
651. Pereyre, M., and J. Valade, *Compt. Rend.*, **260**, 581 (1965); *Bull. Soc. Chim. France* 1928 (1967); Pereyre, M., P. Bellegarde, J. Mendelsohn, and J. Valade, *J. Organomet. Chem.*, **11**, 97 (1968).
652. Permachem Corp., Brit. Pat. 838,722 (June 22, 1960), *C.A.*, **54**, 25540 (1960).
653. Petersen, E., Examination Thesis, Univ. of Giessen, 1966.
654. Petrov, A. A., and V. S. Zavgorodnii, *Zh. Obshch. Khim.*, **34**, 2806 (1964).
655. Petrov, A. D., E. A. Cernysev, and T. L. Krasnova, *Doklady Akad. Nauk SSSR*, **140**, 837 (1961).
656. Petrov, A. D., S. I. Sadykh-Zade, and E. I. Filatova, *Zh. Obshch. Khim.*, **29**, 2936 (1959).
657. Pfeiffer, P., *Ber.*, **35**, 3303 (1902).
658. Pfeiffer, P., and O. Brack, *Z. Anorg. Allg. Chem.*, **87**, 229 (1914).
659. Pfeiffer, P., and R. Lehnardt, *Ber.*, **36**, 1054 (1903).
660. Pfeiffer, P., and R. Lehnardt, *Ber.*, **36**, 3027 (1903).
661. Pfeiffer, P., R. Lehnhardt, R. Prade, K. Schurmann, and P. Truskier, *Z. Anorg. Allg. Chem.*, **68**, 102 (1910).
662. Pfeiffer, P., and K. Schnurmann, *Ber.*, **37**, 319 (1904).
663. Pikina, E. J., T. V. Talalaeva, and K. A. Kocheshkov, *Zh. Obshch. Khim.*, **8**, 1844 (1938).
663a. Plazzogna, G., and G. Pilloni, *Anal. Chim. Acta*, **37**, 260 (1967).
664. Pollard, F. H., G. Nickless, and D. J. Cooke, *J. Chromatog.*, **13**, 48 (1964); **17**, 472 (1965). Further references there.
665. Pollard, F. H., G. Nickless, and P. C. Uden, *J. Chromatog.*, **19**, 28 (1965).
666. Poller, R. C., *J. Inorg. Nucl. Chem.*, **24**, 593 (1962).
667. Poller, R. C., *Proc. Chem. Soc.*, 312 (1963). For further work, see ref. 669.
668. Poller, R. C., *J. Chem. Soc.*, 706 (1963).

669. Poller, R. C., *J. Organomet. Chem.*, **3**, 321 (1965); survey giving further references.
670. Poller, R. C., *Spectrochim. Acta*, **22**, 935 (1966).
671. Polster, R., *Liebigs Ann. Chem.*, **654**, 20 (1962).
671a. Pommier, J. C., Dissertation (Second Thesis), Univ. of Bordeaux, 1966.
672. Pommier, J. C., M. Pereyre, and J. Valade, *Compt. Rend.*, **260**, 6397 (1965).
673. Pommier, J. C., and J. Valade, *Compt. Rend.*, **260**, 4549 (1965).
673a. Pommier, J. C., and J. Valade, *Bull. Soc. Chim. France*, 2697 (1965).
674. Pommier, J. C., and J. Valade, *Bull. Soc. Chim. France*, 975 (1965).
674a. Pommier, J. C., and J. Valade, *J. Organomet. Chem.*, **12**, 433 (1968).
674b. Pope, W. J., and S. J. Peachey, *Proc. Chem. Soc.*, 4244 (1900).
675. Potter, P. E., L. Pratt, and G. Wilkinson, *J. Chem. Soc.*, 524 (1964).
676. Pourcelot, G., M. Le Quan, W. Chodkiewicz, and P. Cadiot, XIXth. IUPAC Congr. London, 1963, Abstr. A, 177.
676a. Price, S. J. W., and A. F. Trotman-Dickenson, *Trans. Faraday Soc.*, **53**, 939, 1208 (1957); **54**, 1630 (1958).
677. Prince, R. H., *J. Chem. Soc.*, 1783 (1959).
677a. Pritchard, H. O., and H. A. Skinner, *Chem. Rev.*, **55**, 745 (1955).
677b. Pryor, W. A., *Free Radicals*, McGraw-Hill Book Co., New York, 1966.
678. Ptitsyna, O. A., O. A. Reutov, and M. F. Turchinskii, *Nauch. Doklady Vysshei Shkoly. Khim. i Khim. Tekhnol.*, 138 (1959).
679. Le Quan, M., and P. Cadiot, *Bull. Soc. Chim. France*, 45 (1965).
680. Le Quan, M., and P. Cadiot, *Bull. Soc. Chim. France*, 35 (1965). Further references there.
681. Quattlebaum, W. M., Jr., and C. A. Noffsinger (Carbide and Carbon Chem. Corp.), U.S. Pat. 2,307,157 (January 5, 1943), *C.A.*, **37**, 3533 (1943).
681a. Rabinovich, J. B., V. J. Telnoi, P. N. Nikolaev, and G. A. Razuvaev, *Dokl. Akad. Nauk SSSR*, **138**, 852 (1961).
682. Ramsden, H. E. (Metal and Thermit Corp.), Brit. Pat. 824,944 (December 9, 1959), *C.A.*, **54**, 17238 (1960), Brit. Pat. 825,039 (December 9, 1959), *C.A.*, **54**, 18438 (1960).
683. Ramsden, H. E., and H. Davidson (Metal and Thermit Corp.), U.S. Pat. 2,675,398 (April 13, 1954), *C.A.*, **48**, 12790 (1954).
684. Rand, L., B. Thir, S. L. Reegen, and K. C. Frisch, *J. Appl. Polymer Sci.*, **9**, 1787 (1965).
684a. Randall, E. W., C. H. Yoder, and J. J. Zuckerman, *Inorg. Nucl. Chem. Letters*, **1**, 105 (1965).
685. Razuvaev, G. A., Yu. I. Dergunov, and N. S. Vyazankin, *Zh. Obshch. Khim.*, **32**, 2515 (1962).
685a. Razuvaev, G. A., O. S. Dyackovskaya, and V. I. Finov, *Dokl. Akad. Nauk SSSR*, **117**, 1113 (1967).
685b. Razuvaev, G. A., O. A. Kruglaya, and G. S. Semchikova, *J. Organomet. Chem.*, **6**, 474 (1966).
686. Razuvaev, G. A., O. A. Shchepetkova, and N. S. Vyazankin, *Zh. Obshch. Khim.*, **31**, 1401, 3762 (1961).
687. Razuvaev, G. A., O. A. Shchepetkova, and N. S. Vyazankin, *Zh. Obshch. Khim.*, **32**, 2152 (1962).
688. Razuvaev, G. A., N. S. Vyazankin, and Yu. I. Dergunov, *Zh. Obshch. Khim.*, **30**, 1310 (1960).
689. Razuvaev, G. A., N. S. Vyazankin, Yu. I. Dergunov, and E. N. Gladysev, *Izvest. Akad. Nauk SSSR*, 848 (1964).

690. Razuvaev, G. A., N. S. Vyazankin, Yu. I. Dergunov, and N. N. Vyshinskii, *Zh. Obshch. Khim.*, **31**, 1712 (1961).
691. Razuvaev, G. A., N. S. Vyazankin, O. S. D'yachkovskaya, I. G. Kiseleva, and Yu. I. Dergunov, *Zh. Obshch. Khim.*, **31**, 4056 (1961).
692. Razuvaev, G. A., N. S. Vyazankin, and O. S. D'yachkovskaya, *Zh. Obshch. Khim.*, **32**, 2161 (1962).
693. Razuvaev, G. A., N. S. Vyazankin, and O. A. Shchepetkova, *Zh. Obshch. Khim.*, **30**, 2498 (1960).
694. Razuvaev, G. A., N. S. Vyazankin, and O. A. Shchepetkova, *Zh. Obshch. Khim.*, **31**, 3762 (1961).
695. Razuvaev, G. A., N. S. Vyazankin, and O. A. Shchepetkova, *Tetrahedron*, **18**, 667 (1962). Further references there.
696. Rees, G., *South African Med. J.*, **36**, No. 9 (1962); reprinted in *Tin and its Uses*, **60**, 1 (1963). Further references there. See also, *Tin and its Uses*, **61**, 16 (1964).
696a. Rees, R. G., and A. F. Webb, *J. Organomet. Chem.*, **12**, 239 (1968).
697. Reichle, W. T., *J. Polymer Sci.*, **49**, 521 (1961).
698. Reichle, W. T., *J. Org. Chem.*, **26**, 4634 (1961).
699. Reichle, W. T., *Inorg. Chem.*, **1**, 650 (1962).
700. Reindl, E., and H. Gelbert (Farbwerke Hoechst AG.), D.B. Pat. 1,100,630 (March 2, 1961), *C.A.*, **55**, 24679 (1961).
701. Reutov, O. A., O. A. Ptitsyna and M. F. Turchinskii, *Doklady Akad. Nauk SSSR*, **139**, 146 (1961).
702. Reverchon, R., *Tin and its Uses*, **65**, 10 (1964).
703. Richardson, B. A., *Wood*, **29**, 57 (1964); cited in *Tin and its Uses*, **64**, 5 (1964). Further references there.
703a. Richter, M., Dr. rer. nat. Thesis, Univ. of Giessen, 1969.
704. Rieche, A., and T. Bertz, *Angew. Chem.* **70**, 507 (1958).
705. Rieche, A., and J. Dahlmann, *Liebigs Ann. Chem.*, **675**, 19 (1964).
706. Rijkens, F., M. J. Janssen, W. Drenth and G. J. M. van der Kerk, *J. Organomet. Chem.*, **2**, 347 (1964). Further references there.
707. Rijkens, F., and G. J. M. van der Kerk, *Rec. Trav. Chim.*, **83**, 723 (1964). Further references there.
707a. Roberts, R. M. G., and F. el Kaissi, *J. Organomet. Chem.*, **12**, 79 (1968).
708. Rochow, E. G., D. T. Hurd and R. N. Lewis, *The Chemistry of Organometallic Compounds*, John Wiley, New York, 1957.
709. Rochow, E. G., D. Seyferth and A. C. Smith, Jr., *J. Amer. Chem. Soc.*, **75**, 3099 (1953).
709a. Rose, M. S. and W. N. Aldridge, *Biochem. J.*, **106**, 821 (1968).
710. Rosenberg, S. D., E. Debreczeni, and E. L. Weinberg, *J. Amer. Chem. Soc.*, **81**, 972 (1959). Further references there.
711. Rosenberg, S. D., and A. J. Gibbons, Jr., *J. Amer. Chem. Soc.*, **79**, 2138 (1957).
712. Rosenberg, S. D., A. J. Gibbons, Jr., and H. E. Ramsden, *J. Amer. Chem. Soc.*, **79**, 2137 (1957).
712a. Ross, A., *Ann. New York Acad. Sci.*, **125**, Art. 1, 107 (1965).
713. Rubincik, G. J., and Z. M. Manulkin, *Zh. Obshch. Khim.*, **34**, 949 (1964).
713a. Rübsamen, K., W. P. Neumann, Ra. Sommer and U. Frommer, *Chem. Ber.*, **102**, 1290 (1969).
714. Rundle, R. E., and D. H. Olson, *Inorg. Chem.* **3**, 596 (1964).
715. Rybakova, N. A., N. K. Taikova and E. N. Silberman, *Trudy Khim. i Khim. Tekhnol.* **2**, 183 (1959).
715a. Sadtler-NMR-Atlas (1968), Spectra 5155M, 5156M.

716. Saitow, A., E. G. Rochow, and D. Seyferth, *J. Org. Chem.*, **23**, 116 (1958).
716a. Sandel, V. R., and H. H. Freedman, *J. Amer. Chem. Soc.*, **90**, 2059 (1968).
717. Sanderson, R. T., *J. Chem. Phys.*, **23**, 2467 (1955).
718. Santo, J. E. (Metal and Thermit Corp.), Belg. Pat. 607,414 (August 22, 1961).
719. Sasin, G. S., *J. Org. Chem.*, **18**, 1142 (1953).
720. Sasin, G. S., A. L. Borror, and R. Sasin, *J. Org. Chem.*, **23**, 1366 (1958).
721. Sasin, G. S., and R. Sasin, *J. Org. Chem.*, **20**, 387 (1955).
722. Sasin, G. S., and R. Sasin, *J. Org. Chem.*, **20**, 770 (1955).
723. Savidan, L., *Bull. Soc. Chim. France*, 411 (1953).
724. Sawyer, A. K., *J. Amer. Chem. Soc.*, **87**, 537 (1965).
724a. Sawyer, A. K., and J. E. Brown, *J. Organomet. Chem.*, **5**, 438 (1966).
725. Sawyer, A. K., J. E. Brown, and E. L. Hanson, *J. Organomet. Chem.*, **3**, 464 (1965).
725a. Sawyer, A. K., J. E. Brown, and G. S. May, *J. Organomet. Chem.*, **11**, 192 (1968).
726. Sawyer, A. K., and H. G. Kuivila, *J. Amer. Chem. Soc.*, **82**, 5958 (1960).
727. Sawyer, A. K., and H. G. Kuivila, *Chem. and Ind.*, 260 (1961).
728. Sawyer, A. K., and H. G. Kuivila, *J. Org. Chem.*, **27**, 610 (1962).
729. Sawyer, A. K., and H. G. Kuivila, *J. Org. Chem.*, **27**, 837 (1962).
730. Sawyer, A. K., and H. G. Kuivila, *J. Amer. Chem. Soc.*, **85**, 1010 (1963).
731. Schenck, G. O., E. Koerner v. Gustorf, and H. Köller, *Angew. Chem.*, **73**, 707 (1961).
732. Scherer, O. J., J. F. Schmidt, and M. Schmidt, *Z. Naturforsch.*, **19b**, 447 (1964).
733. Scherer, O. J., J. F. Schmidt, J. Wokulat, and M. Schmidt, *Z. Naturforsch.*, **20b**, 183 (1965).
734. Scherer, O. J., and M. Schmidt, *Angew. Chem.*, **75**, 642 (1963).
735. Scherer, O. J., and M. Schmidt, *J. Organomet. Chem.*, **1**, 490 (1964).
736. Schick, R., Dissertation, Univ. of Giessen, 1965.
737. Schindlbaur, H., and D. Hammer, *Mh. Chem.*, **94**, 644 (1963).
738. Schmidbaur, H., and H. Hussek, *Angew. Chem.*, **75**, 575 (1963).
739. Schmidbaur, H., and I. Ruidisch, *Inorg. Chem.*, **3**, 599 (1964).
740. Schmidbaur, H., and M. Schmidt, *J. Amer. Chem. Soc.*, **83**, 2963 (1961).
741. Schmidbaur, H., and S. Waldmann, *Chem. Ber.*, **97**, 3381 (1964).
741a. Schmidbaur, H., and W. Wolfsberger, *Chem. Ber.*, **101**, 1664 (1968).
742. Schmidt, M., H. J. Dersin, and H. Schumann, *Chem. Ber.*, **95**, 1428 (1962).
743. Schmidt, M., and H. Ruf, *Chem. Ber.*, **96**, 784 (1963).
744. Schmidt, M., and H. Schumann, *Chem. Ber.*, **96**, 780 (1963).
745. Schmidt, M., and H. Schumann, *Z. Naturforsch.*, **19b**, 74 (1964).
746. Schmidt, U., K. Kabitzke, K. Markau, and W. P. Neumann, *Chem. Ber.*, **98**, 3827 (1965).
747. Schmidt, W., Beckmann Report, 1961, No. 4, 6.
748. Schmitz-Dumont, O., *Z. Anorg. Allg. Chem.*, **248**, 289 (1941).
749. Schneider, B., Diploma Thesis, Univ. of Giessen, 1963.
750. Schneider, B., Dissertation, Univ. of Giessen, 1966; W. P. Neumann, and B. Schneider, *Liebigs. Ann. Chem.*, **707**, 20 (1967).
750a. Schumann, H., and H. Benda, *Angew. Chem.*, **80**, 845, 846 (1968).
750b. Schumann, H., P. Jutzi, and M. Schmidt, *Angew. Chem.*, **77**, 812 (1965).
751. Schumann, H., P. Jutzi, and M. Schmidt, *Angew. Chem.*, **77**, 912 (1965).
752. Schumann, H., H. Köpf, and M. Schmidt, *Angew. Chem.*, **75**, 672 (1963); *J. Organomet. Chem.*, **2**, 159 (1964).
753. Schumann, H., H. Köpf, and M. Schmidt, *Chem. Ber.*, **97**, 1458 (1964).
753a. Schumann, H., Th. Östermann, and M. Schmidt, *Chem. Ber.*, **99**, 2057 (1966).
754. Schumann, H., and M. Schmidt, *Angew. Chem.*, **76**, 344 (1964).

755. Schumann, H., and M. Schmidt, *Angew. Chem.*, **77**, 1049 (1965). Gives numerous references.
756. Schumann, H., K. F. Thom, and M. Schmidt, *J. Organomet. Chem.*, **1**, 167 (1963).
757. Schumann, H., K. F. Thom, and M. Schmidt, *J. Organomet. Chem.*, **2**, 97 (1964).
758. Schwartz, W. T., Jr., and H. W. Post, *J. Organomet. Chem.*, **2**, 357 (1964).
759. Schwartz, W. T., Jr., and H. W. Post, *J. Organomet. Chem.*, **2**, 425 (1964).
760. Seyferth, D., *J. Org. Chem.*, **22**, 1599 (1957).
761. Seyferth, D., *J. Amer. Chem. Soc.*, **79**, 2133 (1957).
762. Seyferth, D., *J. Amer. Chem. Soc.*, **81**, 1844 (1959).
763. Seyferth, D., *Progr. Inorg. Chem.*, **3**, 129 (1962).
764. Seyferth, D., *Record Chem. Progr.*, **26**, 87 (1965). Further references there.
765. Seyferth, D., and D. L. Alleston, *Inorg. Chem.*, **2**, 417 (1963).
765a. Seyferth, D., F. M. Armbrecht, and E. M. Hanson, *J. Organomet. Chem.*, **10**, P25 (1967). Further references there.
765b. Seyferth, D., J. M. Burlitch, H. Dertouzos, and H. D. Simmons, *J. Organomet. Chem.*, **7**, 405 (1967).
766. Seyferth, D., and H. M. Cohen, *Inorg. Chem.*, **2**, 652 (1963).
766a. Seyferth, D., H. Dertouzos, R. Suzuki, and J. Yick-Pui Mui, *J. Org. Chem.*, **32**, 2980 (1967).
766b. Seyferth, D., and A. B. Evnin, *J. Amer. Chem. Soc.*, **89**, 1468 (1967).
766c. Seyferth, D., and T. F. Julia, *J. Organomet. Chem.*, **8**, P13 (1967).
767. Seyferth, D., and N. Kahlen, *J. Org. Chem.*, **25**, 809 (1960); *J. Amer. Chem. Soc.*, **82**, 1080 (1960).
768. Seyferth, D., and R. B. King, *Annual Surveys of Organometallic Chemistry*, **1** (1964); **2** (1965); **3** (1966); From 1967, continued as Section *B* of *Organometallic Chemistry Reviews*.
769. Seyferth, D., G. Raab, and K. A. Brändle, *J. Org. Chem.*, **26**, 2934 (1961).
770. Seyferth, D., and E. G. Rochow, *J. Org. Chem.*, **20**, 250 (1955).
771. Seyferth, D., C. Sarafidis, and A. B. Evnin, *J. Organomet. Chem.*, **2**, 417 (1964).
772. Seyferth, D., and F. G. A. Stone, *J. Amer. Chem. Soc.*, **79**, 515 (1957).
773. Seyferth, D., R. Suzuki, C. J. Murphy, and C. R. Sabet, *J. Organomet. Chem.*, **2**, 431 (1964).
774. Seyferth, D., R. Suzuki, and L. G. Vaughan, *J. Amer. Chem. Soc.*, **88**, 286 (1966).
775. Seyferth, D., and M. Takamizawa, unpublished; cited in ref. 8.
776. Seyferth, D., and L. G. Vaughan, *J. Organomet. Chem.*, **1**, 138 (1963).
777. Seyferth, D., and M. A. Weiner, *J. Org. Chem.*, **24**, 1395 (1959).
778. Seyferth, D., and M. A. Weiner, *Chem. and Ind.*, 402 (1959).
779. Seyferth, D., and M. A. Weiner, *J. Org. Chem.*, **26**, 4797 (1961).
780. Seyferth, D., and M. A. Weiner, *J. Amer. Chem. Soc.*, **83**, 3583 (1961); **84**, 361 (1962).
781. Seyferth, D., M. A. Weiner, L. G. Vaughan, G. Raab, D. E. Welch, H. M. Cohen and D. L. Alleston, *Bull. Soc. Chim. France*, 1364 (1963).
782. Seyferth, D., H. Yamazaki, and D. L. Alleston, *J. Org. Chem.*, **28**, 703 (1963).
783. Shapiro, Ph. J., and E. I. Becker, *J. Org. Chem.*, **27**, 4668 (1962).
784. Sheaffer, B. L., and F. H. Conaty, cited in *Tin and its Uses*, **69**, 9 (1966).
785. Sheverdina, N. I., L. V. Abramova, J. E. Paleeva, and K. A. Kocheshkov, *Chim. Prom.*, 707 (1962).
786. Shimanouchi, T., I. Nakagawa, H. Kyogoku, and M. Ishii, XIXth. IUPAC Congr., London, 1963, Abstr. A., 172.
787. Shostakovskii, M. F., N. V. Komarov, I. S. Guseva, and V. K. Misyunas, *Doklady Akad. Nauk SSSR*, **158**, 918 (1964); Shostakovskii, M. F., V. M. Vlasov, R. G. Mirskov, and I. E. Loginova, *Zh. Obshch. Khim.*, **34**, 3178 (1964).

788. Shostakovskii, M. F., N. V. Komarov, and V. K. Misyunas, *Izvest. Akad. Nauk SSSR*, 368 (1962).
789. Shostakovskii, M. F., V. N. Kotrelev, S. P. Kalinina, G. I. Kuznetsova, L. V. Laine, and A. I. Borisova, *Vysokomol. Soed.*, **3**, 1128, 1131 (1961).
790. Shostakovskii, M. F., V. N. Kotrelev, D. A. Kochkin, G. I. Kuznetsova, S. P. Kalinina, and V. V. Borisenko, *Zh. Obshch. Khim.,* **31**, 1434 (1958).
791. Shostakovskii, M. F., V. M. Vlasov, and R. G. Mirskov, *Zh. Obshch. Khim.*, **33**, 324 (1963).
792. Shostakovskii, M. F., V. M. Vlasov, and R. G. Mirskov, *Zh. Obshch. Khim.*, **34**, 1354 (1964).
793. Shostakovskii, M. F., V. M. Vlasov, R. G. Mirskov, and I. M. Korstaeva, *Zh. Obshch. Khim.*, **35**, 401 (1965). Further references there.
794. Shostakovskii, M. F., V. M. Vlasov, R. G. Mirskov, and I. E. Loginova, *Zh. Obshch. Khim.*, **34**, 3178 (1964).
Shushunov, see Susunov.
795. Siebert, H., *Z. Anorg. Allg. Chem.*, **268**, 177 (1952).
796. Silbert, F. C., and W. R. Kirner, *Ind. Engng. Chem., Anal. Edit.*, **8**, 353 (1936).
796a. Simons, P. B., and W. A. G. Graham, *J. Organomet. Chem.*, **10**, 457 (1967).
797. Sisido, K., and S. Kozima, *J. Org. Chem.*, **27**, 4051 (1962).
797a. Sisido, K., and S. Kozima, *J. Organomet. Chem.*, **11**, 503 (1968).
798. Sisido, K., and S. Kozima, *J. Org. Chem.*, **29**, 907 (1964).
798a. Sisido, K., S. Kozima, and T. Hanada, *J. Organomet. Chem.*, **9**, 99 (1967).
798b. Sisido, K., S. Kozima, and T. Tuzi, *J. Organomet. Chem.*, **9**, 109 (1967).
798c. Sisido, K., T. Miyanisi, K. Nabika, and S. Kozima, *J. Organomet. Chem.*, **11**, 281 (1968).
799. Sisido, K., and Y. Takeda, *J. Org. Chem.*, **26**, 2301 (1961).
800. Sisido, K., Y. Takeda, and Z. Kinugawa, *J. Amer. Chem. Soc.*, **83**, 538 (1961).
801. Skeel, R. T., C. E. Bricker, W. N. Aldridge, and J. E. Cremer, *Z. Anal. Chem.*, **191**, 463 (1962).
802. Skinner, H. A., *Adv. Organomet. Chem.*, **2**, 49 (1964).
803. Skinner, H. A., and L. E. Sutton, *Trans. Faraday Soc.*, **40**, 164 (1944).
804. Smith, A. C., Jr., and E. G. Rochow, *J. Amer. Chem. Soc.*, **75**, 4103, 4105 (1953).
805. Smith, H. V., *The Development of the Organotin Stabilizers*, Tin Research Institute, Greenford, England 1959. Gives references to 109 patents up to early 1958. Also, original paper, *Rubber J. and Internat. Plastics*, **136**, 655 (1959).
806. Smith, T. A., and F. S. Kipping, *J. Chem. Soc.*, **101**, 2553 (1912).
807. Smith, T. A., and F. S. Kipping, *J. Chem. Soc.*, **103**, 2034 (1913).
808. Smith, T. D., *Nature*, **199**, 374 (1963).
809. Smolin, E. M., *Tetrahedron Letters*, 143 (1961).
809a. Sokolov, V. J., and O. A. Reutov, *Uspekhi Khimii*, **34**, 1 (1965).
810. Solerio, A., *Gazz. Chim. Ital.*, **81**, 664 (1951).
811. Sommer, R., Diploma Thesis, Univ. of Giessen, 1961.
812. Sommer, R., Dissertation, Univ. of Giessen, 1964.
812a. Sommer, R., and H. G. Kuivila, *J. Org. Chem.*, **33**, 802 (1968).
812b. Sommer, R., E. Müller, and W. P. Neumann, *Liebigs Ann. Chem.*, **718**, 11 (1968); **721**, 1 (1969).
813. Sommer, R., and W. P. Neumann, *Angew. Chem.*, **78**, 546 (1966).
814. Sommer, R., W. P. Neumann, and M. Scholz, unpublished.
815. Sommer, R., W. P. Neumann, and B. Schneider, *Tetrahedron Letters*, 3875 (1964).
816. Sommer, R., B. Schneider, and W. P. Neumann, *Liebigs Ann. Chem.*, **692**, 12 (1966).

817. Spector, W. S. (Ed.), *Handbook of Toxicology*, Carpenter Co., Springfield, Ohio, 1955.
818. Srivastava, T. N., and M. Onyszchuk, *Canad. J. Chem.*, **41**, 1244 (1963).
819. Srivastava, T. N., and S. K. Tandon, *Indian J. Appl. Chem.*, **26**, 171 (1963).
820. Staab, H. A., *Einführung in die theoret. organ. Chemie*, Verlag Chemie, Weinheim/Bergstr., 1962, p. 212.
821. Stamm, W., *J. Org. Chem.*, **30**, 693 (1965).
822. Stamm, W. A., A. W. Breindel, and A. H. Freiberg (Stauffer Chem.), U.S. Pat. 3,095,434 (June 25, 1963), *C.A*, **59**, 14023 (1963).
823. Stefl, E. P., and Ch. E. Best (Firestone Tire and Rubber Comp.), U.S. Pat. 2,731,482 (January 17, 1956), *C.A.*, **51**, 13918 (1957).
824. Stegmann, H. B., and K. Scheffler, *Tetrahedron Letters*, 3387 (1964).
825. Stern, A., and E. I. Becker, *J. Org. Chem.*, **27**, 4052 (1962).
826. Stern, A., and E. I. Becker, *J. Org. Chem.*, **29**, 3221 (1964).
827. Stolberg, U. Graf zu, *Angew. Chem.*, **75**, 206 (1963); C. A. Burkhard, *J. Amer. Chem. Soc.*, **71**, 963 (1949); H. Gilman and R. A. Tomasi, *J. Org. Chem.*, **28**, 1651 (1963).
828. Stone, F. G. A., *Hydrogen Compounds of the Group IV Elements*, Prentice-Hall, Englewood Cliffs, N.J., 1962.
828a. Stoner, H. B., *Brit. J. Industr. Med.*, **23**, 222 (1966).
829. Strecker, A., *Liebigs Ann. Chem.*, **105**, 306 (1858).
830. Strecker, A., *Liebigs Ann. Chem.*, **123**, 365 (1862).
831. Sundermeyer, W., and W. Verbeek, *Angew. Chem.*, **78**, 107 (1966).
832. Susunov, V. A., and T. G. Brilkina, *Doklady Akad. Nauk SSSR*, **141**, 1391 (1961).
833. Swiger, E. D., and J. D. Graybeal, *J. Amer. Chem. Soc.*, **87**, 1464 (1965).
834. Tagliavini, G., U. Belluco, and G. Pilloni, *Ric. Sci. Rend.*, *A*, **3**, 889 (1963).
835. Tagliavini, G., S. Faleschini, G. Pilloni, and G. Plazzogna, *J. Organomet. Chem.*, **5**, 136 (1966).
836. Taimsalu, P., and J. L. Wood, *Spectrochim. Acta*, **20**, 1043 (1964).
837. Takami, Y., *Rep. Govt. Chem. Ind. Res. Inst. Tokyo*, **57**, 229, 234 (1962).
838. Tamborski, C., F. E. Ford, W. L. Lehn, G. J. Moore, and E. J. Soloski, *J. Org. Chem.*, **27**, 619 (1962).
839. Tamborski, C., F. E. Ford, and E. J. Soloski, *J. Org. Chem.*, **28**, 181 (1963).
840. Tamborski, C., F. E. Ford, and E. J. Soloski, *J. Org. Chem.*, **28**, 237 (1963).
841. Tamborski, C., and H. Gilman (U.S. Dept. of the Air Forces), U.S. Pat. 3,079,414 (February 26, 1963), *C.A.*, **59**, 5196 (1963).
842. Tamborski, C., and E. J. Soloski, *J. Amer. Chem. Soc.*, **83**, 3734 (1961).
843. Tanaka, T., M. Komura, Y. Kawasaki, and R. Okawara, *J. Organomet. Chem.*, **1**, 484 (1964).
844. Tanaka, T., R. Ue-eda, M. Wada, and R. Okawara, *Bull. Chem. Soc. Japan*, **37**, 1554 (1964).
845. Tauberger, G., *Med. Exp.*, **9**, 393 (1963).
846. Tauberger, G., and O. R. Klimmer, *Arch. Exp. Path. Pharmakol.*, **242**, 370 (1961). Gives numerous references.
847. Tchakirian, A., and P. Berillard, *Bull. Soc. Chim. France*, 1300 (1950).
847a. Thayer, J. S., *Organomet. Chem. Rev.*, **1**, 157 (1966).
848. Thayer, J. S., and R. West, *Inorg. Chem.*, **3**, 889 (1964).
849. Thies, C., and J. B. Kisinger, *Inorg. Chem.*, **3**, 551 (1964).
850. Thomas, A. B., and E. G. Rochow, *J. Amer. Chem. Soc.*, **79**, 1843 (1957).
851. Thomas, I. M., *Canad. J. Chem.*, **39**, 1386 (1961).
852. Throndsen, H. P., P. J. Wheatley, and H. Zeiss, *Proc. Chem. Soc.*, 357 (1964).

853. Tobias, R. S., I. Ogrins, and B. A. Nevett, *Inorg. Chem.*, **1**, 638 (1962).
854. Tomilov, A. P., Yu. D. Smirnov, and S. L. Varsavskij, *Zh. Obshch. Khim.*, **35**, 391 (1965).
855. Türler, M., and O. Högl, *Mitt. Lebensmitt. Hyg.* (Bern), **52**, 123 (1961).
856. Tsai, Tsu Tzu, A. Cutler, and W. L. Lehn, *J. Org. Chem.*, **30**, 3049 (1965).
856a. Tsai, Tsu Tzu, and W. L. Lehn, *J. Org. Chem.*, **31**, 2981 (1966).
856b. Tzschach, A., and E. Reiss, *J. Organomet. Chem.*, **8**, 255 (1967).
857. Valade, J., and M. Pereyre, *Compt. Rend.*, 245, 3693 (1962); *Bull. Soc. Chim. France*, 1533 (1962).
858. Valade, J., M. Pereye, and R. Calas, *Compt. Rend.*, **253**, 1216 (1961).
859. Valade, J., and J. C. Pommier, *Bull. Soc. Chim. France*, 199, 951 (1963).
860. Varvill, M. (F. A. Hughes and Co., Ltd.), Brit. Pat. 851,902 (October 19, 1960), *C.A.*, **55**, 15960 (1961).
860a. Vaughan, L. G., and D. Seyferth, *J. Organomet. Chem.*, **5**, 295 (1966).
861. Verdonck, L., and G. P. van der Kelen, *Bull. Soc. Chim. Belges*, **74**, 361 (1965).
861a. Verdonck, L., G. P. van der Kelen, and Z. Eeckhaut, *J. Organomet. Chem.*, **11**, 487 (1968).
862. Vijayaraghavan, K. V., *J. Indian Chem. Soc.*, **22**, 135 (1945).
863. Vind, H. P., and H. Hochman, *Proc. Amer. Wood Preserv. Ass.*, **58**, 1 (1962); cited in *Tin and its Uses*, **57**, 10 (1963).
863a. Voronkov, M. G., and I. Romadane, *Khim. Geterotsikl. Soed.*, **1966**, 892.
864. Vyazankin, N. S., and V. T. Bychkov, *Zh. Obshch. Khim.*, **35**, 684 (1965).
864a. Vyazankin, N. S., E. N. Gladysev, and S. P. Korneva, *Zh. Obshch. Khim.*, **37**, 1736 (1967).
864b. Vyazankin, N. S., and O. A. Kruglaya, *Russ. Chem. Rev.*, **35**, 593 (1966). Review.
865. Vyazankin, N. S., G. A. Razuvaev, and T. N. Brevnova, *Doklady Akad. Nauk SSSR*, **163**, 1389 (1965); *Zh. Obshch. Khim.*, **34**, 1005 (1964); **35**, 2033 (1965).
866. Vyazankin, N. S., G. A. Razuvaev, and V. T. Bychkov, *Zh. Obshch. Khim.*, **35**, 395 (1965).
867. Vyazankin, N. S., G. A. Razuvaev, O. S. D'yachkovskaya, and O. A. Shchepetkova, *Doklady Akad. Nauk SSSR*, **143**, 1348 (1962).
868. Vyazankin, N. S., G. A. Razuvaev, and S. P. Korneva, *Zh. Obshch. Khim.*, **33**, 1041 (1963).
869. Vyazankin, N. S., G. A. Razuvaev, and S. P. Korneva, *Zh. Obshch. Khim.*, **34**, 2787 (1964).
870. Vyazankin, N. S., G. A. Razuvaev, and O. A. Kruglaya, *Izvest. Akad. Nauk SSSR*, 2008 (1962).
870a. Vyazankin, N. S., G. A. Razuvaev, and O. A. Kruglaya, *Organomet. Chem. Rev., Ser. A*, **3**, 323 (1968).
871. Wada, M., M. Nishino, and R. Okawara, *J. Organomet. Chem.*, **3**, 70 (1965).
872. Wada, M., M. Shindo, and R. Okawara, *J. Organomet. Chem.*, **1**, 95 (1963).
873. Walsh, E. J., unpublished; cited in ref. 429. See also ref. 441.
874. Washburn, R. M., and R. A. Baldwin (Amer. Potash and Chemical Corp.), U.S. Pat. 3,112,331 (November 26, 1963), *C.A.*, **60**, 5554 (1964).
875. Watt, G. W., *Chem. Rev.*, **46**, 317 (1950).
875a. Wazer, J. R. van, *Ann. New York Acad. Sci.*, **159**, 5 (1969).
875b. Weber, H. P., and R. F. Bryan, *Chem. Comm.*, 443 (1966).
876. Weber, S., and E. I. Becker, *J. Org. Chem.*, **27**, 1258 (1962).
877. Weinberg, E. L., and L. A. Tomka (Metal and Thermit Corp.), U.S. Pat. 2,858,325 (October 28, 1958), *C.A.*, **53**, 3758 (1959).

878. Wells, A. F., *J. Chem. Soc.*, 55 (1949).
879. Werner, A., and P. Pfeiffer, *Z. Anorg. Allg. Chem.*, **17**, 82 (1898).
880. Westheimer, F. H., and W. A. Jones, *J. Amer. Chem. Soc.*, **63**, 3283 (1941).
881. Wheatley, P. J., *J. Chem. Soc.*, 5027 (1961).
882. Wiberg, E., and H. Behringer, *Z. Anorg. Allg. Chem.*, **329**, 290 (1964).
883. Wiberg, E., and R. Rieger, D.B. Pat. 1,121,050 (December 9, 1960), *C.A.*, **56**, 14328 (1962).
884. Wieber, M., and M. Schmidt, *Z. Naturforsch.*, **18b**, 846 (1963).
885. Wieber, M., and M. Schmidt, *J. Organomet. Chem.*, **1**, 336 (1964). Further references there.
886. Wilkinson, G. R., and M. K. Wilson, *J. Chem. Phys.*, **25**, 784 (1956).
887. Willemsens, L. C., *Organolead Chemistry*, 1964.
888. Willemsens, L. C., and G. J. M. van der Kerk, *Investigations in the Field of Organolead Chemistry*, Internat. Lead Zinc Research Organ. Inc., Schotanus and Jens, Utrecht, Netherlands, 1965.
889. Willemsens, L. C., and G. J. M. van der Kerk, *J. Organomet. Chem.*, **2**, 260 (1964).
890. Williams, D. J., and J. W. Price, *The Analyst*, **89**, 220 (1964). Further references there.
890a. Wirth, H. O., lecture given at the Organotin Symposium, Frankfurt/Main, November, 1968.
891. Wittenberg, D., *Liebigs Ann. Chem.*, **654**, 23 (1962).
892. Wittenberg, D., (Badische Anilin- und Soda-Fabrik), D.B. Pat. 1,124,947 (March 8, 1962), *C.A.*, **57**, 7309 (1962).
893. Wittenberg, D., and H. Gilman, *Quart. Rev.*, **13**, 116 (1959).
894. Wittig, G., and G. Geissler, *Liebigs Ann. Chem.*, **580**, 44 (1953).
895. Wittig, G., F. J. Meyer, and G. Lange, *Liebigs Ann. Chem.*, **571**, 167 (1951).
896. Yakubovich, A. Ya., S. P. Makarov, V. A. Ginsburg, G. I. Gavrilov, and E. N. Merkulova, *Doklady Akad. Nauk SSSR*, **72**, 69 (1950).
897. Yasuda, K., Y. Kawasaki, N. Kasai, and T. Tanaka, *Bull. Chem. Soc. Japan*, **38**, 1216 (1965).
898. Yasuda, K., and R. Okawara, *J. Organomet. Chem.*, **3**, 76 (1965).
899. Yatagai, T., S. Matsuda, and H. Matsuda (Nippon Shokubai Kagaku Kogyo Co.), U.S. Pat. 3,085,102 (April 9, 1963), *C.A.*, **59**, 11560 (1963).
899a. Yergey, A. L., and F. W. Lampe, *J. Amer. Chem. Soc.*, **87**, 4204 (1965).
900. Yngve, V. (Carbide and Carbon Chem. Corp.), U.S. Pat. 2,307,092 (January 5, 1943), *C.A.*, **37**, 3532 (1943).
901. Yoshitomi Phatm. Ind. Ltd., Jap. Pat. 2337 (February 6, 1965).
902. Zakharkin, L. I., and O. Yu. Okhlobystin, *Doklady Akad. Nauk SSSR*, **116**, 236 (1957).
903. Zakharkin, L. I., and O. Yu. Okhlobystin, *Izvest. Akad Nauk SSSR*, 1942 (1959).
904. Zakharkin, L. I., and O. Yu. Okhlobystin, *Zh. Obshch. Khim.*, **31**, 3662 (1961).
905. Zakharkin, L. I., and O. Yu. Okhlobystin, *Izvest. Akad. Nauk SSSR*, 2202 (1963).
906. Zakharkin, L. I., O. Yu. Okhlobystin, and B. N. Strunin, *Izvest Akad. Nauk SSSR*, 2002 (1962).
907. Zakharkin, L. I., O. Yu. Okhlobystin, and B. N. Strunin, *Zh. Prikl. Khim.*, **36**, 2034 (1963).
908. Zavgorodnii, V. S., and A. A. Petrov, *Zh. Obshch. Khim.*, **32**, 3527 (1962); **35**, 760 (1965).
909. Zavgorodnii, V. S., and A. A. Petrov, *Doklady Akad. Nauk SSSR*. **149**, 846 (1963).
910. Zavgorodnii, V. S., and A. A. Petrov, *Zh. Obshch. Khim.*, **34**, 1931 (1964).
911. Zavgorodnii, V. S., and A. A. Petrov, *Zh. Obshch. Khim.*, **35**, 1313 (1965).

912. Zedler, R. J., *Tin and its Uses*, **53**, 7 (1961).
913. Zemlyanskii, N. N., I. P. Goldstein, E. N. Guryanova, E. M. Panov, N. A. Slovokhotova, and K. A. Kocheshkov, *Doklady Akad. Nauk SSSR*, **156**, 131 (1964).
914. Zemlyanskii, N. N., E. P. Panov, and K. A. Kocheshkov, *Zh. Obshch. Khim.*, **32**, 291 (1962).
915. Zemlyanskii, N. N., E. P. Panov, and K. A. Kocheshkov, *Doklady Akad. Nauk SSSR*, **146**, 1335 (1962).
916. Zemlyanskii, N. N., E. M. Panov, O. P. Samagina, and K. A. Kocheshkov, *Zh. Obshch. Khim.*, **35**, 1029 (1965).
917. Zemlyanskii, N. N., E. H. Panov, N. A. Slovokhotova, O. P. Shamagina, and K. A. Kocheshkov, *Doklady Akad. Nauk SSSR*, **149**, 312 (1963).
918. Zhivukhin, S. M., E. D. Dudikova, and V. V. Kireev, *Zh. Obshch. Khim.*, **31**, 3106 (1961).
919. Zhivukhin, S. M., E. D. Dudikova, and A. M. Kotov, *Zh. Obshch. Khim.*, **33**, 3274 (1963).
920. Zhivukhin, S. M., E. D. Dudikova, and N. B. Psijalkovskaja, *Zh. Obshch. Khim.*, **33**, 2958 (1963).
921. Zhivukhin, S. M., E. D. Dudikova, and E. M. Ter-Sarkisjan, *Zh. Obshch. Khim.*, **32**, 3059 (1962).
922. Zhdanov, G. S., and J. G. Ismailzade, *Zh. Fiz. Khim.*, **24**, 1495 (1950).
923. Ziegler, K., "Organo-Aluminium Compounds," in *Organometallic Chemistry*, Ed. H. Zeiss, Reinhold Publ. Corp., New York, 1960.
924. Ziegler, K., W. Deparade, and W. Meye, *Liebigs Ann. Chem.*, **567**, 141 (1950); C. G. Overberger, M. T. O'Shaughnessy, and H. Shalit, *J. Amer. Chem. Soc.*, **71**, 2661 (1949).
925. Ziegler, K., H. G. Gellert, H. Lehmkuhl, W. Pfohl, and K. Zosel, *Angew. Chem.*, **67**, 424 (1955); *Liebigs Ann. Chem.* **629**, 1 (1960).
926. Ziegler, K., H. G. Gellert, H. Martin, K. Nagel, and J. Schneider, *Liebigs Ann. Chem.*, **589**, 91 (1954).
927. Ziegler, K., and H. Lehmkuhl, *Chem.-Ing.-Tech.*, **35**, 325 (1963).
928. Ziegler, K., *et al.*, *Liebigs Ann. Chem.*, **629** (1960).
928a. Zimmer, H., C. W. Blewett, and A. Brakas, *Tetrahedron Letters*, 1615 (1968).
929. Zimmer, H., I. Hechenbleikner, O. A. Homberg, and M. Danzik, *J. Org. Chem.*, **29**, 2632 (1964).
930. Zimmer, H., and H. W. Sparmann, *Chem. Ber.*, **87**, 645 (1954).

Index

Acetyl peroxide, degradation by organotin hydrides, 75
Acyl halides, reduction of, 116
Aldehydes, 95, 99, 110
  as radical scavengers, 116
  from acyl halides, 116
  reduction of, 108–113
Alkoxy-alkylaluminums, 23
Alkyl halides, reduction of, 113–115
Alkyltin hydrides, *see*: Organotin hydrides
Alkynes, 27–29, 96
Alkynyltins, 27
  preparation of, 27
  reaction of, 29
Allenes, 95
Allyltin compounds, 26, 206
Aluminum alkyls, 23, 24, 41, 92
  displacement reactions with, 201
  as hydrostannation catalysts, 92
  by transmetalation, 41
Analysis of organotin compounds, 209–213
Anils, 117
Appearance potentials, 9
Arylformamides, 117
Aryl halides, reduction of, 113–115
Asymmetric tin atom, 5
Azobisisobutyronitrile, 88, 89, 96, 201, 207
  reaction with organotin hydrides, 75
Azo-compounds, 75, 101
  substituted, induced degradation by stannyl radicals, 75
Azomethines, 100

Back-donation to tin atom ($d\pi$–$p\pi$ bonds), 12
Benzoyl peroxide, 39
  degradation by ditins, 133
  degradation by organotin hydrides, 74
Biurets, substituted, 189
Bond-dissociation energies, 6, 8, 9, 72, 131
Bond-lengths, 6–11, 16, 72, 142
  to carbon, 8
Butadiene and its derivatives, 93
t-Butoxy radicals, 39

Carbene intermediates, claimed, 83
Carbodiimides, 103
Carbon dioxide, 101, 103
Carbon disulfide, 103
Carbon–tin bonds, alkylation by, 40
  attack by acids and bases, 34
  cleavage by alkyl halides, 40
  cleavage by free radicals, 38
  cleavage by halogens, 36
  cleavage by halogens, mechanism of, 37
  Grignard-like reactions of, 43
  hydrolyses, 34
  polarity of, 33
  reactions of, 33–44
  reaction with nucleophilic reagents, 35
  reaction with oxygen, 39, 40
Carbonyl compounds, 99–100, 108–113
  hydrogenation of, 108–113
Carboxylic acids, stannyl esters of, 61
Carboxylic esters, reduction of, 117
Catalytic activity of organotins, 187, 245

Catenated compounds of tin, 128
Chiral organotins, 5, 6
Commercial applications of organotins, 238–246
Complexes, with hexa-coordinate tin, 18
  with penta-coordinate tin, 16
Comproportionation, 25, 41, 42, 53, 54, 71, 239
  equilibrium of, 55
  mechanism of, 56
Conductance, 7, 16, 121
Conformation, 9, 10
Cortisone, 112
Covalent radius of tin, 6
Cyclic compounds of Group IV elements (table), 146
Cyclic ethers, preparation with stannoxanes, 184
Cyclic geminal dihalides, 113
Cyclic polytins, 71, 81, 84, 140–147
Cyclohexatin, 5
Cyclopropane, 29

Decarboxylation of stannyl esters, 28
Determination of functional groups, 212
C-Deuterated compounds, preparation of, 110, 115
O-Deuterated compounds, preparation of, 110
Dialkylalkoxytin hydride, 111
Dialkyltins, polymeric, 111, 141
Diazomethane, 29, 61, 183, 200
Dicarboxy-dihalides, 117
Dicyclopentadienyltin, 135
Diels–Alder reaction, 29
Diethyltin, supposed, 134
1,2-Dihalides, 114
α,ω-Dihalo-polytins, 139, 146
1,4-Dilithiumtetraphenylbutadiene, 193
Diorganotin dihydroxides, 161
Diorganotin oxides, 160–163, 205
  modifications of, 162
Diorganotin sulfides, 171
Diphenyltin, supposed, 135
Dipole moments, 8, 10, 11, 13
Dissociation constants of substituted benzoic acids, 13
Dissociation energy; *see*: Bond-dissociation
1,4-Distannacyclohexane, 196
1,6-Distanna-2,5,7,10-tetraoxacyclodecane, 197
Distannthianes, 170
Distannyl acetals, 186
Distannyl-ureas, 180, 189
Di-*t*-butyl peroxide, 39
Ditins, 128–134
  supposed dissociation of, 131
  autoxidation of, 129
  reaction with free radicals, 39
  substituted, 130, 133
  unsymmetrical, 125
  unsymmetrically substituted, 80
Dodecaarylcyclohexatins, 143
Donor–acceptor bond, 15–19
Double-bonding ($d\pi$–$p\pi$), 13, 151

Electrochemical methods, 24
Electron density at tin atom, change of, 12
Electron-donor character, 13–19
Electronegativities, 7, 8, 11, 12, 18, 35, 72
  table of, 12
Electronegativity of tin atom, 33
Electronegativity series of organic groups, 37
Electrophilic aromatic substitutions, 12
Electrophilic substitution at the saturated carbon atom, 34, 37, 42, 56
Elemental analysis, 211
Enthalpies of combustion and formation, 8

Force constants for Sn–Cl bonds, 13
Fungicides, 242

Galvinoxyl, 73, 86, 89
Gas–liquid chromatography, 210
Geminal polyhalides, 113
Grignard reagents, 22, 24, 26, 239
Grignard-like reactions of carbon–tin bonds, 43

α-Halo-ketones, 114
Hammett σ-constants, 35
Heterocycles, tin-containing, 191–198
  with one tin atom, 191–195
  with two or more heteroatoms, 195–198
Hexa-coordinate tin, 18, 34
Hexaorganoditins, 128–133
  reaction with free radicals, 39
  structure of, 131

Hexaorganostannoxanes; *see*: Stannoxanes
Hexatins, linear, 137
Historical aspects of organotin chemistry, 1
Hydrocortisone, 112
Hydrogenations by organotin hydrides, 107–119
Hydrostannation, by use of radical catalysts, 86
of aldehydes and ketones, 99
of alkynes, 96
of allenes, 95
of conjugated dienes, 93
of internal olefins, 89
of olefins, 85
of polar conjugated systems, 103
of unsaturated compounds, 85–106
of unsaturated nitrogen compounds, 100
use of organometallic catalysts, 92
Hydroxybenzaldehydes, 100

Imidazolyl compounds, penta-coordination in, 14
Inductive effect, 12, 37
of tin, 11
Industrial preparation of organotins, 239
Infrared spectroscopy, 10, 68, 215–216, 218–219
Insecticides, 242
Intoxication by organotins, therapy, 235
Ionic dissociation, 7
Isocyanates, 102
Isomorphism of tetraphenyltin, 21
Isothiocyanates, 102

Ketones, 99, 108–113

Lewis bases, 15–17
Ligand exchange reactions, 42, 81, 128, 136, 137, 140, 152, 165, 173
Literature sources, 3, 4
Lithium alkenyls, 26
Lithium alkyls, 22, 26, 27, 43, 121
by transmetalation, 42

Macromolecular organotins, 199–208
from olefinic organotin compounds, 206
preparation by hydrostannation, 200
with stannyl substituents, 206
with tin atoms in the main chain, 199

MAC values, 231
Malonic ester derivatives, 103, 105, 182
Malononitrile derivatives, 103, 104, 181
Mass spectrometry, 9, 227
Metal carbonyls containing covalent tin–metal bonds, 177
Metal halides, reduction of, 119
Metal ketyls, 124
Metal oxides, reduction of, 119
Mössbauer spectra, 8, 226
Molecular compounds, 16

Network polymers, 201
formed by tin–oxygen bonds, 162, 164
Nitro-compounds, reduction of, 117
α,β-unsaturated, 106
Nitroso-compounds, 101
Nomenclature, 4
Nuclear magnetic resonance spectroscopy, 216–226
Nucleophilic attack at the tin atom, 34, 79

Octaphenylcyclooctatetraene, 194
Olefins, hydrostannation of, 85–96
isomerization by stannyl radicals, 88, 94
Organohalotin hydrides, 70, 84, 93, 100, 142, 144
Organolead hydrides, 85
Organotin alkoxides, 110, 165–166
addition to polar multiple bonds, 184
condensation reactions, 179–184
cyclic, 196
transesterification with carboxylic esters, 167
condensation reactions with proton donors, 179
Organotin azides, 59
addition to acetylenic compounds, 59
Organotin cadmium compounds, 176
Organotin cyanides, 59
structure of, 60
Organotin deuterides, 67, 110, 115, 123
Organotin dicyanides, 182
Organotin fulminates, 59
Organotin halides, 7, 45–58
conductivity of, 7, 45
hydrolysis of, 45
preparation of, 49–58
preparation of, by cleavage of C–Sn or Sn–Sn bonds, 57

Organotin halides—*contd.*
preparation of, by comproportionation, 53
preparation of, by partial alkylation of inorganic tin compounds, 52
preparation of, from other organotin halides, 58
preparation of, starting from tin metal, 49
Organotin hydrides, 66–119
as radical scavengers, 74
as reducing agents, 107–119
condensations leading to Sn–Sn bonds, 78
decomposition of, 69
estimation of, 76
exchange reactions with other stannyl compounds, 77
hydrostannation reactions of, 85–106
polarity of, 72
preparation of, 66–71
reactions of, 72
reaction with bases, 83
reaction with diazo-compounds, 82
reactions with metal alkyls, 81
reactions with peroxides, 74, 75
reactions with proton donors, 76
reactions with radical sources and short-lived radicals, 73
Organotin isocyanates, 59
Organotin isothiocyanates, 59
Organotin-mercury compounds, 176
Organotin oxides, potential of, in organic synthesis, 179
Organotin peroxides, 129, 167–169
Organotin phenoxides, 100, 165–166
Organotin pseudohalides, 59–60
Organotins as insecticides and fungicides, 242
Organotins as polymer stabilizers, 240
Organotins, toxicity of, 230
Organotins, with functional groups, 85
Organotin-selenium compounds, 174
Organotin-sulfur compounds, 169–174
Organotin-tellurium compounds, 174
Organotin-zinc compounds, 176
Oxygen–oxygen bond, cleavage by stannyl radicals, 74, 75

Paper chromatography, 209
Penta-coordinate tin, 14–16, 18, 62, 64, 77, 104, 105, 148, 160, 161, 164, 166, 168, 170
Pentatins, linear, 137, 138
Peroxides, induced decomposition by stannyl radicals, 74, 75
Phenoxy radicals, 73; *see also*: Galvinoxyl
Phenyl isocyanate, 102
Phenylaluminum compounds, 24
Planar trialkylstannyl group, 15, 18
Plumbyl esters, 15
Polyadditions, 201
cis-1,4-Polybutadiene, stannyl-substituted, 206
Polyethers, containing stannoxane groups, 205
Polyethers, organotin-substituted, 208
Polymer stabilizers, 240
Polymerizations, 88
catalyzed by organotin compounds, 187
Polytins, 81, 134–147
acyclic, 136–140
cyclic, 140–147
cyclic, aliphatic, 144
cyclic, aromatic, 141
cyclic, color of, 147
Polyurethanes, 245
formation of, 187
Progesterone, 112
Publications, number of, 2
Pyrazoline derivatives, 29
Pyridine, as catalyst, 141–145

Radical reactions, 8, 38, 73
Radical scavengers, 86; *see also*: Galvinoxyl
Radical sources, 10, 73, 86, 87, 93, 96, 98, 101, 102, 115, 117, 154, 201; *see also*: Azobisisobutyronitrile
Raman spectroscopy, 215, 218–221
Redistribution reactions, 25, 40–42, 166
Rotation isomerism, 10, 20

Silicones, 69, 162, 245
Sodium-naphthalene, 121, 143
Sodium tetraalkylaluminates, 23
Sodium tin hydrides, alkylation of, 67
1,1′-Spirobisstannacyclohexane, 193
1,1′-Spirobistannole, 194
Stabilizers of polymers, 240

Stabilizers of stannoxanes, 240
Stalinon affair, 235
Stannacyclodecadienes, 195
Stanna-2,6-cycloheptadienes, 194
Stannacyclohexanes, 191, 192
Stannacyclopentadiene, 193
Stannacyclopentane, 192–194
  dioxa-, 197
9-Stannafluorene, 194
1-Stanna-4-heterocyclohexanes, 196
Stannazanes, 151
Stannazine, pheno-, 198
Stannepines, 194
Stannepine, benzo-, 195
  dibenzo-, 195
  dibenzodioxa-, 197
Stannoles, 193, 194
  dibenzo-, 194
Stannonic acids, 163–165, 205
  structure of, 163
Stannosiloxanes, 205
Stannoxane hydrate, 157
Stannoxanes, 155–161, 164, 240
  addition to polar multiple bonds, 184
  condensation reactions, with proton donors, 179
  condensations with other compounds, 183
  disubstituted, 17, 158
  polymeric, 204
Stannthianes, polymeric, 204, 205
Stannylacetylenes, 27–29, 181
Stannyl-alkali compounds, 120–124
  ligand exchange of, 123
Stannyl-alkaline earth compounds, 124
Stannyl alkyl selenides, 175
Stannyl alkyl sulfides, 173
Stannylamines, 100, 101, 148–151
  addition to polar multiple bonds, 184
  condensation reactions, with proton donors, 179
  condensation with organotin hydrides, 79, 183
  disproportionation of, 150
  penta-coordination in, 148
  reaction with metal hydrides, 126
Stannylarsines, 152–154
Stannyl aryl sulfides, 173
Stannylation of polymers, 206
Stannylbismuthines, 152–154
Stannylboron compounds, 125
Stannyl carbamates, 180, 189
N-Stannylcarbamic acid, 188
N-Stannylcarbamic ester, 187
Stannyl enol ethers, 105
Stannyl esters, 61–65, 76
  association, 15
  comproportionation with, 63
  decarboxylation of, 28
  of carboxylic acids, 61
  of formic acid, elimination of CO_2, 69
  of inorganic acids, 64
  of thioacids, 174
  penta-coordination in, 14, 62, 64
  of malonic acid, 44, 180
  polymeric, 204–205
Stannyl formamidines, 103
Stannyl-germanium compounds, 126–127
Stannyl hydroperoxides, 167
O-Stannylhydroxylamine, 101
O-Stannyl keten acetals, 103, 105, 182
N-Stannyl ketenimines, 103, 104, 181
α-Stannyl ketones, 105
Stannyl-lead compounds, 126–127
Stannyl-lithium compounds, conductivity of, 121
  constitution of, 121
Stannylmetallics, 120–124
O-Stannyl nitronic acid esters, 106, 182
Stannyl phosphines, 152–154
Stannylpyridine, 35
Stannyl radicals, 73–76, 88, 93–96
Stannyl-silicon compounds, 126–127
Stannyl stibines, 152–154
Structure of organotin compounds, 5–19
Substitution at saturated carbon atoms, 34, 37, 42, 56
Substitution at unsaturated carbon atoms, 34, 37
Succinyl (di)chloride, 117

Technology of organotins, 238
Telomerization, 88
Tetraalkylditin dihalides, 71, 84, 133
Tetraalkylditin dihydrides, 71
Tetraallyltin, 206
Tetraaryltins, 21
Tetraethyllead, modifications of, 20
Tetraethyltin, modifications of, 20
Tetrahedral angle, 5, 9
Tetrahedral model, 5

Tetraorganotins, 20–44
  physical properties of, 21
  preparation by electrochemical methods, 24
  preparation of, 21, 24, 239
  redistribution reactions of, 25
  unsymmetrical, 24
  with functional groups, 52
Tetratins, linear, 71, 137, 138
Thallium alkyls, by transmetalation, 41, 42
Thin-layer chromatography, 210
Thiostannonic anhydrides, 172
Tin alloys, reaction with alkyl halides, 22
Tin dialkyls, supposed, 5
Tin halides, reaction with alkylaluminum compounds, 23
  reaction with alkyllithium, 22
  reaction with Grignard reagents, 22
Tin–halogen bond, partial double-bond character of, 11
  contraction of, 7
Tin heterocycles, 191–193
Tin-119 magnetic resonance spectroscopy, 226
Tin–nitrogen bonds, compounds with; *see*: Stannylamines
Tin–mercury bonds, 82
Tin–sulfur heterocycles, 174
Tin tetrachloride, etherate of, 23
Tin tetrahydride, 66
Tin–tin bond, cleavage by halogens, 57
Tin–tin bonds, by condensation of organotin hydrides, 78
  compounds with; *see*: Ditins, Polytins
Toxicity of organotin compounds, 230–237
Transamination, 150
Transesterification, 167, 173, 182
Transition-metal chemistry, 178
Transmetalation, 26, 40, 41, 42
Trialkylaluminum, exchange reaction with, 81
Trialkylphosphines, 60
Trialkylstannyl acyl peroxides, 168
  alkyl peroxides, 167
  Bis(trialkylstannyl) peroxides, 169
Trialkyltin chlorides, 16
Trialkyltin sulfides, 103
Triazoles, 59
2-Tributylstannylbutadiene, 206
"Tributyltin oxide"; *see*: Stannoxanes, 158
1,3,5-Tristannacyclohexane, 196
Tritins, linear, 136, 138
  Bis(triorganostannyl) sulfides, 170
Triorganotin hydrosulfides, 169
Triorganotin hydroxides, 155, 159

"Ultradissociation", supposed, 131
Ultraviolet spectroscopy, 214–215
Unsymmetrical tetraorganotins, 24

Valence-bond calculations, 29
Valence electrons, 5
Vinyltin compounds, 25, 26, 206
  reactions of, 26
Vinyl ketones, 106

Ziegler catalysts, 206